The Human Brain

Edited by Paul F Kisak

Contents

1 Human brain   1
- 1.1 Structure 1
  - 1.1.1 General features 2
  - 1.1.2 Comparative anatomy 2
  - 1.1.3 Cerebral cortex 3
  - 1.1.4 Cortical divisions 3
  - 1.1.5 Functional divisions 4
  - 1.1.6 Development 6
- 1.2 Function 6
  - 1.2.1 Cognition 6
  - 1.2.2 Lateralization 6
  - 1.2.3 Language 7
  - 1.2.4 Metabolism 8
- 1.3 Clinical significance 8
  - 1.3.1 Effects of brain damage 9
  - 1.3.2 Electrodes and magnetic fields 9
  - 1.3.3 Imaging 9
  - 1.3.4 Structural and functional imaging 10
- 1.4 Evolution 10
- 1.5 See also 11
- 1.6 References 11
- 1.7 Bibliography 13
- 1.8 External links 14

2 Cerebral cortex   15
- 2.1 Structure 15
  - 2.1.1 Layered structure 15
  - 2.1.2 Areas 17
  - 2.1.3 Development 17
  - 2.1.4 Thickness 18

		2.1.5 Blood supply	19

- 2.2 Function 19
  - 2.2.1 Connections 19
  - 2.2.2 Cortical areas 19
- 2.3 Clinical significance 20
- 2.4 Other animals 21
- 2.5 Additional images 21
- 2.6 See also 21
- 2.7 References 21
- 2.8 External links 23

3 Frontal lobe 24
- 3.1 Structure 24
- 3.2 Function 25
- 3.3 Clinical significance 25
  - 3.3.1 Damage 25
- 3.4 History 25
  - 3.4.1 Psychosurgery 25
  - 3.4.2 Theories of function 26
- 3.5 In other animals 26
- 3.6 Additional images 26
- 3.7 See also 27
- 3.8 References 27
- 3.9 External links 27

4 Cerebral hemisphere 28
- 4.1 Structure 28
  - 4.1.1 Development 28
- 4.2 Function 28
  - 4.2.1 Hemisphere lateralization 28
- 4.3 See also 29
- 4.4 References 29

5 Lobe (anatomy) 30
- 5.1 Examples of lobes 30
  - 5.1.1 Examples of lobules 30
- 5.2 References 30

6 Parietal lobe 31
- 6.1 Structure 31

	6.2	Function	31
	6.3	Clinical significance	32
	6.4	Additional images	33
	6.5	See also	33
	6.6	References	33

7 Temporal lobe — 35

- 7.1 Structure — 35
  - 7.1.1 Medial temporal lobe — 35
- 7.2 Function — 35
  - 7.2.1 Visual memories — 35
  - 7.2.2 Processing sensory input — 35
  - 7.2.3 Language recognition — 35
  - 7.2.4 New memories — 36
- 7.3 Clinical significance — 36
  - 7.3.1 Unilateral temporal lesion — 36
  - 7.3.2 Dominant hemisphere — 36
  - 7.3.3 Non-dominant hemisphere — 36
  - 7.3.4 Bitemporal lesions (additional features) — 36
  - 7.3.5 Damage — 36
  - 7.3.6 Disorders — 36
- 7.4 References — 37
- 7.5 External links — 37

8 Occipital lobe — 38

- 8.1 Structure — 38
- 8.2 Function — 38
- 8.3 Clinical significance — 39
  - 8.3.1 Epilepsy — 39
- 8.4 Additional images — 39
- 8.5 See also — 40
- 8.6 References — 40

9 Limbic lobe — 41

- 9.1 History — 41
- 9.2 Gallery — 41
- 9.3 References — 41
- 9.4 External links — 41

10 Insular cortex — 42

- 10.1 Structure 42
  - 10.1.1 Connections 42
  - 10.1.2 Cytoarchitecture 42
  - 10.1.3 Development 42
- 10.2 Function 43
  - 10.2.1 Interoceptive awareness 43
  - 10.2.2 Motor control 43
  - 10.2.3 Homeostasis 43
  - 10.2.4 Self 43
  - 10.2.5 Social emotions 43
  - 10.2.6 Emotions 43
  - 10.2.7 Salience 44
- 10.3 Clinical significance 44
  - 10.3.1 Progressive non-fluent aphasia 44
  - 10.3.2 Addiction 44
  - 10.3.3 Subjective certainty in ecstatic seizures 45
  - 10.3.4 Other clinical conditions 45
- 10.4 History 45
- 10.5 Additional images 45
- 10.6 See also 46
- 10.7 References 46
- 10.8 External links 49

11 Cerebrum 50

- 11.1 Structure 50
  - 11.1.1 Cerebral cortex 50
  - 11.1.2 Cerebral hemispheres 51
- 11.2 Development 51
- 11.3 Functions 51
  - 11.3.1 Movement 51
  - 11.3.2 Sensory processing 51
  - 11.3.3 Olfaction 51
  - 11.3.4 Language and communication 51
  - 11.3.5 Learning and memory 52
- 11.4 Other animals 52
- 11.5 Additional Images 52
- 11.6 See also 52
- 11.7 Notes 52
- 11.8 References 52

11.9 External links 53

12 White matter   54
    12.1 Structure 54
        12.1.1 Location 54
        12.1.2 Myelinated axon length 54
    12.2 Function 55
    12.3 Clinical significance 55
    12.4 References 56
    12.5 External links 56

13 Grey matter   57
    13.1 Structure 57
    13.2 Function 57
    13.3 Clinical significance 58
        13.3.1 Research 58
    13.4 History 58
        13.4.1 Etymology 58
    13.5 Additional images 59
    13.6 See also 59
    13.7 References 59
    13.8 External links 59

14 Forebrain   60
    14.1 See also 60
    14.2 External links 60

15 Midbrain   61
    15.1 Structure 61
        15.1.1 Corpora quadrigemina 61
        15.1.2 Cerebral peduncle 61
        15.1.3 Anatomical features of cross-sections through the midbrain 62
        15.1.4 Development 62
    15.2 Function 63
    15.3 See also 63
    15.4 References 63

16 Hindbrain   64
    16.1 Myelencephalon 64
    16.2 Metencephalon 64

16.3 Evolution 64

16.4 Additional images 65

16.5 References 65

16.6 External links 65

17 Ventricular system 66

17.1 Structure 66

    17.1.1 Ventricles 66

    17.1.2 Development 66

17.2 Function 67

    17.2.1 Flow of cerebrospinal fluid 67

    17.2.2 Protection of the brain 68

17.3 Clinical significance 68

17.4 Additional images 69

17.5 See also 69

17.6 References 69

17.7 External links 69

18 Medulla oblongata 70

18.1 Anatomy 70

    18.1.1 External surfaces 71

    18.1.2 Blood supply 71

    18.1.3 Development 71

18.2 Function 72

18.3 Clinical significance 72

18.4 Other animals 72

18.5 Additional images 72

18.6 References 72

18.7 External links 73

19 Pons 74

19.1 Structure 74

    19.1.1 Development 74

    19.1.2 Nucleus 74

19.2 Function 74

19.3 Clinical significance 75

19.4 Other animals 75

    19.4.1 Evolution 75

19.5 Additional images 75

19.6 References ... 75
19.7 External links ... 75

20 Brainstem   76
20.1 Structure ... 76
20.1.1 Midbrain ... 76
20.1.2 Pons ... 77
20.1.3 Medulla oblongata ... 77
20.1.4 Ventral view of medulla and pons ... 77
20.1.5 Dorsal view of medulla and pons ... 78
20.1.6 Development ... 79
20.2 Function ... 79
20.3 Clinical significance ... 79
20.4 Additional images ... 80
20.5 See also ... 80
20.6 References ... 80
20.7 External links ... 80

21 Superior colliculus   81
21.1 Structure ... 81
21.1.1 Neural circuit ... 81
21.1.2 Mosaic structure ... 83
21.1.3 Related structures ... 83
21.2 Function ... 83
21.2.1 Eye movements ... 84
21.3 Other animals ... 84
21.3.1 Primates ... 85
21.3.2 Other vertebrates ... 85
21.3.3 Lamprey ... 85
21.3.4 Bats ... 86
21.4 See also ... 86
21.5 Additional images ... 86
21.6 Notes ... 86
21.7 External links ... 87
21.8 References ... 87

22 Thalamus   89
22.1 Anatomy ... 89
22.1.1 Morphology ... 89

		22.1.2 Blood supply	89

- 22.1.2 Blood supply 89
- 22.1.3 Thalamic nuclei 89
- 22.1.4 Connections 90
- 22.2 Function 90
- 22.3 Development 91
  - 22.3.1 Early brain development 91
  - 22.3.2 The formation of the mid-diencephalic organiser (MDO) 91
  - 22.3.3 Maturation and parcellation of the thalamus 91
- 22.4 Clinical significance 92
- 22.5 Additional images 92
- 22.6 See also 92
- 22.7 References 92
- 22.8 External links 94

23 Hypothalamus   95

- 23.1 Structure 95
  - 23.1.1 Nuclei 95
  - 23.1.2 Neural connections 95
  - 23.1.3 Sexual dimorphism 96
  - 23.1.4 Development 96
- 23.2 Function 97
  - 23.2.1 Hormone release 97
  - 23.2.2 Stimulation 98
  - 23.2.3 Control of food intake 99
  - 23.2.4 Fear processing 99
  - 23.2.5 Sexual orientation 100
- 23.3 See also 100
- 23.4 Additional images 101
- 23.5 References 101
- 23.6 Further reading 102
- 23.7 External links 102

24 Basal ganglia   103

- 24.1 Structure 103
  - 24.1.1 Striatum 104
  - 24.1.2 Pallidum 105
  - 24.1.3 Substantia nigra 105
  - 24.1.4 Subthalamic nucleus 105
  - 24.1.5 Circuit connections 105

	24.2 Function	107
	24.2.1 Eye movements	107
	24.2.2 Role in motivation	107
	24.2.3 Neurotransmitters	107
	24.3 Clinical significance	108
	24.4 History	108
	24.4.1 Terminology	109
	24.5 In other animals	109
	24.6 See also	109
	24.7 References	109
	24.8 External links	110
25	**Olfactory bulb**	**112**
	25.1 Structure	112
	25.1.1 Layers	113
	25.2 Function	113
	25.2.1 Lateral inhibition	113
	25.2.2 Accessory Olfactory Bulb	114
	25.2.3 Further processing	115
	25.2.4 Adult neurogenesis	116
	25.3 Clinical significance	116
	25.4 Other animals	116
	25.4.1 Evolution	116
	25.5 See also	117
	25.6 References	117
	25.7 Further reading	118
	25.8 External links	118
26	**Hippocampus**	**119**
	26.1 Name	119
	26.2 Anatomy	120
	26.3 Functions	121
	26.3.1 Role in memory	122
	26.3.2 Role in spatial memory and navigation	123
	26.3.3 Hippocampal formation	124
	26.4 Physiology	124
	26.4.1 Theta rhythm	125
	26.4.2 Sharp waves	125
	26.4.3 Long-term potentiation	125

26.5 Pathology 126

    26.5.1 Aging 126

    26.5.2 Stress 126

    26.5.3 Epilepsy 126

    26.5.4 Schizophrenia 127

    26.5.5 Transient global amnesia 127

26.6 Evolution 127

26.7 See also 128

26.8 Notes 128

26.9 References 130

26.10 Further reading 135

    26.10.1 Journals 135

    26.10.2 Books 135

26.11 External links 135

27 Amygdala   136

27.1 Structure 136

    27.1.1 Hemispheric specializations 137

    27.1.2 Amygdalar development 137

    27.1.3 Gender distinction 137

27.2 Function 138

    27.2.1 Connections 138

    27.2.2 Emotional learning 138

    27.2.3 Memory modulation 139

27.3 Neuropsychological correlates of amygdala activity 139

    27.3.1 Sexual orientation 140

    27.3.2 Social interaction 140

    27.3.3 Aggression 140

    27.3.4 Fear 141

    27.3.5 Alcoholism and binge drinking 141

    27.3.6 Anxiety 141

    27.3.7 Posttraumatic stress disorder 141

    27.3.8 Bipolar disorder 141

    27.3.9 Political orientation 142

27.4 See also 142

27.5 Further reading 142

27.6 References 142

27.7 External links 145

28 Pallium (neuroanatomy) — 146
- 28.1 Structure — 146
- 28.2 Evolution — 147
- 28.3 In humans — 148
- 28.4 In amphibians and other anamniotes — 148
- 28.5 In reptiles and birds — 148
- 28.6 See also — 148

29 Gyrus — 149
- 29.1 Structure — 149
    - 29.1.1 Development — 149
- 29.2 Clinical significance — 149
- 29.3 Notable gyri — 150
- 29.4 References — 150
- 29.5 See also — 150

30 Sulcus (neuroanatomy) — 152
- 30.1 Structure — 152
    - 30.1.1 Importance of expanded surface area — 152
    - 30.1.2 Variation — 153
    - 30.1.3 Development — 153
- 30.2 Notable sulci — 153
- 30.3 Other animals — 153
    - 30.3.1 Macaque — 154
- 30.4 See also — 154
- 30.5 References — 154
- 30.6 External links — 154

31 Development of the nervous system in humans — 155
- 31.1 Embryonic stage — 155
    - 31.1.1 Neurulation — 155
    - 31.1.2 Formation of the spinal cord — 156
    - 31.1.3 Formation of the brain — 156
    - 31.1.4 Evolution of Nervous System — 156
- 31.2 Human brain development — 156
- 31.3 Neuronal migration — 157
    - 31.3.1 Radial migration — 157
    - 31.3.2 Axophilic migration — 157
    - 31.3.3 Tangential migration — 157

         31.3.4 Others 157
   31.4 Neurotrophic factors 157
   31.5 Adult neurogenesis 157
   31.6 Adult neural development 158
   31.7 See also 158
   31.8 References 158

32 **Lateralization of brain function**   **159**
   32.1 Interaction and Role 159
         32.1.1 Theme 160
         32.1.2 Specifics 160
   32.2 Failures of lateralization 161
   32.3 History of research on lateralization 161
         32.3.1 Broca 161
         32.3.2 Wernicke 161
         32.3.3 Advance in imaging technique 161
         32.3.4 Movement and sensation 161
         32.3.5 Split-brain patients 161
         32.3.6 Pop psychology 162
   32.4 Sex differences 162
   32.5 Handedness 162
   32.6 Self-harm 162
   32.7 Lateralized cognitive processes 162
         32.7.1 Lateralization of language processes 163
         32.7.2 Handedness and language 163
         32.7.3 Methods of study 164
   32.8 Pathology 164
         32.8.1 Hemisphere damage 164
         32.8.2 Plasticity 164
         32.8.3 Broca's aphasia 164
         32.8.4 Wernicke's aphasia 164
   32.9 Misapplication of concept 165
   32.10 Advantages of brain lateralization 165
   32.11 Additional images 165
   32.12 See also 165
   32.13 References 165
   32.14 Further reading 168

33 **Corpus callosum**   **169**

| | 33.1 Structure . . . . | 169 |

- 33.1 Structure 169
  - 33.1.1 Variation 170
  - 33.1.2 Sexual dimorphism 170
- 33.2 Other correlations 170
- 33.3 Clinical significance 170
  - 33.3.1 Epilepsy 170
  - 33.3.2 Other disease 171
  - 33.3.3 Brain split procedure 171
- 33.4 History 171
- 33.5 In other animals 172
- 33.6 Additional images 172
- 33.7 References 172
- 33.8 External links 174

34 Brain mapping — 175

- 34.1 Overview 175
- 34.2 History 175
- 34.3 Current Atlas tools 176
- 34.4 See also 176
- 34.5 References 176
- 34.6 Further reading 177
- 34.7 External links 177

35 Outline of brain mapping — 179

- 35.1 Broad scope 179
  - 35.1.1 The neuron doctrine 179
  - 35.1.2 Map, atlas, and database projects 180
- 35.2 Imaging and recording systems 180
  - 35.2.1 General 181
  - 35.2.2 Specific systems 181
  - 35.2.3 Imaging and recording componentry 181
- 35.3 Scientists, academics and researchers 183
- 35.4 Research Institutions 183
- 35.5 Journals 184
- 35.6 See also 184
- 35.7 Notes and references 184
- 35.8 Text and image sources, contributors, and licenses 185
  - 35.8.1 Text 185
  - 35.8.2 Images 193

    35.8.3 Content license 199

Chapter 1

Human brain

The **human brain** is the main organ of the human central nervous system. It is located in the head, protected by the skull. It has the same general structure as the brains of other mammals, but with a more developed cerebral cortex. Large animals such as whales and elephants have larger brains in absolute terms, but when measured using a measure of relative brain size, which compensates for body size, the quotient for the human brain is almost twice as large as that of a bottlenose dolphin, and three times as large as that of a chimpanzee, though the quotient for a treeshrew's brain is larger than that of a human's.[3] Much of the size of the human brain comes from the cerebral cortex, especially the frontal lobes, which are associated with executive functions such as self-control, planning, reasoning, and abstract thought. The area of the cerebral cortex devoted to vision, the visual cortex, is also greatly enlarged in humans compared to other animals.

The human cerebral cortex is a thick layer of neural tissue that covers the two cerebral hemispheres that make up most of the brain. This layer is folded in a way that increases the amount of surface area that can fit into the volume available. The pattern of folds is similar across individuals but shows many small variations. The cortex is divided into four lobes – the frontal lobe, parietal lobe, temporal lobe, and occipital lobe. (Some classification systems also include a limbic lobe and treat the insular cortex as a lobe.) Within each lobe are numerous cortical areas, each associated with a particular function, including vision, motor control, and language. The left and right hemispheres are broadly similar in shape, and most cortical areas are replicated on both sides. Some areas, though, show strong lateralization, particularly areas that are involved in language. In most people, the left hemisphere is dominant for language, with the right hemisphere playing only a minor role. There are other functions, such as visual-spatial ability, for which the right hemisphere is usually dominant.

Despite being protected by the thick bones of the skull, suspended in cerebrospinal fluid, and isolated from the bloodstream by the blood–brain barrier, the human brain is susceptible to damage and disease. The most common forms of physical damage are closed head injuries such as a blow to the head or other trauma, a stroke, or poisoning by a number of chemicals that can act as neurotoxins, such as alcohol. Infection of the brain, though serious, is rare because of the protective blood-to brain and blood-to cerebral fluid barriers. The human brain is also susceptible to degenerative disorders, such as Parkinson's disease, forms of dementia including Alzheimer's disease, (mostly as the result of aging) and multiple sclerosis. A number of psychiatric conditions, such as schizophrenia and clinical depression, are thought to be associated with brain dysfunctions, although the nature of these is not well understood. The brain can also be the site of brain tumors and these can be benign or malignant.

There are some techniques for studying the brain that are used in other animals that are not suitable for use in humans and vice versa; it is easier to obtain individual brain cells taken from other animals, for study. It is also possible to use invasive techniques in other animals such as inserting electrodes into the brain or disabling certains parts of the brain in order to examine the effects on behaviour – techniques that are not possible to be used in humans. However, only humans can respond to complex verbal instructions or be of use in the study of important brain functions such as language and other complex cognitive tasks, but studies from humans and from other animals, can be of mutual help. Medical imaging technologies such as functional neuroimaging and EEG recordings are important techniques in studying the brain. The complete functional understanding of the human brain is an ongoing challenge for neuroscience.

1.1 Structure

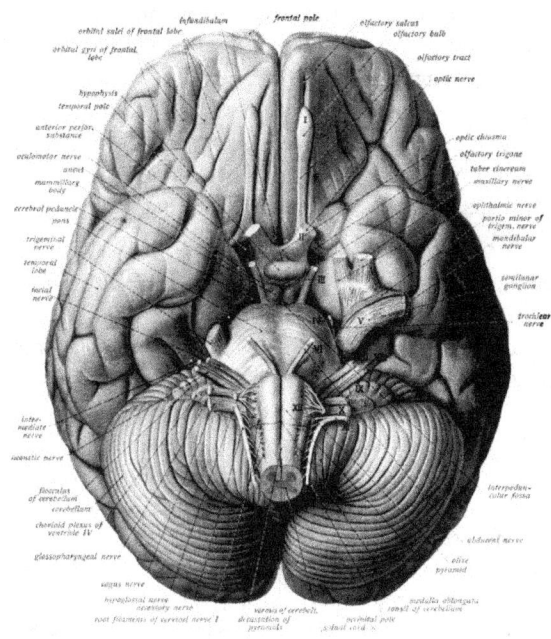

Human brain viewed from below

1.1.1 General features

The adult human brain weighs on average about 1.2–1.4 kg (2.6–3.1 lb), or about 2% of total body weight,[4][5] with a volume of around 1260 cm^3 in men and 1130 cm^3 in women, although there is substantial individual variation.[6] Neurological differences between the sexes have not been shown to correlate in any simple way with IQ or other measures of cognitive performance.[7]

The human brain is composed of neurons, glial cells, and blood vessels. The number of neurons is estimated at roughly 100 billion.[8] The adult human brain is estimated to contain 86±8 billion neurons, with a roughly equal number (85±10 billion) of non-neuronal cells. Out of these, 16 billion (or 19% of all brain neurons) are located in the cerebral cortex (including subcortical white matter), 69 billion (or 80% of all brain neurons) are in the cerebellum.[5][9]

The cerebral hemispheres (the cerebrum) form the largest part of the human brain and are situated above other brain structures. They are covered with a cortical layer (the cerebral cortex) which has a convoluted topography.[10] Underneath the cerebrum lies the brainstem, resembling a stalk on which the cerebrum is attached. At the rear of the brain, beneath the cerebrum and behind the brainstem, is the cerebellum, a structure with a horizontally furrowed surface, the cerebellar cortex, that makes it look different from any other brain area. The same structures are present in other mammals, although they vary considerably in relative size. As a rule, the smaller the cerebrum, the less convoluted the cortex. The cortex of a rat or mouse is almost perfectly smooth. The cortex of a dolphin or whale, on the other hand, is more convoluted than the cortex of a human.

The living brain is very soft, having a gel-like consistency similar to soft tofu. Although referred to as grey matter, the live cortex is pinkish-beige in color and slightly off-white in the interior.

1.1.2 Comparative anatomy

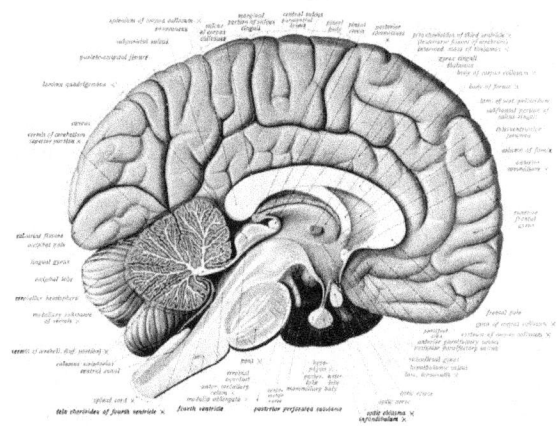

Human brain viewed through a mid-line incision

The human brain has many properties that are common to all vertebrate brains, including a basic division into three parts called the forebrain, midbrain, and hindbrain, with interconnected fluid-filled ventricles, and a set of generic vertebrate brain structures including the medulla oblongata and pons of the brainstem, the cerebellum, optic tectum, thalamus, hypothalamus, basal ganglia, olfactory bulb, and many others.

As a mammalian brain, the human brain has special features that are common to all mammalian brains, most notably a six-layered cerebral cortex and a set of associated structures, including the hippocampus and amygdala. The upper surface of the forebrain of other vertebrates is covered in a layer of neural tissue called the pallium. The pallium is a relatively simple three-layered cell structure. The hippocampus and the amygdala originate from the pallium but in mammals they are much more complex.

As a primate brain, the human brain has a much larger cerebral cortex, in proportion to body size, than most mammals, and a very highly developed visual system. The shape of the brain within the skull is also altered somewhat as a consequence of the upright position in which primates hold their heads.

As a hominid brain, the human brain is substantially enlarged even in comparison to the brain of a typical monkey.

The sequence of evolution from *Australopithecus* (four million years ago) to *Homo sapiens* (modern man) was marked by a steady increase in brain size, particularly in the frontal lobes, which are associated with a variety of high-level cognitive functions.

Humans and other primates have some differences in gene sequence, and genes are differentially expressed in many brain regions. The functional differences between the human brain and the brains of other animals also arise from many gene–environment interactions.[11]

The neuroimmune system of the brain is structurally distinct from the peripheral immune system which protects the rest of the body; in particular, the immune system is composed primarily of hematopoietic cells and anatomical barriers, while the neuroimmune system is composed of glia, mast cells, and various brain barriers (e.g., blood–brain barrier and blood-cerebrospinal fluid barrier).

1.1.3 Cerebral cortex

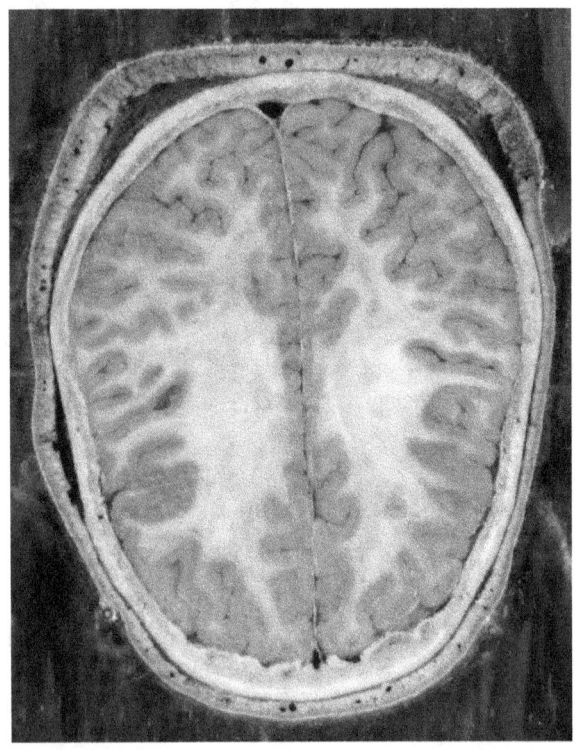

Bisection of the head of an adult female, showing the cerebral cortex, with its extensive folding, and the underlying white matter[12]

The dominant feature of the human brain is corticalization, or wrinkling of the cortex. The cerebral cortex in humans is so large that it overshadows every other part of the brain. A few subcortical structures show alterations reflecting this trend. The cerebellum, for example, has a medial zone connected mainly to subcortical motor areas, and a lateral zone connected primarily to the cortex. In humans the lateral zone takes up a much larger fraction of the cerebellum than in most other mammalian species. Corticalization is reflected in function as well as structure. In a rat, surgical removal of the entire cerebral cortex leaves an animal that is still capable of walking around and interacting with the environment.[13]

In a human, comparable cerebral cortex damage produces a permanent state of coma. The amount of association cortex, relative to the other two categories of sensory and motor, increases dramatically as one goes from simpler mammals, such as the rat and the cat, to more complex ones, such as the chimpanzee and the human.[14] A gene present in the human genome but not in the chimpanzee (ArhGAP11B) seems to play a major role in corticalization and human encephalisation. The cerebral cortex is essentially a sheet of neural tissue, folded in a way that allows a large surface area to fit within the confines of the skull. When unfolded, each cerebral hemisphere has a total surface area of about 1.3 square feet $(0.12\ m^2)$.[15] Each cortical ridge is called a gyrus, and each groove or fissure separating one gyrus from another is called a sulcus.

1.1.4 Cortical divisions

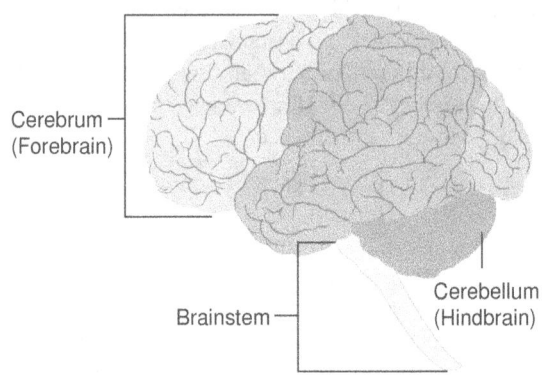

Regions of the lateral surface of the brain, and particularly the lobes of the forebrain:
Beige – frontal lobe
Blue – parietal lobe
Green – occipital lobe
Pink – temporal lobe

The cerebral cortex is nearly symmetrical with left and right hemispheres that are approximate mirror images of each other. Each hemisphere is conventionally divided into four "lobes", the frontal lobe, parietal lobe, occipital lobe, and temporal lobe. With one exception, this division into lobes does not derive from the structure of the cortex, though the lobes are named after the bones of the skull that overlie them, the frontal bone, parietal bone, temporal bone, and occipital bone. The borders between lobes lie beneath the sutures that link the skull bones together. The exception is the border between the frontal and parietal lobes, which lies behind the corresponding suture; instead it follows the anatomical boundary of the central sulcus, a deep fold in the brain's structure where the primary somatosensory cortex and primary motor cortex meet.

Because of the arbitrary way most of the borders between lobes are demarcated, they have little functional significance. With the exception of the occipital lobe, a small area that is entirely dedicated to vision, each of the lobes contains a variety of brain areas that have minimal functional relationship. The parietal lobe, for example, contains areas involved in somatosensation, hearing, language, attention, and spatial cognition. In spite of this heterogeneity, the division into lobes is convenient for reference. The main functions of the frontal lobe are to control attention, abstract thinking, behavior, problem solving tasks, and physical reactions and personality. The occipital lobe is the smallest lobe; its main functions are visual reception, visual-spatial processing, movement, and color recognition. The temporal lobe controls auditory and visual memories, language, and some hearing and speech.

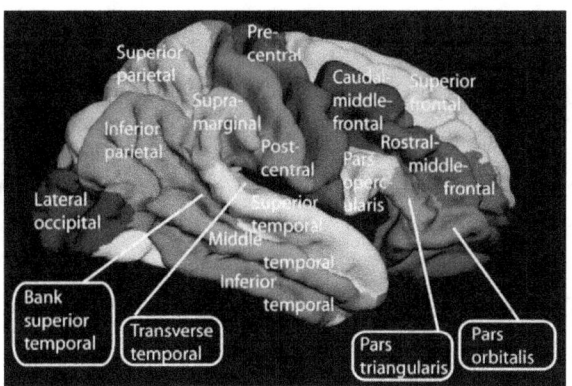

Lateral surface of the cerebral cortex

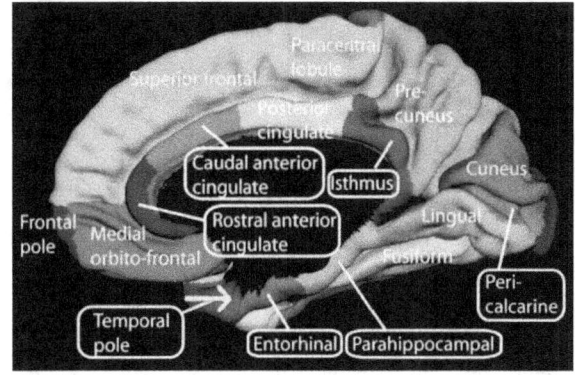

Medial surface of the cerebral cortex

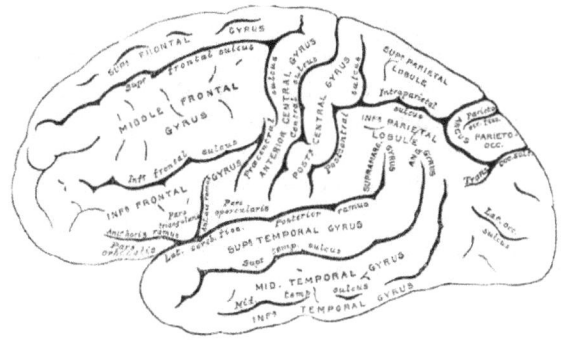

Major gyri and sulci on the lateral surface of the cortex

Although there are enough variations in the shape and placement of gyri and sulci (cortical folds) to make every brain unique, most human brains show sufficiently consistent patterns of folding that allow them to be named. Many of the gyri and sulci are named according to the location on the lobes or other major folds on the cortex. These include:

- *Superior, Middle, Inferior frontal gyrus*: in reference to the frontal lobe

- *Medial longitudinal fissure*, which separates the left and right cerebral hemispheres

- *Precentral and Postcentral sulcus*: in reference to the central sulcus, which separates the frontal lobe from the parietal lobe

- *Lateral sulcus*, which divides the frontal lobe and parietal lobe above from the temporal lobe below

- *Parieto-occipital sulcus*, which separates the parietal lobes from the occipital lobes, is seen to some small extent on the lateral surface of the hemisphere, but mainly on the medial surface.

- *Trans-occipital sulcus*: in reference to the occipital lobe

1.1.5 Functional divisions

Functions of the cortex are divided it into three categories of regions: One consists of the primary sensory areas, which receive signals from the sensory nerves and tracts by way of relay nuclei in the thalamus. Primary sensory areas include the visual area of the occipital lobe, the auditory area in parts of the temporal lobe and insular cortex, and the somatosensory cortex in the parietal lobe. A second category is the primary motor cortex, which sends axons down to motor neurons in the brainstem and spinal cord.

1.1. STRUCTURE

This area occupies the rear portion of the frontal lobe, directly in front of the somatosensory area. The third category consists of the remaining parts of the cortex, which are called the association areas. These areas receive input from the sensory areas and lower parts of the brain and are involved in the complex processes of perception, thought, and decision-making.[16]

Cytoarchitecture

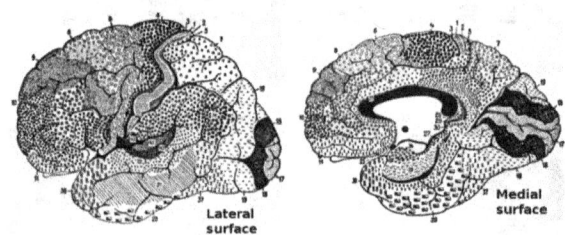

Brodmann's classification of areas of the cortex

Different parts of the cerebral cortex are involved in different cognitive and behavioral functions. The differences show up in a number of ways: the effects of localized brain damage, regional activity patterns exposed when the brain is examined using functional imaging techniques, connectivity with subcortical areas, and regional differences in the cellular architecture of the cortex. Neuroscientists describe most of the cortex—the part they call the neocortex—as having six layers, but not all layers are apparent in all areas, and even when a layer is present, its thickness and cellular organization may vary. Scientists have constructed maps of cortical areas on the basis of variations in the appearance of the layers as seen with a microscope. One of the most widely used schemes came from Korbinian Brodmann, who split the cortex into 51 different areas and assigned each a number (many of these Brodmann areas have since been subdivided). For example, Brodmann area 1 is the primary somatosensory cortex, Brodmann area 17 is the primary visual cortex, and Brodmann area 25 is the anterior cingulate cortex.[17]

Topography

Many of the brain areas Brodmann defined have their own complex internal structures. In a number of cases, brain areas are organized into topographic maps, where adjoining bits of the cortex correspond to adjoining parts of the body, or of some more abstract entity. A simple example of this type of correspondence is the primary motor cortex, a strip of tissue running along the anterior edge of the central sulcus, shown in the image to the right. Motor areas innervating each part of the body arise from a distinct

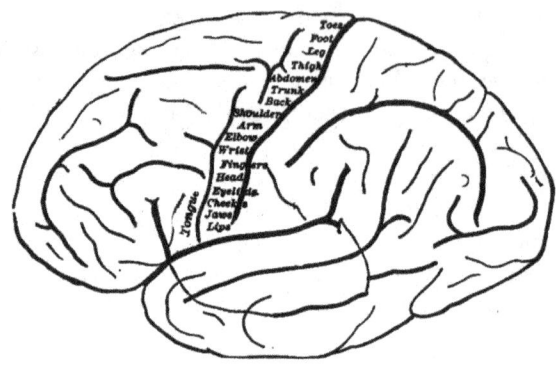

Topography of the primary motor cortex, showing which body part is controlled by each zone

zone, with neighboring body parts represented by neighboring zones. Electrical stimulation of the cortex at any point causes a muscle-contraction in the represented body part. This "somatotopic" representation is not evenly distributed, however. The head, for example, is represented by a region about three times as large as the zone for the entire back and trunk. The size of any zone correlates to the precision of motor control and sensory discrimination possible. The areas for the lips, fingers, and tongue are particularly large, considering the proportional size of their represented body parts.

In visual areas, the maps are retinotopic—that is, they reflect the topography of the retina, the layer of light-activated neurons lining the back of the eye. In this case too the representation is uneven: the fovea—the area at the center of the visual field—is greatly overrepresented compared to the periphery. The visual circuitry in the human cerebral cortex contains several dozen distinct retinotopic maps, each devoted to analyzing the visual input stream in a particular way. The primary visual cortex (Brodmann area 17), which is the main recipient of direct input from the visual part of the thalamus, contains many neurons that are most easily activated by edges with a particular orientation moving across a particular point in the visual field. Visual areas farther downstream extract features such as color, motion, and shape.

In auditory areas, the primary map is tonotopic. Sounds are parsed according to frequency (i.e., high pitch vs. low pitch) by subcortical auditory areas, and this parsing is reflected by the primary auditory zone of the cortex. As with the visual system, there are a number of tonotopic cortical maps, each devoted to analyzing sound in a particular way.

Within a topographic map there can sometimes be finer levels of spatial structure. In the primary visual cortex, for example, where the main organization is retinotopic and the main responses are to moving edges, cells that respond to different edge-orientations are spatially segregated from one another.

1.1.6 Development

Main article: Neural development in humans
Further information: Human brain development timeline

During the first three weeks of gestation, the human embryo's ectoderm forms a thickened strip called the neural plate. The neural plate then folds and closes to form the neural tube. This tube flexes as it grows, forming the crescent-shaped cerebral hemispheres at the head, and the cerebellum and pons towards the tail.

1.2 Function

1.2.1 Cognition

Main articles: Cognition and Mind

Understanding the mind–body problem – the relationship between the brain and the mind – is a significant challenge both philosophically and scientifically. This is because of the difficulty reconciling how mental activities, such as thoughts and emotions, can be implemented by physical structures such as neurons and synapses, or by any other type of physical mechanism. This difficulty was expressed by Gottfried Leibniz in an analogy known as *Leibniz's Mill*:

> One is obliged to admit that perception and what depends upon it is inexplicable on mechanical principles, that is, by figures and motions. In imagining that there is a machine whose construction would enable it to think, to sense, and to have perception, one could conceive it enlarged while retaining the same proportions, so that one could enter into it, just like into a windmill. Supposing this, one should, when visiting within it, find only parts pushing one another, and never anything by which to explain a perception.
>
> — Leibniz, Monadology[18]

Doubt about the possibility of a mechanistic explanation of thought drove René Descartes, and most of humankind along with him, to dualism: the belief that the mind exists independently of the brain.[19] There has always, however, been a strong argument in the opposite direction. There is clear empirical evidence that physical manipulations of, or injuries to, the brain (for example by drugs or by lesions, respectively) can affect the mind in potent and intimate ways.[20] For example, a person suffering from Alzheimer's disease – a condition that causes physical damage to the brain – also experiences a compromised mind. Similarly, someone who has taken a psychedelic drug may temporarily lose their sense of personal identity (ego death) or experience profound changes to their perception and thought processes. Likewise, a patient with epilepsy who undergoes cortical stimulation mapping with electrical brain stimulation would also, upon stimulation of his or her brain, experience various complex feelings, hallucinations, memory flashbacks, and other complex cognitive, emotional, or behavioral phenomena.[21] Following this line of thinking, a large body of empirical evidence for a close relationship between brain activity and mental activity has led most neuroscientists and contemporary philosophers to be materialists, believing that mental phenomena are ultimately the result of, or reducible to, physical phenomena.[22]

1.2.2 Lateralization

Main article: Lateralization of brain function
Each hemisphere of the brain interacts primarily with one

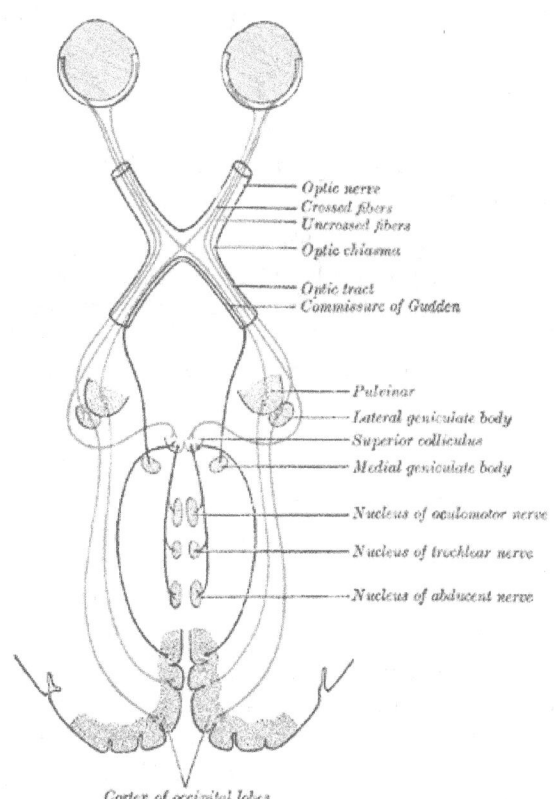

Routing of neural signals from the two eyes to the brain

half of the body, but for reasons that are unclear, the connections are crossed: the left side of the brain interacts with the right side of the body, and vice versa. Motor connections from the brain to the spinal cord, and sensory connections from the spinal cord to the brain, both cross the midline at the level of the brainstem. Visual input follows a more complex rule: the optic nerves from the two eyes come together at a point called the optic chiasm, and half of the fibers from each nerve split off to join the other. The result is that connections from the left half of the retina, in both eyes, go to the left side of the brain, whereas connections from the right half of the retina go to the right side of the brain. Because each half of the retina receives light coming from the opposite half of the visual field, the functional consequence is that visual input from the left side of the world goes to the right side of the brain, and vice versa. Thus, the right side of the brain receives somatosensory input from the left side of the body, and visual input from the left side of the visual field—an arrangement that presumably is helpful for visuomotor coordination.

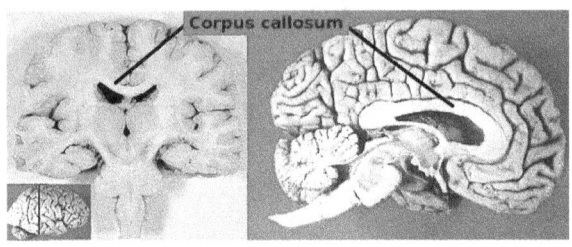

The corpus callosum, a nerve bundle connecting the two cerebral hemispheres, with the lateral ventricles directly below

The two cerebral hemispheres are connected by a very large nerve bundle (the largest white matter structure in the brain) called the corpus callosum, which crosses the midline above the level of the thalamus.[23] There are also two much smaller connections, the anterior commissure and hippocampal commissure, as well as many subcortical connections that cross the midline. The corpus callosum is the main avenue of communication between the two hemispheres, though. It connects each point on the cortex to the mirror-image point in the opposite hemisphere, and also connects to functionally related points in different cortical areas.

In most respects, the left and right sides of the brain are symmetrical in terms of function. For example, the counterpart of the left-hemisphere motor area controlling the right hand is the right-hemisphere area controlling the left hand. There are, however, several very important exceptions, involving language and spatial cognition. In most people, the left hemisphere is "dominant" for language: a stroke that damages a key language area in the left hemisphere can leave the victim unable to speak or understand, whereas equivalent damage to the right hemisphere would cause only minor impairment to language skills.

A substantial part of current understanding of the interactions between the two hemispheres has come from the study of "split-brain patients"—people who underwent surgical transection of the corpus callosum in an attempt to reduce the severity of epileptic seizures. These patients do not show unusual behavior that is immediately obvious, but in some cases can behave almost like two different people in the same body, with the right hand taking an action and then the left hand undoing it. Most of these patients, when briefly shown a picture on the right side of the point of visual fixation, are able to describe it verbally, but when the picture is shown on the left, are unable to describe it, but may be able to give an indication with the left hand of the nature of the object shown.

1.2.3 Language

Main article: Language processing in the brain

The study of how language is represented, processed,

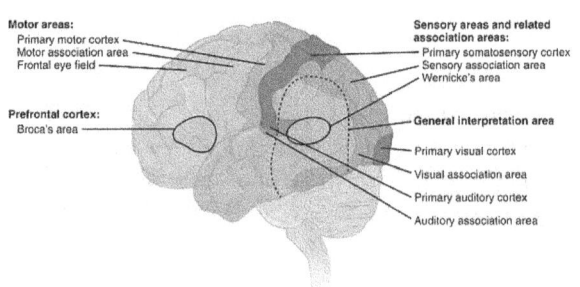

Locations of two brain areas historically associated with language processing, Broca's area and Wernicke's area, and associated regions of sound processing and speech.
(Associated cortical regions involved in vision, touch sensation, and non-speech movement are also shown.)

and acquired by the brain is neurolinguistics, which is a large multidisciplinary field drawing from cognitive neuroscience, cognitive linguistics, and psycholinguistics. This field originated from the 19th-century discovery that damage to different parts of the brain appeared to cause different symptoms: physicians noticed that individuals with damage to a portion of the left inferior frontal gyrus now known as Broca's area had difficulty in producing language (aphasia of speech), whereas those with damage to a region in the left superior temporal gyrus, now known as Wernicke's area, had difficulty in understanding it.[24]

Since then, there has been substantial debate over what linguistic processes these and other parts of the brain subserve,[25] and although Broca's and Wernicke's areas have traditionally been associated with language functions, they may also be involved in certain non-speech functions. There is also debate over whether or not there even is a strong one-to-one relationship between brain regions and language functions that emerges during neocortical development.[26] More recently, research on language has increasingly used more modern methods including electrophysiology and functional neuroimaging, to examine how language processing occurs.

1.2.4 Metabolism

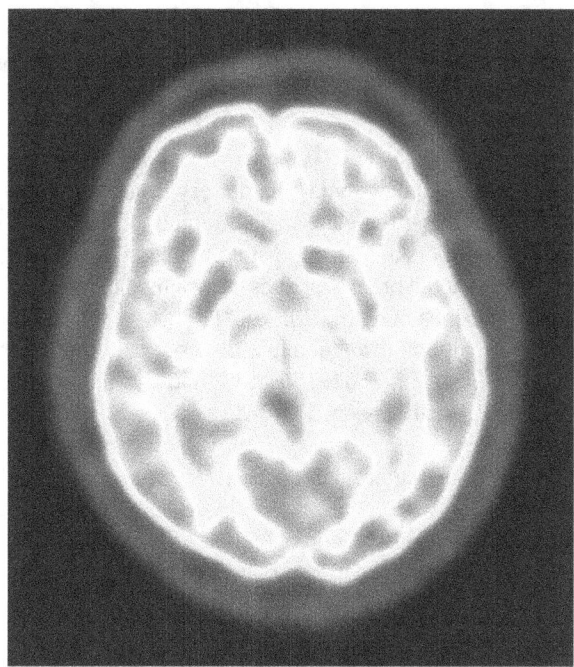

PET image of the human brain showing energy consumption

The brain consumes up to twenty percent of the energy used by the human body, more than any other organ.[27] Brain metabolism normally relies upon blood glucose as an energy source, but during times of low glucose (such as fasting, exercise, or limited carbohydrate intake), the brain will use ketone bodies for fuel with a smaller need for glucose. The brain can also utilize lactate during exercise.[28] Long-chain fatty acids cannot cross the blood–brain barrier, but the liver can break these down to produce ketones. However, the medium-chain fatty acids octanoic and heptanoic acids can cross the barrier and be used by the brain.[29][30][31] The brain stores glucose in the form of glycogen, albeit in significantly smaller amounts than that found in the liver or skeletal muscle.[32]

Although the human brain represents only 2% of the body weight, it receives 15% of the cardiac output, 20% of total body oxygen consumption, and 25% of total body glucose utilization.[33] The need to limit body weight has led to selection for a reduction of brain size in some species, such as bats, who need to be able to fly.[34] The brain mostly uses glucose for energy, and deprivation of glucose, as can happen in hypoglycemia, can result in loss of consciousness. The energy consumption of the brain does not vary greatly over time, but active regions of the cortex consume somewhat more energy than inactive regions: this fact forms the basis for the functional brain imaging methods PET and fMRI.[35] These are nuclear medicine, functional imaging techniques which produce a three-dimensional image of metabolic activity.

1.3 Clinical significance

Clinically, death is defined as an absence of brain activity as measured by EEG. Injuries to the brain tend to affect large areas of the organ, sometimes causing major deficits in intelligence, memory, personality, and movement. Head trauma caused, for example, by vehicular or industrial accidents, is a leading cause of death in youth and middle age. In many cases, more damage is caused by resultant edema than by the impact itself. Stroke, caused by the blockage or rupturing of blood vessels in the brain, is another major cause of death from brain damage.

Other problems in the brain can be more accurately classified as diseases. Neurodegenerative diseases, such as Alzheimer's disease, Parkinson's disease, Huntington's disease and motor neuron diseases are caused by the gradual death of individual neurons, leading to diminution in movement control, memory, and cognition. These are mostly the result of the aging brain which has shown enlarged ventricles and decreased cortical regions on scanning.[36] There are five motor neuron diseases, the most common of which is amyotrophic lateral sclerosis (ALS).

Some infectious diseases affecting the brain are caused by viruses and bacteria. Infection of the meninges, the membranes that cover the brain, can lead to meningitis. Bovine spongiform encephalopathy (also known as "mad cow disease") is deadly in cattle and humans and is linked to prions. Kuru is a similar prion-borne degenerative brain disease affecting humans, (endemic only to Papua New Guinea tribes). Both are linked to the ingestion of neural tissue, and may explain the tendency in human and some non-human species to avoid cannibalism. Viral or bacterial causes have been reported in multiple sclerosis, and are established causes of encephalopathy, and encephalomyelitis.

Brain tumors both benign and malignant can form. These can either originate in the cerebral tissue or in the meninges. The most common are those growths that affect the glial cells known as gliomas. (This term has been extended to include all primary brain tumors.)[37] Secondary cancers can form in the brain as a result of brain metastasis.

Mental disorders, such as clinical depression,

1.3. CLINICAL SIGNIFICANCE

schizophrenia, bipolar disorder and post-traumatic stress disorder may involve particular patterns of neuropsychological functioning related to various aspects of mental and somatic function. These disorders may be treated by psychotherapy, psychiatric medication, social intervention and personal recovery work or cognitive behavioural therapy; the underlying issues and associated prognoses vary significantly between individuals.

Many brain disorders are congenital, occurring during development. Tay-Sachs disease, fragile X syndrome, and Down syndrome are all linked to genetic and chromosomal errors. Many other syndromes, such as the intrinsic circadian rhythm disorders, are suspected to be congenital as well. Normal development of the brain can be altered by genetic factors, drug use, nutritional deficiencies, and infectious diseases during pregnancy.

Epileptic, and non-epileptic seizures can cause cognitive impairment when the seizures become widespread, occur repeatedly in the same brain area or last for too long. Seizures can be assessed using EEG and various medical imaging techniques. They can sometimes be treated using anticonvulsant drugs and certain neurosurgical procedures and auxiliary treatments may also be used.

1.3.1 Effects of brain damage

A key source of information about the function of brain regions is the effects of damage to them.[38] In humans, strokes have long provided a "natural laboratory" for studying the effects of brain damage. Most strokes result from a blood clot lodging in the brain and blocking the local blood supply, causing damage or destruction of nearby brain tissue: the range of possible blockages is very wide, leading to a great diversity of stroke symptoms. Analysis of strokes is limited by the fact that damage often crosses into multiple regions of the brain, not along clear-cut borders, making it difficult to draw firm conclusions.

Transient ischemic attacks (TIAs) are mini-strokes that can cause sudden dimming or loss of vision (including amaurosis fugax), speech impairment ranging from slurring to dysarthria or aphasia, and mental confusion. But unlike a stroke, the symptoms of a TIA can resolve within a few minutes or 24 hours. Brain injury may still occur in a TIA lasting only a few minutes.[39][40] A silent stroke or silent cerebral infarct (SCI) differs from a TIA in that there are no immediately observable symptoms. An SCI may still cause long lasting neurological dysfunction affecting such areas as mood, personality, and cognition. An SCI often occurs before or after a TIA or major stroke.[41]

1.3.2 Electrodes and magnetic fields

By placing electrodes on the scalp, it is possible to record the summed electrical activity of the cortex using a methodology known as electroencephalography (EEG).[42] EEG records average neuronal activity from the cerebral cortex and can detect changes in activity over large areas but with low sensitivity for sub-cortical activity. EEG recordings are sensitive enough to detect tiny electrical impulses lasting only a few milliseconds. Most EEG devices have good temporal resolution, but low spatial resolution.

Electrodes can also be placed directly on the surface of the brain (usually during surgical procedures that require removal of part of the skull). This technique, called electrocorticography (ECoG), offers finer spatial resolution than electroencephalography, but is very invasive. In addition to measuring the electric field directly via electrodes placed over the skull, it is possible to measure the magnetic field that the brain generates using a method known as magnetoencephalography (MEG).[43] This technique also has good temporal resolution like EEG but with much better spatial resolution. The greatest disadvantage of MEG is that, because the magnetic fields generated by neural activity are very subtle, the neural activity must be relatively close to the surface of the brain to detect its magnetic field. MEGs can only detect the magnetic signatures of neurons located in the depths of cortical folds (*sulci*) that have dendrites oriented in a way that produces a field.

1.3.3 Imaging

Further information: Brain mapping and Outline of brain mapping

Neuroscientists, along with researchers from allied disci-

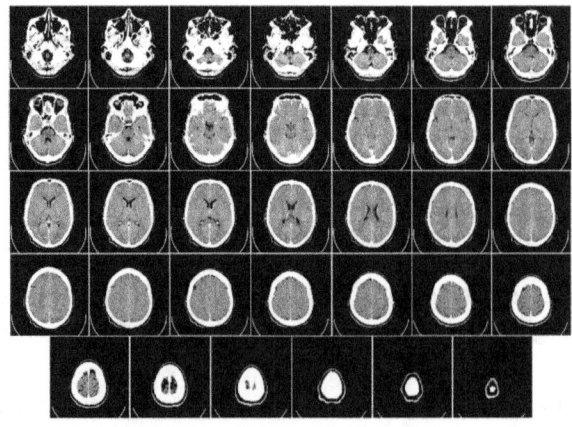

Computed tomography of human brain, from base of the skull to top, taken with intravenous contrast medium

plines, study how the human brain works. Such research

has expanded considerably in recent decades. The "Decade of the Brain", an initiative of the United States Government in the 1990s, is considered to have marked much of this increase in research,[44] and was followed in 2013 by the BRAIN Initiative.

Information about the structure and function of the human brain comes from a variety of experimental methods. Most information about the cellular components of the brain and how they work comes from studies of animal subjects, using a variety of techniques. Some techniques, however, are used mainly on humans.

1.3.4 Structural and functional imaging

Main article: Neuroimaging

There are several methods for detecting brain activity

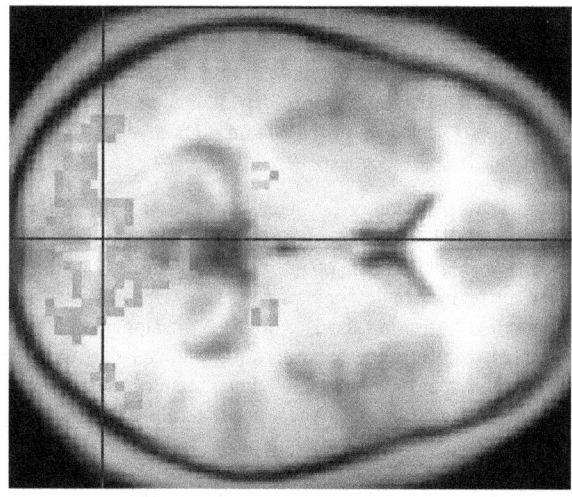

A scan of the brain using fMRI

changes using three-dimensional imaging of local changes in blood flow. The older methods are SPECT and PET, which depend on injection of radioactive tracers into the bloodstream. A newer method, functional magnetic resonance imaging (fMRI), has considerably better spatial resolution and involves no radioactivity.[45] Using the most powerful magnets currently available, fMRI can localize brain activity changes to regions as small as one cubic millimeter. The downside is that the temporal resolution is poor: when brain activity increases, the blood flow response is delayed by 1–5 seconds and lasts for at least 10 seconds. Thus, fMRI is a very useful tool for learning which brain regions are involved in a given behavior, but gives little information about the temporal dynamics of their responses. A major advantage for fMRI is that, because it is non-invasive, it can readily be used on human subjects. Another new non-invasive functional imaging method is functional near-infrared spectroscopy.

1.4 Evolution

See also: Brain size, Human evolution, and Encephalization
In the course of evolution of the Homininae, the human

A reconstruction of Homo habilis

brain has grown in volume from about 600 cm^3 in *Homo habilis* to about 1500 cm^3 in *Homo sapiens neanderthalensis*. Subsequently, there has been a shrinking over the past 28,000 years. The male brain has decreased from 1,500 cm^3 to 1,350 cm^3 while the female brain has shrunk by the same relative proportion.[46] For comparison, *Homo erectus*, a relative of humans, had a brain size of 1,100 cm^3. However, the little *Homo floresiensis*, with a brain size of 380 cm^3, a third of that of their proposed ancestor *H. erectus*, used fire, hunted, and made stone tools at least as sophisticated as those of *H. erectus*.[47] In spite of significant changes in social capacity, there has been very little change in brain size from Neanderthals to the present day.[48] The notion "As large as you need and as small as you can" has been used to summarize the opposite evolutionary constraints on human brain size.[49][50] Changes in the size of the human brain during evolution have been reflected in changes in the ASPM and microcephalin genes.[51]

Studies tend to indicate small to moderate correlations (averaging around 0.3 to 0.4) between brain volume and IQ.[52] The most consistent associations are observed within the frontal, temporal, and parietal lobes, the hippocampi, and the cerebellum, but these only account for a relatively small amount of variance in IQ, which itself has only a partial relationship to general intelligence and real-world performance.[53][54] One study indicated that in humans, fertility and intelligence tend to be negatively correlated— that is to say, the more intelligent, as measured by IQ, exhibit a lower total fertility rate than the less intelligent. According to the model, the present rate of decline is predicted to be 1.34 IQ points per decade.[55]

1.5 See also

- Cephalic disorder
- Cephalization
- Common misconceptions about the brain
- Enchanted loom
- Functional specialization (brain)
- History of neuroscience
- Lateralization of brain function
- List of neuroscience databases
- List of regions in the human brain
- Neural development in humans
- Neuroanatomy
- Neuroanthropology
- Outline of the human brain
- Philosophy of mind
- Ten percent of the brain myth

1.6 References

[1] "*Cerebrum* Etymology". *dictionary.com*. Retrieved 24 October 2015.

[2] "*Encephalo-* Etymology". *Online Etymology Dictionary*. Retrieved 24 October 2015.

[3] "Tupaia belangeri". The Genome Institute, Washington University. Retrieved January 2016. Check date values in: |access-date= (help)

[4] Parent, A; Carpenter MB (1995). "Ch. 1". *Carpenter's Human Neuroanatomy*. Williams & Wilkins. ISBN 978-0-683-06752-1.

[5] Kristin L. Bigos, Ahmad R. Hariri, Daniel R. Weinberger (2015). *Neuroimaging Genetics: Principles and Practices*. Oxford University Press. p. 157. ISBN 0199920222. Retrieved January 2, 2016.

[6] Cosgrove, KP; Mazure CM; Staley JK (2007). "Evolving knowledge of sex differences in brain structure, function, and chemistry". *Biol Psychiat.* **62** (8): 847–55. doi:10.1016/j.biopsych.2007.03.001. PMC 2711771. PMID 17544382.

[7] Gur RC, Turetsky BI, Matsui M, Yan M, Bilker W, Hughett P, Gur RE (1999). "Sex differences in brain gray and white matter in healthy young adults: correlations with cognitive performance". *The Journal of Neuroscience.* **19** (10): 4065–72. PMID 10234034.

[8] "there was, to our knowledge, no actual, direct estimate of numbers of cells or of neurons in the entire human brain to be cited until 2009. A reasonable approximation was provided by Williams and Herrup (1988), from the compilation of partial numbers in the literature. These authors estimated the number of neurons in the human brain at about 85 billion [...] With more recent estimates of 21–26 billion neurons in the cerebral cortex (Pelvig et al., 2008) and 101 billion neurons in the cerebellum (Andersen et al., 1992), however, the total number of neurons in the human brain would increase to over 120 billion neurons." Herculano-Houzel, Suzana. "The human brain in numbers: a linearly scaled-up primate brain". *Front. Hum. Neurosci.* **3**. doi:10.3389/neuro.09.031.2009.

[9] "despite the widespread quotes that the human brain contains 100 billion neurons and ten times more glial cells, the absolute number of neurons and glial cells in the human brain remains unknown. Here we determine these numbers by using the isotropic fractionator and compare them with the expected values for a human-sized primate. We find that the adult male human brain contains on average 86.1 ± 8.1 billion NeuN-positive cells ("neurons") and 84.6 ± 9.8 billion NeuN-negative ("nonneuronal") cells." Azevedo, F.A.C., Carvalho, L.R.B., Grinberg, L.T., Farfel, J.M., Ferretti, R.E.L., Leite, R.E.P., Filho, W.J., Lent, R., Herculano-Houzel, S. (2009). "Equal numbers of neuronal and nonneuronal cells make the human brain an isometrically scaled-up primate brain.". *Journal of Comparative Neurology.* **513** (5): 532–541. doi:10.1002/cne.21974. PMID 19226510.

[10] Kandel, ER; Schwartz JH; Jessel TM (2000). *Principles of Neural Science*. McGraw-Hill Professional. p. 324. ISBN 978-0-8385-7701-1.

[11] Jones R (2012). "Neurogenetics: What makes a human brain?". *Nature Reviews Neuroscience.* **13** (10): 655. doi:10.1038/nrn3355. PMID 22992645.

[12] From the National Library of Medicine's Visible Human Project. In this project, two human cadavers (from a man and a woman) were frozen and then sliced into thin sections, which were individually photographed and digitized. The slice here is taken from a small distance below the top of the brain, and shows the cerebral cortex (the convoluted cellular layer on the outside) and the underlying white matter, which consists of myelinated fiber tracts traveling to and from the cerebral cortex.

[13] Vanderwolf et al., 1978

[14] Gray *Psychology* 2002

[15] Toro et al., 2008

[16] Principles of Anatomy and Physiology 12th Edition - Tortora,Page 519.

[17] Principles of Anatomy and Physiology 12th Edition - Tortora,Page 519-fig. (14.15)

[18] Rescher N (1992). *G. W. Leibniz's Monadology*. Psychology Press. p. 83. ISBN 978-0-415-07284-7.

[19] Hart, WD (1996). Guttenplan S, ed. *A Companion to the Philosophy of Mind*. Blackwell. pp. 265–267.

[20] Churchland, PS (1989). "Ch. 8". *Neurophilosophy*. MIT Press. ISBN 978-0-262-53085-9.

[21] Aslihan Selimbeyoglu, Josef Parvizi. "Electrical stimulation of the human brain: perceptual and behavioral phenomena reported in the old and new literature" (2010). Frontiers in Human Neuroscience.

[22] James H. Schwartz. *Appendix D: Consciousness and the Neurobiology of the Twenty-First Century*. In Kandel, ER; Schwartz JH; Jessell TM. (2000). *Principles of Neural Science, 4th Edition*.

[23] Eric Mooshagian. "Anatomy of the Corpus Callosum Reveals Its Function". Jneurosci.org. Retrieved 2014-03-05.

[24] Damasio, H. (2001). Neural basis of language disorders. In R. Chapey (Ed.), Language intervention strategies in adult aphasia. 4th edition (pp. 18–36). Baltimore: Williams & Wilkins.

[25] Regarding the function of Broca's region, see for example the following:

- Grodzinsky, Y. 2000. The neurology of syntax: language use without Broca's area. Behavioral and Brain Sciences, 23.1, pp. 1–71.
- Hagoort, P. 2013. MUC (Memory, Unification, Control) and beyond. Frontiers in Language Sciences.

[26] Caplan, Waters; Dede, Michaud; Reddy (2007). "A study of syntactic processing in aphasia I: Behavioral (psycholinguistic) aspects". *Brain and Language*. **101** (2): 103–150. doi:10.1016/j.bandl.2006.06.225.

[27] Swaminathan, Nikhil (29 April 2008). "Why Does the Brain Need So Much Power?". *Scientific American*. Scientific American, a Division of Nature America, Inc. Retrieved 19 November 2010.

[28] Quistorff, Bjørn; Secher, Niels; Van Lieshout, Johanne (July 24, 2008). "Lactate fuels the human brain during exercise". *The FASEB Journal*. **22** (10): 3443–3449. doi:10.1096/fj.08-106104. Retrieved May 9, 2011.

[29] "Energy Contribution of Octanoate to Intact Rat Brain Metabolism Measured by 13C Nuclear Magnetic Resonance Spectroscopy". Jneurosci.org. 2003-07-02. Retrieved 2014-03-05.

[30] Journal of Cerebral Blood Flow & Metabolism (2012-10-17). "Journal of Cerebral Blood Flow & Metabolism - Abstract of article: Heptanoate as a neural fuel: energetic and neurotransmitter precursors in normal and glucose transporter I-deficient (G1D) brain". Nature.com. Retrieved 2014-03-05.

[31] MedBio.info > Integration of Metabolism Professor em. Robert S. Horn, Oslo, Norway. Retrieved on May 1, 2010.

[32] Obel, LF; Müller, MS; Walls, AB; Sickmann, HM; Bak, LK; Waagepetersen, HS; Schousboe, A (2012). "Brain glycogen-new perspectives on its metabolic function and regulation at the subcellular level.". *Frontiers in neuroenergetics*. **4**: 3. doi:10.3389/fnene.2012.00003. PMC 3291878. PMID 22403540.

[33] Clark, DD; Sokoloff L (1999). Siegel GJ, Agranoff BW, Albers RW, Fisher SK, Uhler MD, eds. *Basic Neurochemistry: Molecular, Cellular and Medical Aspects*. Philadelphia: Lippincott. pp. 637–670. ISBN 978-0-397-51820-3.

[34] Safi, K; Seid, MA; Dechmann, DK (2005). "Bigger is not always better: when brains get smaller". *Biol Lett*. **1** (3): 283–286. doi:10.1098/rsbl.2005.0333. PMC 1617168. PMID 17148188.

[35] Raichle, M; Gusnard, DA (2002). "Appraising the brain's energy budget". *Proc. Natl. Acad. Sci. U.S.A.* **99** (16): 10237–10239. doi:10.1073/pnas.172399499. PMC 124895. PMID 12149485.

[36] Craik, F.; Salthouse, T. (2000). *The Handbook of Aging and Cognition* (2nd ed.). Mahwah, NJ: Lawrence Erlbaum. ISBN 0-8058-2966-0. OCLC 44957002.

[37] Dorland's (2012). *Dorland's Illustrated Medical Dictionary* (32nd ed.). Elsevier. p. 784. ISBN 978-1-4160-6257-8.

[38] Andrews, *Neuropsychology*

[39] Ferro, J. M. Rodrigues; et al. (1996). "Diagnosis of transient ischemic attack by the nonneurologist. A validation study". *Stroke*. **27** (12): 2225–2229. doi:10.1161/01.STR.27.12.2225. PMID 8969785.

[40] Easton, J. D. Albers; et al. (2009). "Definition and evaluation of transient ischemic attack: a scientific statement for healthcare professionals from the American Heart Association/American Stroke Association Stroke Council; Council on Cardiovascular Surgery and Anesthesia; Council on Cardiovascular Radiology and Intervention; Council on Cardiovascular Nursing; and the Interdisciplinary Council on Peripheral Vascular Disease. The American Academy of Neurology affirms the value of this statement as an educational tool for neurologists". *Stroke*. **40** (6): 2276–2293. doi:10.1161/STROKEAHA.108.192218. PMID 19423857.

[41] Coutts, S. B.; Simon, J. E.; et al. (2005). "Silent ischemia in minor stroke and TIA patients identified on MR imaging". *Neurology*. **65** (4): 513–517. doi:10.1212/01.WNL.0000169031.39264.ff. PMID 16116107.

[42] *Fisch and Spehlmann's EEG primer*

[43] Preissl, *Magnetoencephalography*

[44] Jones, Edward G.; Mendell, Lorne M. (April 30, 1999). "Assessing the Decade of the Brain". *Science*. American Association for the Advancement of Science. **284** (5415): 739. doi:10.1126/science.284.5415.739. PMID 10336393. Retrieved 2010-04-05.

[45] Buxton, *Introduction to Functional Magnetic Resonance Imaging*

[46] "If Modern Humans Are So Smart, Why Are Our Brains Shrinking?". DiscoverMagazine.com. 2011-01-20. Retrieved 2014-03-05.

[47] Brown P, Sutikna T, Morwood MJ, et al. (2004). "A new small-bodied hominin from the Late Pleistocene of Flores, Indonesia". *Nature*. **431** (7012): 1055–61. doi:10.1038/nature02999. PMID 15514638.

[48] Viegas, Jennifer (March 12, 2013). "Brain comparison suggests that Neanderthals lacked social skills". NBC News. Retrieved December 7, 2013.

[49] Davidson, Iain. "As large as you need and as small as you can'--implications of the brain size of Homo floresiensis, (Iain Davidson)". Une-au.academia.edu. Retrieved 2011-10-30.

[50] P. Thomas Schoenemann (2006). "Evolution of the Size and Functional Areas of the Human Brain". *Annu. Rev. Anthropol.* **35**: 379–406. doi:10.1146/annurev.anthro.35.081705.123210.

[51] http://www.uchospitals.edu/news/2005/20050908-humanbrain.html

[52] McDaniel, Michael (2005). "Big-brained people are smarter" (PDF). *Intelligence*. **33**: 337–346. doi:10.1016/j.intell.2004.11.005.

[53] Luders et al., 2008

[54] Hoppe & Stojanovic, 2008

[55] Meisenberg, G. (2009). "Wealth, Intelligence, Politics and Global Fertility Differentials". *Journal of Biosocial Science*. **41** (4): 519–535. doi:10.1017/S0021932009003344. PMID 19323856.

1.7 Bibliography

- Andrews, DG (2001). *Neuropsychology*. Psychology Press. ISBN 978-1-84169-103-9.

- Buxton, RB (2002). *An Introduction to Functional Magnetic Resonance Imaging: Principles and Techniques*. Cambridge University Press. ISBN 978-0-521-58113-4.

- Campbell, Neil A. and Jane B. Reece. (2005). *Biology*. Benjamin Cummings. ISBN 0-8053-7171-0

- Cosgrove, KP; Mazure CM; Staley JK (2007). "Evolving knowledge of sex differences in brain structure, function, and chemistry". *Biol Psychiat*. **62** (8): 847–55. doi:10.1016/j.biopsych.2007.03.001. PMC 2711771. PMID 17544382.

- Fisch, BJ; Spehlmann R (1999). *Fisch and Spehlmann's EEG Primer: Basic Principles of Digital and Analog EEG*. Elsevier Health Sciences. ISBN 978-0-444-82148-5.

- Gray, Peter (2002). *Psychology* (4th ed.). Worth Publishers. ISBN 0-7167-5162-3.

- Kandel, ER; Schwartz JH; Jessel TM (2000). *Principles of Neural Science*. McGraw-Hill Professional. ISBN 978-0-8385-7701-1.

- McGilchrist, Iain (2009). *The Master and His Emissary: The Divided Brain and the Making of the Western World*. USA: Yale University Press. ISBN 0-300-14878-X.

- Parent, A; Carpenter MB (1995). *Carpenter's Human Neuroanatomy*. Williams & Wilkins. ISBN 978-0-683-06752-1.

- Preissl, H (2005). *Magnetoencephalography*. Academic Press. ISBN 978-0-12-366869-1.

- Ramachanandran, V S (2011), *The Tell-Tale Brain: A Neuroscientist's Quest for What Makes Us Human*. W. W. Norton & Company.

- Simon, Seymour (1999). *The Brain*. HarperTrophy. ISBN 0-688-17060-9

- Thompson, Richard F. (2000). *The Brain: An Introduction to Neuroscience*. Worth Publishers. ISBN 0-7167-3226-2

- Toro, R; Perron M; Pike B; Richer L. Veillette S; Pausova Z; Paus T (2008). "Brain size and folding of the human cerebral cortex". *Cerebral cortex (New York, N.Y. : 1991)*. **18** (10): 2352–7. doi:10.1093/cercor/bhm261. PMID 18267953.

- Vanderwolf, C. H.; Kolb, B.; Cooley, R. K. (Feb 1978). "Behavior of the rat after removal of the neocortex and hippocampal formation". *Journal of comparative and physiological psychology.* **92** (1): 156–175. doi:10.1037/h0077447. ISSN 0021-9940. PMID 564358.

1.8 External links

- Atlas of the Human Brain
- The Whole Brain Atlas
- High-Resolution Cytoarchitectural Primate Brain Atlases
- Brain Facts and Figures
- Interactive Human Brain 3D Tool

Chapter 2

Cerebral cortex

For the scientific journal, see Cerebral Cortex (journal).

The **cerebral cortex** is the cerebrum's (brain) outer layer of neural tissue in humans and other mammals. It is divided into two cortices, along the sagittal plane: the left and right cerebral hemispheres divided by the medial longitudinal fissure. The cerebral cortex plays a key role in memory, attention, perception, awareness, thought, language, and consciousness. The human cerebral cortex is 2 to 4 millimetres (0.079 to 0.157 in) thick.[1]

In large mammals, the cerebral cortex is folded, giving a much greater surface area in the confined volume of the skull. A fold or ridge in the cortex is termed a gyrus (plural gyri) and a groove or fissure is termed a sulcus (plural sulci). In the human brain more than two-thirds of the cerebral cortex is buried in the sulci.

The cerebral cortex is composed of gray matter, consisting mainly of cell bodies (with astrocytes being the most abundant cell type in the cortex as well as the human brain in general) and capillaries. It contrasts with the underlying white matter, consisting mainly of the white myelinated sheaths of neuronal axons. The most recent part of the cerebral cortex to develop in the evolutionary history of mammals is the neocortex (also called isocortex), which differentiated into six horizontal layers; the more ancient part of the cerebral cortex, the hippocampus, has at most three cellular layers. Neurons in various layers connect vertically to form small microcircuits, called cortical columns. Different neocortical regions known as Brodmann areas are distinguished by variations in their cytoarchitectonics (histological structure) and functional roles in sensation, cognition and behavior.

2.1 Structure

2.1.1 Layered structure

The different cortical layers each contain a characteristic distribution of neuronal cell types and connections with

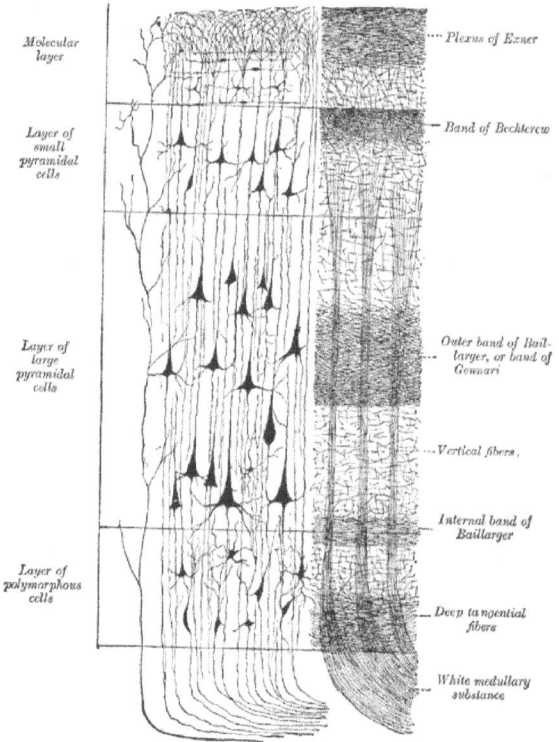

Cerebral cortex. (Poirier.) To the left, the groups of cells; to the right, the systems of fibers. Quite to the left of the figure a sensory nerve fiber is shown. Cell body layers are labeled on the left, and fiber layers are labeled on the right.

other cortical and subcortical regions. There are direct connections between different cortical areas and indirect connections via the thalamus, for example. One of the clearest examples of cortical layering is the Stria of Gennari in the primary visual cortex. This is a band of whiter tissue that can be observed with the naked eye in the fundus of the calcarine sulcus of the occipital lobe. The Stria of Gennari is composed of axons bringing visual information from the thalamus into layer four of the visual cortex.

Staining cross-sections of the cortex to reveal the position of neuronal cell bodies and the intracortical axon tracts al-

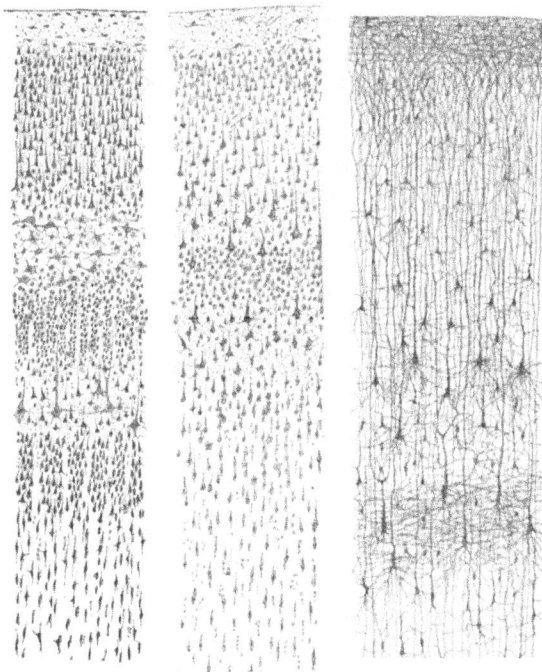

Three drawings of cortical lamination by Santiago Ramon y Cajal, each showing a vertical cross-section, with the surface of the cortex at the top. Left: Nissl-stained visual cortex of a human adult. Middle: Nissl-stained motor cortex of a human adult. Right: Golgi-stained cortex of a 1½ month old infant. The Nissl stain shows the cell bodies of neurons; the Golgi stain shows the dendrites and axons of a random subset of neurons.

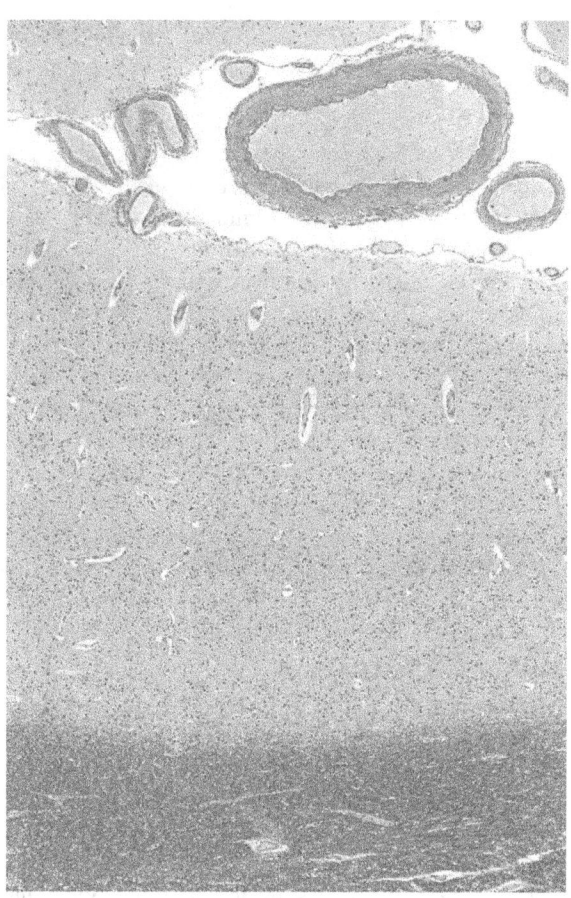

Micrograph showing the visual cortex (predominantly pink). Subcortical white matter (predominantly blue) is seen at the bottom of the image. HE-LFB stain.

lowed neuroanatomists in the early 20th century to produce a detailed description of the *laminar structure of the cortex* in different species. After the work of Korbinian Brodmann (1909) the neurons of the cerebral cortex are grouped into six main layers, from outside (pial surface) to inside (white matter):

1. Layer I, the molecular layer, contains few scattered neurons and consists mainly of extensions of apical dendritic tufts of pyramidal neurons and horizontally oriented axons, as well as glial cells.[2] During development Cajal-Retzius[3] and subpial granular layer cells[4] are present in this layer. Also, some spiny stellate cells can be found here. Inputs to the apical tufts are thought to be crucial for the "feedback" interactions in the cerebral cortex involved in associative learning and attention.[5] While it was once thought that the input to layer I came from the cortex itself,[6] it is now realized that layer I across the cerebral cortex mantle receives substantial input from "matrix" or M-type thalamus cells[7] (in contrast to "core" or C-type that go to layer IV).[8]

2. Layer II, the external granular layer, contains small pyramidal neurons and numerous stellate neurons.

3. Layer III, the external pyramidal layer, contains predominantly small and medium-size pyramidal neurons, as well as non-pyramidal neurons with vertically oriented intracortical axons; layers I through III are the main target of interhemispheric corticocortical afferents, and layer III is the principal source of corticocortical efferents.

4. Layer IV, the internal granular layer, contains different types of stellate and pyramidal neurons, and is the main target of thalamocortical afferents from thalamus type C neurons[8] as well as intra-hemispheric corticocortical afferents.

5. Layer V, the internal pyramidal layer, contains large pyramidal neurons which give rise to axons leaving the cortex and running down to subcortical structures (such as the basal ganglia). In the primary motor cortex of the frontal lobe, layer V contains Betz cells, whose axons travel through the internal capsule,

the brain stem and the spinal cord forming the corticospinal tract, which is the main pathway for voluntary motor control.

6. Layer VI, the polymorphic or multiform layer, contains few large pyramidal neurons and many small spindle-like pyramidal and multiform neurons; layer VI sends efferent fibers to the thalamus, establishing a very precise reciprocal interconnection between the cortex and the thalamus.[9] That is, layer VI neurons from one cortical column connect with thalamus neurons that provide input to the same cortical column. These connections are both excitatory and inhibitory. Neurons send excitatory fibers to neurons in the thalamus and also send collaterals to the thalamic reticular nucleus that inhibit these same thalamus neurons or ones adjacent to them.[10] One theory is that because the inhibitory output is reduced by cholinergic input to the cerebral cortex, this provides the brainstem with adjustable "gain control for the relay of lemniscal inputs".[10]

The cortical layers are not simply stacked one over the other; there exist characteristic connections between different layers and neuronal types, which span all the thickness of the cortex. These cortical microcircuits are grouped into cortical columns and minicolumns. It has been proposed that the minicolumns are the basic functional units of the cortex.[11] In 1957, Vernon Mountcastle showed that the functional properties of the cortex change abruptly between laterally adjacent points; however, they are continuous in the direction perpendicular to the surface. Later works have provided evidence of the presence of functionally distinct cortical columns in the visual cortex (Hubel and Wiesel, 1959),[12] auditory cortex, and associative cortex.

Cortical areas that lack a layer IV are called agranular. Cortical areas that have only a rudimentary layer IV are called dysgranular.[13] Information processing within each layer is determined by different temporal dynamics with that in the layers II/III having a slow 2 Hz oscillation while that in layer V having a fast 10–15 Hz one.[14]

2.1.2 Areas

Based on the differences in lamination the cerebral cortex can be classified into two parts, the large area of **neocortex** and the much smaller area of **allocortex**:

- The neocortex (also known as the isocortex or neopallium) is the part of the mature cerebral cortex with six distinct layers. Examples of neocortical areas include the granular primary motor cortex, also known as Brodmann area 4, and the striate primary visual cortex, or Brodmann area 17. The neocortex has two types of cortex, the **true isocortex** and the **proisocortex**. The proisocortex contains Brodmann areas 24, 25, and 32

- The allocortex is the part of the cerebral cortex with less than six layers and has three regions, the archicortex with three cortical laminae and the paleocortex which has four or five, and a transitional area adjacent to the allocortex, the **periallocortex**. Examples of allocortex are the olfactory cortex and the hippocampus.

There is a transitional area between the neocortex and the allocortex called the **paralimbic cortex**, where layers 2, 3 and 4 are merged. This area incorporates the proisocortex of the neocortex and the periallocortex of the allocortex. In addition, the cerebral cortex may be classified on the basis of gross topographical conventions into four lobes: the temporal lobe, the occipital lobe, the parietal lobe, and the frontal lobe.

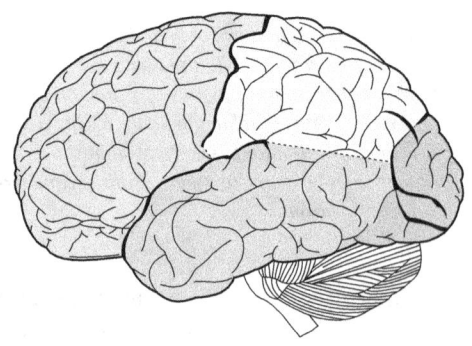

Frontal lobe
Temporal lobe
Parietal lobe
Occipital lobe

2.1.3 Development

The ontongenic development of the cerebral cortex is a complex and finely tuned process influenced by the interplay between genes and environment.[15] The cerebral cortex develops from the most anterior part of the neural plate, a specialized part of the embryonic ectoderm.[16] The neural plate folds and closes to form the neural tube. From the cavity inside the neural tube develops the ventricular system, and, from the epithelial cells of its walls, the neurons and glia of the nervous system. The most anterior (front, or cranial) part of the neural plate, the prosencephalon, which is evident before neurulation begins, gives rise to the cerebral hemispheres and its later cortex.[17]

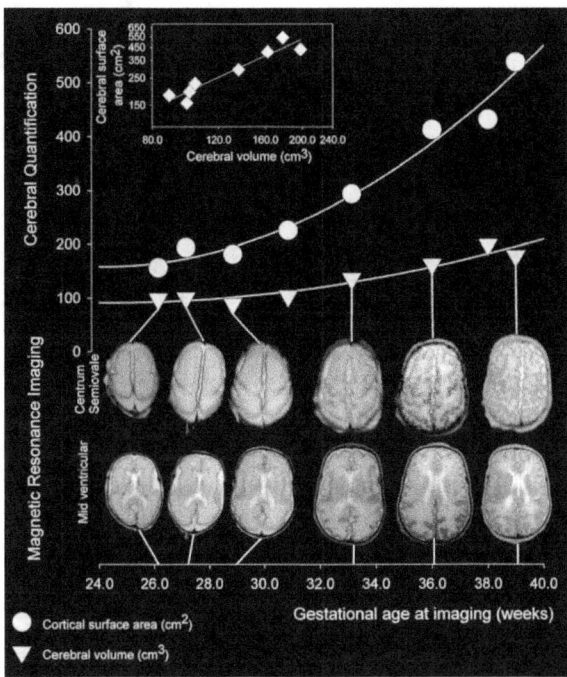

Human cortical development between 26 and 39 week gestational age

Cortical neurons are generated within the ventricular zone, next to the ventricles. At first, this zone contains progenitor cells, which divide to produce glial cells and neurons.[18] The glial fibers produced in the first divisions of the progenitor cells are radially oriented, spanning the thickness of the cortex from the ventricular zone to the outer, pial surface, and provide scaffolding for the migration of neurons outwards from the ventricular zone.[19][20] The first divisions of the progenitor cells are symmetric, which duplicates the total number of progenitor cells at each mitotic cycle. Then, some progenitor cells begin to divide asymmetrically, producing one postmitotic cell that migrates along the radial glial fibers, leaving the ventricular zone, and one progenitor cell, which continues to divide until the end of development, when it differentiates into a glial cell or an ependymal cell. As the G1 phase of mitosis is elongated, in what is seen as selective cell-cycle lengthening, the newly-born neurons migrate to more superficial layers of the cortex.[21] The migrating daughter cells become the pyramidal cells of the cerebral cortex.[22] The development process is time ordered and regulated by hundreds of genes and epigenetic regulatory mechanisms.[23]

The layered structure of the mature cerebral cortex is formed during development. The first pyramidal neurons generated migrate out of the ventricular zone and subventricular zone, together with reelin producing Cajal–Retzius neurons, from the preplate. Next, a cohort of neurons migrating into the middle of the preplate divides this transient layer into the superficial marginal zone, which will become layer one of the mature neocortex, and the subplate,[24] forming a middle layer called the cortical plate. These cells will form the deep layers of the mature cortex, layers five and six. Later born neurons migrate radially into the cortical plate past the deep layer neurons, and become the upper layers (two to four). Thus, the layers of the cortex are created in an inside-out order.[25] The only exception to this inside-out sequence of neurogenesis occurs in the layer I of primates, in which, in contrast to rodents, neurogenesis continues throughout the entire period of corticogenesis.[26]

The map of functional cortical areas, which include primary motor and visual cortex, originates from a 'protomap',[27] which is regulated by molecular signals such as fibroblast growth factor FGF8 early in embryonic development.[28][29] These signals regulate the size, shape, and position of cortical areas on the surface of the cortical primordium, in part by regulating gradients of transcription factor expression, through a process called cortical patterning. Examples of such transcription factors include the genes EMX2 and PAX6.[30] Rapid expansion of the cortical surface area is regulated by the amount of self-renewal of radial glial cells and is partly regulated by FGF and Notch genes.[31] During the period of cortical neurogenesis and layer formation, many higher mammals begin the process of gyrification, which generates the characteristic folds of the cerebral cortex.[32][33] Gyrification is regulated by the gene Trnp1[34] and by FGF and SHH signaling[35][36]

2.1.4 Thickness

For mammals, species with larger brains (in absolute terms, not just in relation to body size) tend to have thicker cortices.[37] The range, however, is not very great; only a factor of 7 differentiates between the thickest and thinnest cortices. The smallest mammals, such as shrews, have a neocortical thickness of about 0.5 mm; the ones with the largest brains, such as humans and fin whales, have thicknesses of 2.3—2.8 mm. There is an approximately logarithmic relationship between brain weight and cortical thickness.[37]

Magnetic resonance imaging (MRI) of the brain makes it possible to get a measure for the thickness of the human cerebral cortex and relate it to other measures. The thickness of different cortical areas varies but in general, sensory cortex is thinner than motor cortex.[38] One study has found some positive association between the cortical thickness and intelligence.[39] Another study has found that the somatosensory cortex is thicker in migraine sufferers, though it is not known if this is the result of migraine attacks or the cause of them.[40][41] A later study using a larger patient population reports no change in the cortical thickness

2.2. FUNCTION

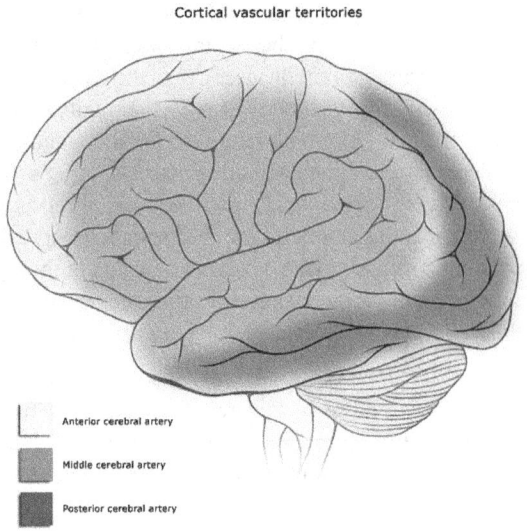

Cortical blood supply

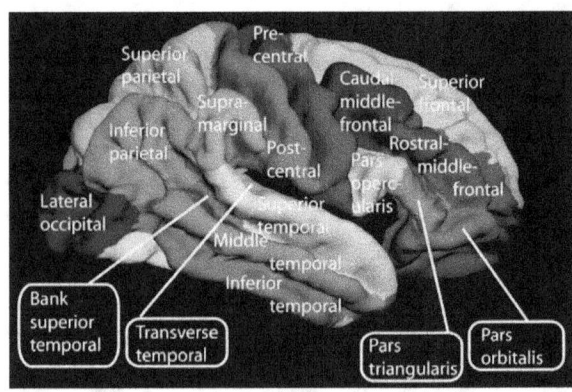

Lateral surface of the human cerebral cortex

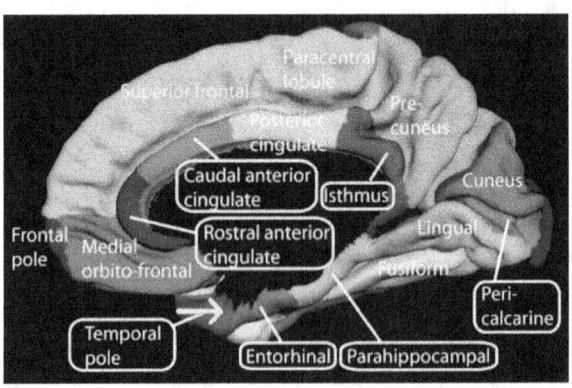

Medial surface of the human cerebral cortex

in migraine sufferers.[42] A genetic disorder of the cerebral cortex, whereby decreased folding in certain areas results in a microgyrus, where there are four layers instead of six, is in some instances seen to be related to dyslexia.[43]

2.1.5 Blood supply

Blood is supplied to the cerebral cortex via the cerebral circulation.

2.2 Function

2.2.1 Connections

The cerebral cortex is connected to various subcortical structures such as the thalamus and the basal ganglia, sending information to them along efferent connections and receiving information from them via afferent connections. Most sensory information is routed to the cerebral cortex via the thalamus. Olfactory information, however, passes through the olfactory bulb to the olfactory cortex (piriform cortex). The vast majority of connections are from one area of the cortex to another, rather than to subcortical areas; Braitenberg and Schüz (1991) put the figure as high as 99%.[44]

2.2.2 Cortical areas

The cortex is commonly described as comprising three parts: sensory, motor, and association areas.

Sensory areas

The sensory areas are the cortical areas that receive and process information from the senses. Parts of the cortex that receive sensory inputs from the thalamus are called primary sensory areas. The senses of vision, audition, and touch are served by the primary visual cortex, primary auditory cortex and primary somatosensory cortex respectively. In general, the two hemispheres receive information from the opposite (contralateral) side of the body. For example, the right primary somatosensory cortex receives information from the left limbs, and the right visual cortex receives information from the left visual field. The organization of sensory maps in the cortex reflects that of the corresponding sensing organ, in what is known as a topographic map. Neighboring points in the primary visual cortex, for example, correspond to neighboring points in the retina. This topographic map is called a retinotopic map. In the same way, there exists a tonotopic map in the primary auditory cortex and a somatotopic map in the primary sensory cortex. This last topographic map of the body onto the posterior central gyrus has been illustrated as a deformed human representation, the somatosensory homunculus, where the size of different body parts reflects the relative density of their innervation. Areas with lots of sensory innervation, such as the fingertips and the lips, require more cortical area to process finer sensation.

Motor areas

The motor areas are located in both hemispheres of the cortex. They are shaped like a pair of headphones stretching from ear to ear. The motor areas are very closely related to the control of voluntary movements, especially fine fragmented movements performed by the hand. The right half of the motor area controls the left side of the body, and vice versa.

Two areas of the cortex are commonly referred to as motor:

- Primary motor cortex, which *executes* voluntary movements
- Supplementary motor areas and premotor cortex, which *select* voluntary movements.

In addition, motor functions have been described for:

- Posterior parietal cortex, which guides voluntary movements in space
- Dorsolateral prefrontal cortex, which decides which voluntary movements to make according to higher-order instructions, rules, and self-generated thoughts.

Just underneath the cerebral cortex are interconnected subcortical masses of grey matter called basal ganglia (or nuclei). The basal ganglia receive input from the substantia nigra of the midbrain and motor areas of the cerebral cortex, and send signals back to both of these locations. They are involved in motor control. They are found lateral to the thalamus. The main components of the basal ganglia are the caudate nucleus, the putamen, the globus pallidus, the substantia nigra, the nucleus accumbens, and the subthalamic nucleus. The putamen and globus pallidus are also collectively known as the lentiform nucleus, because together they form a lens-shaped body. The putamen and caudate nucleus are also collectively called the corpus striatum after their striped appearance.[45][46]

Association areas

The association areas are the parts of the cerebral cortex that do not belong to the primary regions. They function to produce a meaningful perceptual experience of the world, enable us to interact effectively, and support abstract thinking and language. The parietal, temporal, and occipital lobes - all located in the posterior part of the cortex - integrate sensory information and information stored in memory. The frontal lobe or prefrontal association complex is involved in planning actions and movement, as well as abstract thought. Globally, the association areas are organized as distributed networks.[47] Each network connects areas distributed across widely spaced regions of the cortex. Distinct networks are positioned adjacent to one another yielding a complex series of interwoven networks. The specific organization of the association networks is debated with evidence for interactions, hierarchical relationships, and competition between networks.[48] In humans, association networks are particularly important to language function. In the past it was theorized that language abilities are localized in the left hemisphere in areas 44/45, the Broca's area, for language expression and area 22, the Wernicke's area, for language reception. However, language is no longer limited to easily identifiable areas. More recent research suggests that the processes of language expression and reception occur in areas other than just those structures around the lateral sulcus, including the frontal lobe, basal ganglia, cerebellum, and pons.[49]

2.3 Clinical significance

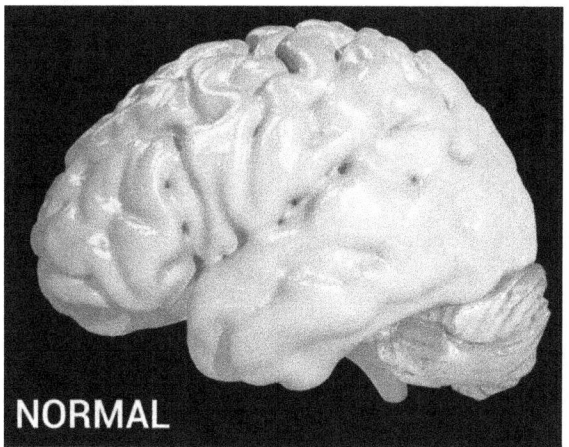

There is marked cortical atrophy in Alzheimer's Disease, associated with loss of gyri and sulci in the temporal lobe and parietal lobe, and parts of the frontal cortex and cingulate gyrus.

Neurodegenerative diseases such as Alzheimer's disease and Lafora disease, show as a marker, an atrophy of the grey matter of the cerebral cortex.[50]

2.4 Other animals

The cerebral cortex is derived from the pallium, a layered structure found in the forebrain of all vertebrates. The basic form of the pallium is a cylindrical layer enclosing fluid-filled ventricles. Around the circumference of the cylinder are four zones, the dorsal pallium, medial pallium, ventral pallium, and lateral pallium, which are thought respectively to give rise to the neocortex, hippocampus, amygdala, and olfactory cortex.

Until recently no counterpart to the cerebral cortex had been recognized in invertebrates. However, a study published in the journal *Cell* in 2010, based on gene expression profiles, reported strong affinities between the cerebral cortex and the mushroom bodies of ragworms.[51] Mushroom bodies are structures in the brains of many types of worms and arthropods that are known to play important roles in learning and memory; the genetic evidence indicates a common evolutionary origin, and therefore indicates that the origins of the earliest precursors of the cerebral cortex date back to the early Precambrian era.

2.5 Additional images

- Motor and Sensory Regions of the Cerebral Cortex
- Motor and Sensory Regions of the Cerebral Cortex
- Cortical areas

2.6 See also

- *Cerebral Cortex (journal)*
- Limbic system
- List of regions in the human brain
- Subplate
- Brain-computer interface
- EMX1
- Gyrification

2.7 References

[1] Kandel, Eric R.; Schwartz, James H.; Jessell, Thomas M. (2000). *Principles of Neural Science* (Fourth ed.). United State of America: McGraw-Hill. p. 324. ISBN 0-8385-7701-6.

[2] Shipp, Stewart (2007-06-17). "Structure and function of the cerebral cortex". *Current Biology*. **17** (12): R443–9. doi:10.1016/j.cub.2007.03.044. PMID 17580069. Retrieved 2009-02-17.

[3] Meyer, Gundela; Goffinet, André M.; Fairén, Alfonso (1999). "Feature Article: What is a Cajal–Retzius cell? A Reassessment of a Classical Cell Type Based on Recent Observations in the Developing Neocortex". *Cereb. Cortex*. **9** (8): 765–775. doi:10.1093/cercor/9.8.765.

[4] Judaš, Miloš; Pletikos, Mihovil (2010). "The discovery of the subpial granular layer in the human cerebral cortex". *Translational Neuroscience*. **1** (3): 255–260. doi:10.2478/v10134-010-0037-4.

[5] Gilbert CD, Sigman M (2007). "Brain states: top-down influences in sensory processing". *Neuron*. **54** (5): 677–96. doi:10.1016/j.neuron.2007.05.019. PMID 17553419.

[6] Cauller L (1995). "Layer I of primary sensory neocortex: where top-down converges upon bottom-up". *Behav Brain Res*. **71** (1–2): 163–70. doi:10.1016/0166-4328(95)00032-1. PMID 8747184.

[7] Rubio-Garrido P, Pérez-de-Manzo F, Porrero C, Galazo MJ, Clascá F (2009). "Thalamic input to distal apical dendrites in neocortical layer 1 is massive and highly convergent". *Cereb Cortex*. **19** (10): 2380–95. doi:10.1093/cercor/bhn259. PMID 19188274.

[8] Jones EG (1998). "Viewpoint: the core and matrix of thalamic organization". *Neuroscience*. **85** (2): 331–45. doi:10.1016/S0306-4522(97)00581-2. PMID 9622234.

[9] Creutzfeldt, O. 1995. *Cortex Cerebri*. Springer-Verlag.

[10] Lam YW, Sherman SM (2010). "Functional Organization of the Somatosensory Cortical Layer 6 Feedback to the Thalamus". *Cereb Cortex*. **20** (1): 13–24. doi:10.1093/cercor/bhp077. PMC 2792186. PMID 19447861.

[11] Mountcastle V (1997). "The columnar organization of the neocortex". *Brain*. **120** (4): 701–722. doi:10.1093/brain/120.4.701. PMID 9153131.

[12] HUBEL DH, WIESEL TN (October 1959). "Receptive fields of single neurones in the cat's striate cortex". *J. Physiol. (Lond.)*. **148** (3): 574–91. doi:10.1113/jphysiol.1959.sp006308. PMC 1363130. PMID 14403679.

[13] S.M. Dombrowski, C.C. Hilgetag, and H. Barbas. Quantitative Architecture Distinguishes Prefrontal Cortical Systems in the Rhesus Monkey.Cereb. *Cortex* 11: 975-988. "...they either lack (agranular) or have only a rudimentary granular layer IV (dysgranular)."

[14] Sun W, Dan Y (2009). "Layer-specific network oscillation and spatiotemporal receptive field in the visual cortex". *Proc Natl Acad Sci U S A*. **106** (42): 17986–17991.

doi:10.1073/pnas.0903962106. PMC 2764922. PMID 19805197.

[15] Pletikos, Mihovil; Sousa, Andre MM; et al. (22 January 2014). "Temporal Specification and Bilaterality of Human Neocortical Topographic Gene Expression". *Neuron.* **81** (2): 321–332. doi:10.1016/j.neuron.2013.11.018.

[16] Natasha Warren; Damira Caric; Thomas Pratt; Julia A. Clausen; Pundit Asavaritikrai; John O. Mason; Robert E. Hill; David J. Price; Oxford Journals (1999). "The transcription factor, Pax6, is required for cell proliferation and differentiation in the developing cerebral cortex". *National Institutes of Health.* pp. 627–635. PMID 10498281.

[17] Larsen, W J. Human Embryology 3rd edition 2001. pp 421-422 ISBN 0-443-06583-7

[18] Stephen C. Noctor; Alexander C. Flint; Tamily A. Weissman; Ryan S. Dammerman & Arnold R. Kriegstein (2001). "Neurons derived from radial glial cells establish radial units in neocortex". *Nature.* **409** (6821): 714–720. doi:10.1038/35055553. PMID 11217860.

[19] Rakic, P (October 2009). "Evolution of the neocortex: a perspective from developmental biology.". *Nature reviews. Neuroscience.* **10** (10): 724–35. doi:10.1038/nrn2719. PMC 2913577. PMID 19763105.

[20] Rakic, P (November 1972). "Extrinsic cytological determinants of basket and stellate cell dendritic pattern in the cerebellar molecular layer.". *The Journal of Comparative Neurology.* **146** (3): 335–54. doi:10.1002/cne.901460304. PMID 4628749.

[21] Calegari, F; Haubensack W; Haffner C; Huttner WB (2005). "Selective lengthening of the cell cycle in the neurogenic subpopulation of neural progenitor cells during mouse brain development.". *J Neurosci.* **25** (28): 6533–8. doi:10.1523/jneurosci.0778-05.2005.

[22] P. Rakic (1988). "Specification of cerebral cortical areas". *Science.* **241** (4862): 170–176. doi:10.1126/science.3291116. PMID 3291116.

[23] Hu, X.L.; Wang,Y.; Shen, Q. (2012). "Epigenetic control on cell fate choice in neural stem cells". *Protein & Cell.* **3** (4): 278–290. doi:10.1007/s13238-012-2916-6. PMID 22549586.

[24] Kostović, Ivica. "Developmental history of the transient subplate zone in the visual and somatosensory cortex of the macaque monkey and human brain". *Journal of Comparative Neurology.* **297** (3): 441–470. doi:10.1002/cne.902970309.

[25] Rakic, P (1 February 1974). "Neurons in rhesus monkey visual cortex: systematic relation between time of origin and eventual disposition.". *Science.* **183** (4123): 425–7. doi:10.1126/science.183.4123.425. PMID 4203022.

[26] Zecevic N, Rakic P (2001). "Development of layer I neurons in the primate cerebral cortex". *J Neurosci.* **21** (15): 5607–19. PMID 11466432.

[27] Rakic, P (8 July 1988). "Specification of cerebral cortical areas.". *Science.* **241** (4862): 170–6. doi:10.1126/science.3291116. PMID 3291116.

[28] Fukuchi-Shimogori, T; Grove, EA (2 November 2001). "Neocortex patterning by the secreted signaling molecule FGF8.". *Science.* **294** (5544): 1071–4. doi:10.1126/science.1064252. PMID 11567107.

[29] Garel, S; Huffman, KJ; Rubenstein, JL (May 2003). "Molecular regionalization of the neocortex is disrupted in Fgf8 hypomorphic mutants.". *Development (Cambridge, England).* **130** (9): 1903–14. doi:10.1242/dev.00416. PMID 12642494.

[30] Bishop, KM; Goudreau, G; O'Leary, DD (14 April 2000). "Regulation of area identity in the mammalian neocortex by Emx2 and Pax6.". *Science.* **288** (5464): 344–9. doi:10.1126/science.288.5464.344. PMID 10764649.

[31] Rash, BG; Lim, HD; Breunig, JJ; Vaccarino, FM (26 October 2011). "FGF signaling expands embryonic cortical surface area by regulating Notch-dependent neurogenesis.". *The Journal of neuroscience : the official journal of the Society for Neuroscience.* **31** (43): 15604–17. doi:10.1523/jneurosci.4439-11.2011. PMID 22031906.

[32] Rajagopalan, V; Scott, J; Habas, PA; Kim, K; Corbett-Detig, J; Rousseau, F; Barkovich, AJ; Glenn, OA; Studholme, C (23 February 2011). "Local tissue growth patterns underlying normal fetal human brain gyrification quantified in utero.". *The Journal of neuroscience : the official journal of the Society for Neuroscience.* **31** (8): 2878–87. doi:10.1523/jneurosci.5458-10.2011. PMID 21414909.

[33] Lui, Jan H.; Hansen, David V.; Kriegstein, Arnold R. (2011-07-08). "Development and evolution of the human neocortex". *Cell.* **146** (1): 18–36. doi:10.1016/j.cell.2011.06.030. ISSN 1097-4172. PMC 3610574. PMID 21729779.

[34] Stahl, Ronny; Walcher, Tessa; De Juan Romero, Camino; Pilz, Gregor Alexander; Cappello, Silvia; Irmler, Martin; Sanz-Aquela, José Miguel; Beckers, Johannes; Blum, Robert (2013-04-25). "Trnp1 regulates expansion and folding of the mammalian cerebral cortex by control of radial glial fate". *Cell.* **153** (3): 535–549. doi:10.1016/j.cell.2013.03.027. ISSN 1097-4172. PMID 23622239.

[35] Wang, Lei; Hou, Shirui; Han, Young-Goo (2016-05-23). "Hedgehog signaling promotes basal progenitor expansion and the growth and folding of the neocortex". *Nature Neuroscience.* **19**: 888–96. doi:10.1038/nn.4307. ISSN 1546-1726. PMID 27214567.

[36] Rash, Brian G.; Tomasi, Simone; Lim, H. David; Suh, Carol Y.; Vaccarino, Flora M. (2013-06-26). "Cortical gyrification induced by fibroblast growth factor 2 in the

mouse brain". *The Journal of Neuroscience: The Official Journal of the Society for Neuroscience.* **33** (26): 10802–10814. doi:10.1523/JNEUROSCI.3621-12.2013. ISSN 1529-2401. PMC 3693057. PMID 23804101.

[37] Nieuwenhuys R, Donkelaar HJ, Nicholson C (1998). *The central nervous system of vertebrates, Volume 1*. Springer. pp. 2011–2012. ISBN 978-3-540-56013-5.

[38] Frithjof Kruggel; Martina K. Brückner; Thomas Arendt; Christopher J. Wiggins; D. Yves von Cramon (2003). "Analyzing the neocortical fine-structure". *Medical Image Analysis.* **7** (3): 251–264. doi:10.1016/S1361-8415(03)00006-9.

[39] Katherine L. Narr; Roger P. Woods; Paul M. Thompson; Philip Szeszko; Delbert Robinson; Teodora Dimtcheva; Mala Gurbani; Arthur W. Toga; Robert M. Bilder (2007). "Relationships between IQ and Regional Cortical Grey Matter Thickness in Healthy Adults". *Cerebral Cortex.* **17** (9): 2163–2171. doi:10.1093/cercor/bhl125. PMID 17118969.

[40] Alexandre F.M. DaSilva; Cristina Granziera; Josh Snyder; Nouchine Hadjikhani (2007). "Thickening in the somatosensory cortex of patients with migraine". *Neurology.* **69** (21): 1990–1995. doi:10.1212/01.wnl.0000291618.32247.2d. PMID 18025393.

[41] Catharine Paddock (2007-11-20). "Migraine Sufferers Have Thicker Brain Cortex". Medical News Today.

[42] Datte R, Detre JA, et al. (Oct 2011). "Absence of changes in cortical thickness in patients with migraine". *Cephalagia.* **31** (14): 1452–8. doi:10.1177/0333102411421025. PMID 21911412.

[43] Habib M (2000). "The neurological basis of developmental dyslexia: an overview and working hypothesis". *Brain.* **123** (12): 2373–99. doi:10.1093/brain/123.12.2373. PMID 11099442.

[44] Braitenberg, V and Schüz, A 1991. "Anatomy of the Cortex: Statistics and Geometry" NY: Springer-Verlag

[45] Saladin, Kenneth. Anatomy and Physiology: The Unity of Form and Function, 5th Ed. New York: McGraw-Hill Companies Inc., 2010. Print.

[46] Dorland's Medical Dictionary for Health Consumers, 2008.

[47] Yeo BT, Krienen FM, Sepulcre J, Sabuncu MR, Lashkari D, Hollinshead M, Roffman JL, Smoller JW, Zöllei L, Polimeni JR, Fischl B, Liu H, Buckner RL (2011). "The organization of the human cerebral cortex estimated by intrinsic functional connectivity". *Journal of Neurophysiology.* **106** (3): 1125–1165. doi:10.1152/jn.00338.2011. PMC 3174820. PMID 21653723.

[48] Rupesh Kumar Srivastava; Jürgen Schmidhuber (2014). "Understanding Locally Competitive Networks". arXiv.org. arXiv:1410.1165.

[49] Cathy J. Price (2000). "The anatomy of language: contributions from functional neuroimaging". *Journal of Anatomy.* **197** (3): 335–359. doi:10.1046/j.1469-7580.2000.19730335.x.

[50] Ortolano S, Vieitez I, et al. (2014). "Loss of cortical neurons underlies the neuropathology of Lafora disease". *Mol Brain.* **7**: 7. doi:10.1186/1756-6606-7-7. PMC 3917365. PMID 24472629.

[51] Tomer, R; Denes, AS; Tessmar-Raible, K; Arendt, D; Tomer R; Denes AS; Tessmar-Raible K; Arendt D (2010). "Profiling by image registration reveals common origin of annelid mushroom bodies and vertebrate pallium". *Cell.* **142** (5): 800–809. doi:10.1016/j.cell.2010.07.043. PMID 20813265.

2.8 External links

- hier-20 at NeuroNames
- Stained brain slice images which include the "cerebral cortex" at the BrainMaps project
- Webvision - The primary visual cortex Comprehensive article about the structure and function of the primary visual cortex.
- Webvision - Basic cell types Image of the basic cell types of the monkey cerebral cortex.
- Development of the Cerebral Cortex Different topics on cortical development in the form of columns written by leading scientists.
- Cerebral Cortex - Cell Centered Database
- NIF Search - Cerebral Cortex via the Neuroscience Information Framework

Chapter 3

Frontal lobe

The **frontal lobe**, located at the front of the brain, is one of the four major lobes of the cerebral cortex in the mammalian brain. The frontal lobe is located at the front of each cerebral hemisphere and positioned in front of the parietal lobe and above and in front of the temporal lobe. It is separated from the parietal lobe by a space between tissues called the central sulcus, and from the temporal lobe by a deep fold called the lateral sulcus also called the Sylvian fissure. The precentral gyrus, forming the posterior border of the frontal lobe, contains the primary motor cortex, which controls voluntary movements of specific body parts.

The frontal lobe contains most of the dopamine-sensitive neurons in the cerebral cortex. The dopamine system is associated with reward, attention, short-term memory tasks, planning, and motivation. Dopamine tends to limit and select sensory information arriving from the thalamus to the forebrain. A report from the National Institute of Mental Health says a gene variant that reduces dopamine activity in the prefrontal cortex is related to poorer performance and inefficient functioning of that brain region during working memory tasks, and to a slightly increased risk for schizophrenia.[1]

3.1 Structure

On the lateral surface of the human brain, the central sulcus separates the frontal lobe from the parietal lobe. The lateral sulcus separates the frontal lobe from the temporal lobe.

The frontal lobe bottom can be divided into a lateral, polar, orbital (above the orbit; also called basal or ventral), and medial part. Each of these parts consists of particular gyri:

- Lateral part: lateral part of the superior frontal gyrus, middle frontal gyrus, inferior frontal gyrus.

- Polar part: Transverse frontopolar gyri, frontomarginal gyrus.

- Orbital part: Lateral orbital gyrus, anterior orbital

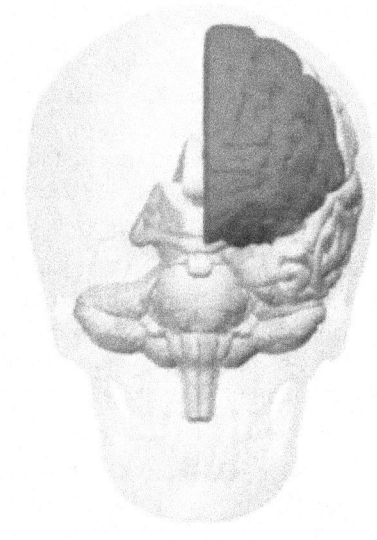

Animation. Frontal lobe (red) of left cerebral hemisphere.

gyrus, posterior orbital gyrus, medial orbital gyrus, gyrus rectus.

- Medial part: Medial part of the superior frontal gyrus, cingulate gyrus.

The gyri are separated by sulci. E.g., the precentral gyrus is in front of the central sulcus, and behind the precentral sulcus. The superior and middle frontal gyri are divided by the superior frontal sulcus. The middle and inferior frontal gyri are divided by the inferior frontal sulcus.

In humans, the frontal lobe reaches full maturity around the late 20s,[2] marking the cognitive maturity associated with adulthood. A small amount of atrophy, however, is normal in the aging person's frontal lobe. Fjell, in 2009, studied atrophy of the brain in people aged 60–91 years. The 142 healthy participants were scanned using MRI. Their results were compared to those of 122 participants with Alzheimer's disease. A follow-up one year later showed there to have been a marked volumetric decline in those with Alzheimer's and a much smaller decline (averaging 0.5%) in the healthy group.[3] These findings corroborate those of Coffey, who in 1992 indicated that the frontal lobe decreases in volume approximately 0.5%–1% per year.[4]

3.2 Function

The frontal lobe plays a large role in voluntary movement. It houses the primary motor cortex which regulates activities like walking.

The function of the frontal lobe involves the ability to project future consequences resulting from current actions, the choice between good and bad actions (or better and best) (also known as conscience), the override and suppression of socially unacceptable responses, and the determination of similarities and differences between things or events.

The frontal lobe also plays an important part in integrating longer non-task based memories stored across the brain. These are often memories associated with emotions derived from input from the brain's limbic system. The frontal lobe modifies those emotions to generally fit socially acceptable norms.

Psychological tests that measure frontal lobe function include finger tapping (as the frontal lobe controls voluntary movement), the Wisconsin Card Sorting Test, and measures of language and numeracy skills.[5]

3.3 Clinical significance

3.3.1 Damage

Stuss, et al. discuss in a review of many studies how damage to the frontal lobe can occur in an assortment of ways and result in many different consequences. Transient ischemic attacks (TIAs) also known as mini-strokes, and strokes are common causes of frontal lobe damage in older adults (65 and over). These strokes and mini-strokes can occur due to the blockage of blood flow to the brain or as a result of the rupturing of an aneurysm in a cerebral artery. Other ways in which injury can occur include head injuries such as traumatic brain injuries incurred following accidents, diagnoses such as Alzheimer's disease or Parkinson's disease (which cause dementia symptoms), and frontal lobe epilepsy (which can occur at any age).[6]

Symptoms

Common effects of damage to the frontal lobe are varied. Patients who have experienced frontal lobe trauma may know the appropriate response to a situation but display inappropriate responses to those same situations in "real life". Similarly, emotions that are felt may not be expressed in the face or voice. For example, someone who is feeling happy would not smile, and his or her voice would be devoid of emotion. Along the same lines, though, the person may also exhibit excessive, unwarranted displays of emotion. Depression is common in stroke patients; it affects a great number of those who have experienced one. Also common along with depression is a loss of or decrease in motivation. Someone might not want to carry out normal daily activities and would not feel "up to it".[6] Those who are close to the person who has experienced the damage may notice that the person no longer behaves like him or herself.[7] This personality change is characteristic of damage to the frontal lobe and was exemplified in the case of Phineas Gage. The frontal lobe is the same part of the brain that is responsible for executive functions such as planning for the future, judgment, decision-making skills, attention span, and inhibition. These functions can decrease drastically in someone whose frontal lobe is damaged.[6]

Consequences that are seen less frequently are also varied. Confabulation may be the most frequently indicated "less common" effect. In the case of confabulation, someone gives false information while maintaining the belief that it is the truth; he or she cannot remember the accurate information. In a small number of patients, uncharacteristic cheerfulness can be noted. This effect is seen mostly in patients with lesions to the right frontal portion of the brain.[6][8]

Another infrequent effect is that of reduplicative paramnesia, in which patients believe that the location in which they currently reside is a replica of one located somewhere else. Similarly, those who experience Capgras syndrome after frontal lobe damage believe that an identical "replacement" has taken the identity of a close friend, relative, or other person and is posing as that person. This last effect is seen mostly in schizophrenic patients who also have a neurological disorder in the frontal lobe.[6][9]

3.4 History

3.4.1 Psychosurgery

In the early 20th century, a medical treatment for mental illness, first developed by Portuguese neurologist Egas Moniz, involved damaging the pathways connecting the frontal lobe to the limbic system. Frontal lobotomy (sometimes called frontal leucotomy) successfully reduced distress but at the cost of often blunting the subject's emotions, volition and personality. The indiscriminate use of this psychosurgical procedure, combined with its severe side effects and a mortality rate of 7.4 to 17 per cent,[10] gained it a bad reputation. The frontal lobotomy has largely died out as a psychiatric treatment. More precise psychosurgical procedures are still used, although rarely. They may include anterior capsulotomy (bilateral thermal lesions of the anterior limbs of the internal capsule) or the bilateral cingulotomy (involving lesions of the anterior cingulate gyri) and might be used to treat otherwise untreatable obsessional disorders or clinical depression.

3.4.2 Theories of function

Theories of frontal lobe function can be separated into four categories:

- Single-process theories, which propose that "damage to a single process or system is responsible for a number of different dysexecutive symptoms" [11]

- Multi-process theories, which propose "that the frontal lobe executive system consists of a number of components that typically work together in everyday actions (heterogeneity of function)" [12]

- Construct-led theories, which propose that "most if not all frontal functions can be explained by one construct (homogeneity of function) such as working memory or inhibition" [13]

- Single-symptom theories, which propose that a specific dysexecutive symptom (e.g., confabulation) is related to the processes and construct of the underlying structures.[14]

Other theories include:

- Stuss (1999) suggests a differentiation into two categories according to homogeneity and heterogeneity of function.

- Grafman's managerial knowledge units (MKU) / structured event complex (SEC) approach (cf. Wood & Grafman, 2003)

- Miller & Cohen's integrative theory of prefrontal functioning (e.g. Miller & Cohen, 2001)

- Rolls's stimulus-reward approach and Stuss's anterior attentional functions (Burgess & Simons, 2005; Burgess, 2003; Burke, 2007).

It may be highlighted that the theories described above differ in their focus on certain processes/systems or constructlets. Stuss (1999) remarks that the question of homogeneity (single construct) or heterogeneity (multiple processes/systems) of function "may represent a problem of semantics and/or incomplete functional analysis rather than an unresolvable dichotomy" (p. 348). However, further research will show if a unified theory of frontal lobe function that fully accounts for the diversity of functions will be available.

3.5 In other animals

For many years, many scientists thought that the frontal lobe was disproportionately enlarged in humans compared to other primates. They thought that this was an important feature of human evolution and was the primary reason why human cognition differs from that of other primates. However, this view has been challenged by newer research. Using magnetic resonance imaging to determine the volume of the frontal cortex in humans, all extant ape species and several monkey species, Semendeferi *et al.* found that the human frontal cortex was not relatively larger than the cortex of other great apes but was relatively larger than the frontal cortex of lesser apes and the monkeys.[15]

3.6 Additional images

- Left frontal lobe (click to view animation)

- Lobes

- Base of brain.

- Human brain showing the four major lobes of the cerebrum. Beneath the cerebral cortex are the cerebellum, pons, olive, and medulla oblongata

- Drawing to illustrate the relations of the brain to the skull.

- Frontal lobe

- Frontal lobe

- Cerebrum.Inferior view.Deep dissection

- Ventricles of brain and basal ganglia.Superior view. Horizontal section.Deep dissection

3.7 See also

This article uses anatomical terminology; for an overview, see Anatomical terminology.

- Broca's area
- Limen insulae
- Regions in the human brain

3.8 References

[1] "Gene Slows Frontal Lobes, Boosts Schizophrenia Risk". National Institute of Mental Health. May 29, 2001. Retrieved 2013-06-20.

[2] Giedd JN, Blumenthal J, Jeffries NO, et al. (October 1999). "Brain development during childhood and adolescence: a longitudinal MRI study". *Nature Neuroscience.* **2** (10): 861–3. doi:10.1038/13158. PMID 10491603.

[3] Fjell AM, Walhovd KB, Fennema-Notestine C, et al. (December 2009). "One-year brain atrophy evident in healthy aging". *The Journal of Neuroscience.* **29** (48): 15223–31. doi:10.1523/JNEUROSCI.3252-09.2009. PMC 2827793. PMID 19955375.

[4] Coffey CE, Wilkinson WE, Parashos IA, et al. (March 1992). "Quantitative cerebral anatomy of the aging human brain: a cross-sectional study using magnetic resonance imaging". *Neurology.* **42** (3 Pt 1): 527–36. doi:10.1212/wnl.42.3.527. PMID 1549213.

[5] Kimberg DY, Farah MJ (December 1993). "A unified account of cognitive impairments following frontal lobe damage: the role of working memory in complex, organized behavior". *Journal of Experimental Psychology. General.* **122** (4): 411–28. doi:10.1037/0096-3445.122.4.411. PMID 8263463.

[6] Stuss DT, Gow CA, Hetherington CR (June 1992). "'No longer Gage': frontal lobe dysfunction and emotional changes". *Journal of Consulting and Clinical Psychology.* **60** (3): 349–59. doi:10.1037/0022-006X.60.3.349. PMID 1619089.

[7] Rowe AD, Bullock PR, Polkey CE, Morris RG (March 2001). "'Theory of mind' impairments and their relationship to executive functioning following frontal lobe excisions". *Brain.* **124** (Pt 3): 600–16. doi:10.1093/brain/124.3.600. PMID 11222459.

[8] Robinson RG, Kubos KL, Starr LB, Rao K, Price TR (March 1984). "Mood disorders in stroke patients. Importance of location of lesion". *Brain.* **107** (1): 81–93. doi:10.1093/brain/107.1.81. PMID 6697163.

[9] Durani, Shiban K.; Ford, Rodney; Sajjad, S. H. (September 1991). "Capgras syndrome associated with a frontal lobe tumour". *Irish Journal of Psychological Medicine.* **8** (2): 135–6. doi:10.1017/S0790966700015093.

[10] Ogren K, Sandlund M (2007). "Lobotomy at a state mental hospital in Sweden. A survey of patients operated on during the period 1947-1958". *Nordic Journal of Psychiatry.* **61** (5): 355–62. doi:10.1080/08039480701643498. PMID 17990197.

[11] (Burgess, 2003, p. 309).

[12] (Burgess, 2003, p. 310).

[13] (Stuss, 1999, p. 348; cf. Burgess & Simons, 2005).

[14] (cf. Burgess & Simons, 2005).

[15] Semendeferi K, Lu A, Schenker N, Damasio H (March 2002). "Humans and great apes share a large frontal cortex". *Nature Neuroscience.* **5** (3): 272–6. doi:10.1038/nn814. PMID 11850633.

3.9 External links

- NIF Search - Frontal Lobe via the Neuroscience Information Framework

Chapter 4

Cerebral hemisphere

The vertebrate cerebrum (brain) is formed by two **cerebral hemispheres** that are separated by a groove, the medial longitudinal fissure. The brain can thus be described as being divided into left and right cerebral hemispheres. Each of these hemispheres has an outer layer of grey matter, the cerebral cortex, that is supported by an inner layer of white matter. In eutherian (placental) mammals, the hemispheres are linked by the corpus callosum, a very large bundle of nerve fibers. Smaller commissures, including the anterior commissure, the posterior commissure and the fornix, also join the hemispheres and these are also present in other vertebrates. These commissures transfer information between the two hemispheres to coordinate localized functions.

The central sulcus is a prominent fissure which separates the parietal lobe from the frontal lobe and the primary motor cortex from the primary somatosensory cortex.

Macroscopically the hemispheres are roughly mirror images of each other, with only subtle differences, such as the Yakovlevian torque seen in the human brain, which is a slight warping of the right side, bringing it just forward of the left side. On a microscopic level, the cytoarchitecture of the cerebral cortex, shows the functions of cells, quantities of neurotransmitter levels and receptor subtypes to be markedly asymmetrical between the hemispheres.[1][2] However, while some of these hemispheric distribution differences are consistent across human beings, or even across some species, many observable distribution differences vary from individual to individual within a given species.

4.1 Structure

Each cerebral hemisphere has an outer layer of cerebral cortex which is of grey matter and an inner layer or core of white matter known as the centrum ovale.[3]

4.1.1 Development

The cerebral hemispheres are derived from the telencephalon. They arise five weeks after conception as bilateral invaginations of the walls. The hemispheres grow round in a C-shape and then back again, pulling all structures internal to the hemispheres (such as the ventricles) with them. The intraventricular foramina (also called the foramina of Monro) allows communication with the lateral ventricles. The choroid plexus is formed from ependymal cells and vascular mesenchyme.

4.2 Function

4.2.1 Hemisphere lateralization

Main article: Lateralization of brain function

Broad generalizations are often made in popular psychology about certain functions (e.g. logic, creativity) being lateralized, that is, located in the right or left side of the brain. These claims are often inaccurate, as most brain functions are actually distributed across both hemispheres. Most scientific evidence for asymmetry relates to low-level perceptual functions rather than the higher-level functions popularly discussed (e.g. subconscious processing of grammar, not "logical thinking" in general).[4][5]

The best example of an established lateralization is that of Broca's and Wernicke's Areas (language) where both are often found exclusively on the left hemisphere. These areas frequently correspond to handedness however, meaning the localization of these areas is regularly found on the hemisphere opposite to the dominant hand. Function lateralization such as semantics, prosodic, intonation, accentuation, prosody, etc has since been called into question and largely been found to have a neuronal basis in both hemispheres.[6][7]

Perceptual information is processed in both hemispheres,

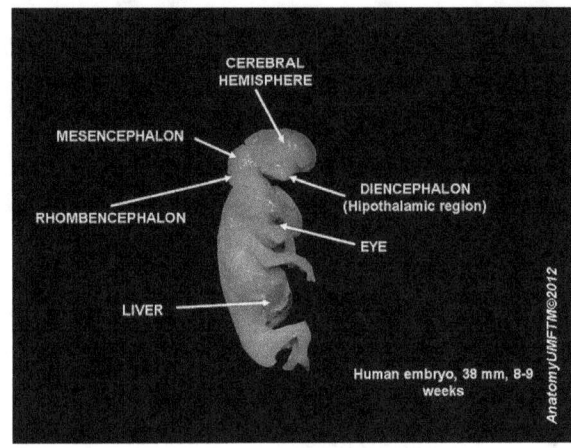

Cerebral hemispheres of a human embryo at 8 weeks.

but is laterally partitioned: information from each side of the body is sent to the opposite hemisphere (visual information is partitioned somewhat differently, but still lateralized). Similarly, motor control signals sent out to the body also come from the hemisphere on the opposite side. Thus, hand preference (which hand someone prefers to use) is also related to hemisphere lateralization.

In some aspects, the hemispheres are asymmetrical; one side is slightly bigger. There are higher levels of the neurotransmitter norepinephrine on the right and higher levels of dopamine on the left. There is more white matter (longer axons) on right and more grey matter (cell bodies) on the left.[8]

Linear reasoning functions of language such as grammar and word production are often lateralized to the left hemisphere of the brain. In contrast, holistic reasoning functions of language such as intonation and emphasis are often lateralized to the right hemisphere of the brain. Other integrative functions such as intuitive or heuristic arithmetic, binaural sound localization, etc. seem to be more bilaterally controlled.[9]

4.3 See also

This article uses anatomical terminology; for an overview, see Anatomical terminology.

- Hemispherectomy

- Corpus callosotomy

4.4 References

[1] Anderson, B.; Rutledge, V. (1996). "Age and hemisphere effects on dendritic structure". *Brain.* **119**: 1983–1990. doi:10.1093/brain/119.6.1983.

[2] Hutsler, J.; Galuske, R.A.W. (2003). "Hemispheric asymmetries in cerebral cortical networks". *Trends in Neurosciences.* **26** (8): 429–435. doi:10.1016/S0166-2236(03)00198-X.

[3] Bogousslavsky, J; Regli, F (October 1992). "Centrum ovale infarcts: subcortical infarction in the superficial territory of the middle cerebral artery.". *Neurology.* **42** (10): 1992–8. PMID 1340771.

[4] Western et al. 2006 "Psychology: Australian and New Zealand edition" John Wiley p.107

[5] "Neuromyth 6" http://www.oecd.org/document/63/0,3746,en_2649_35845581_34555007_1_1_1_1,00.html Retrieved October 15, 2011.

[6] Weiss, Peter H., and Simon D. Ubben. "Where Language Meets Meaningful Action: A Combined Behavior and Lesion." Springer. 29 Oct. 2014. Web. 31 Mar. 2016.

[7] Riès, Stephanie K., and Nina F. Dronkers. "Choosing Words: Left Hemisphere, Right Hemisphere, or Both? Perspective on the Lateralization of Word Retrieval."Wiley Online Library. 14 Jan. 2016. Web. 31 Mar. 2016.

[8] R. Carter, Mapping the Mind, Phoenix, London, 2004, Originally Weidenfeld and Nicolson, 1998.

[9] Dehaene S, Spelke E, Pinel P, Stanescu R, Tsivkin S (1999). "Sources of mathematical thinking: behavioral and brain-imaging evidence". *Science.* **284** (5416): 970–4. doi:10.1126/science.284.5416.970. PMID 10320379.

Chapter 5

Lobe (anatomy)

In anatomy, a **lobe** is a clear anatomical division or extension[1] of an organ (as seen for example in the brain, the lung, liver or the kidney) that can be determined without the use of a microscope at the gross anatomy level. This is in contrast to the much smaller **lobule**, which is a clear division only visible under the microscope.[2]

Interlobar ducts connect lobes and interlobular ducts connect lobules.

5.1 Examples of lobes

- The four lobes of the human cerebral cortex
  - the frontal lobe
  - the parietal lobe
  - the occipital lobe
  - the temporal lobe
- The three lobes of the human cerebellum
  - the flocculonodular lobe
  - the anterior lobe
  - the posterior lobe
- The two lobes of the thymus
- The two and three lobes of the lungs
  - Left lung: superior and inferior
  - Right lung: superior, middle, inferior
- The four lobes of the liver
  - Left lobe of liver
  - Right lobe of liver
  - Quadrate lobe of liver
  - Caudate lobe of liver
- The renal lobes of the kidney

5.1.1 Examples of lobules

- the cortical lobules of the kidney
- the testicular lobules
- the lobules of the mammary gland
- the lobules of the lung
- the lobules of the thymus

5.2 References

[1] lobe at eMedicine Dictionary

[2] SIU SOM Histology GI

Chapter 6

Parietal lobe

The **parietal lobe** is one of the four major lobes of the cerebral cortex in the brain of mammals. The parietal lobe is positioned above the occipital lobe and behind the frontal lobe and central sulcus.

The parietal lobe integrates sensory information among various modalities, including spatial sense and navigation (proprioception), the main sensory receptive area for the sense of touch (mechanoreception) in the somatosensory cortex which is just posterior to the central sulcus in the postcentral gyrus,[1] and the dorsal stream of the visual system. The major sensory inputs from the skin (touch, temperature, and pain receptors), relay through the thalamus to the parietal lobe.

Several areas of the parietal lobe are important in language processing. The somatosensory cortex can be illustrated as a distorted figure — the homunculus (Latin: "little man"), in which the body parts are rendered according to how much of the somatosensory cortex is devoted to them.[2] The superior parietal lobule and inferior parietal lobule are the primary areas of body or spatial awareness. A lesion commonly in the right superior or inferior parietal lobule leads to hemineglect.

The name comes from the overlying parietal bone, which is named from the Latin *paries-*, meaning "wall".

6.1 Structure

The parietal lobe is defined by three anatomical boundaries: The central sulcus separates the parietal lobe from the frontal lobe; the parieto-occipital sulcus separates the parietal and occipital lobes; the lateral sulcus (sylvian fissure) is the most lateral boundary, separating it from the temporal lobe; and the medial longitudinal fissure divides the two hemispheres. Within each hemisphere, the somatosensory cortex represents the skin area on the contralateral surface of the body.[2]

Immediately posterior to the central sulcus, and the most anterior part of the parietal lobe, is the postcentral gyrus

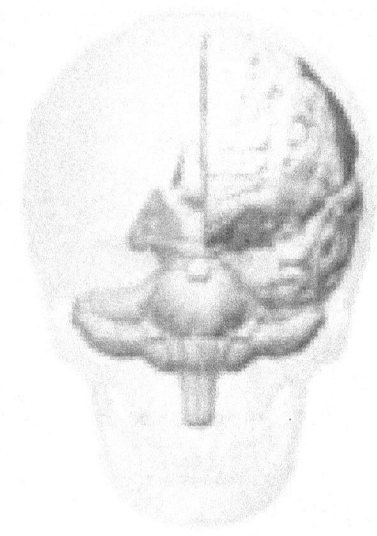

Animation. Parietal lobe (red) of left cerebral hemisphere.

(Brodmann area 3), the primary somatosensory cortical area. Dividing this and the posterior parietal cortex is the postcentral sulcus.

The posterior parietal cortex can be subdivided into the superior parietal lobule (Brodmann areas 5 + 7) and the inferior parietal lobule (39 + 40), separated by the intraparietal sulcus (IPS). The intraparietal sulcus and adjacent gyri are essential in guidance of limb and eye movement, and—based on cytoarchitectural and functional differences—is further divided into medial (MIP), lateral (LIP), ventral (VIP), and anterior (AIP) areas.

6.2 Function

Cortical functions of the parietal lobe are:

- Two point discrimination

through touch alone without other sensory input (i.e., visual)

- Graphesthesia - recognizing writing on skin by touch alone
- Touch localization (bilateral simultaneous stimulation)

The parietal lobe plays important roles in integrating sensory information from various parts of the body, knowledge of numbers and their relations,[3] and in the manipulation of objects. Its function also includes processing information relating to the sense of touch.[4] Portions of the parietal lobe are involved with visuospatial processing. Although multisensory in nature, the posterior parietal cortex is often referred to by vision scientists as the dorsal stream of vision (as opposed to the ventral stream in the temporal lobe). This dorsal stream has been called both the "where" stream (as in spatial vision)[5] and the "how" stream (as in vision for action).[6] The posterior parietal cortex (PPC) receives somatosensory and/or visual input, which then, through motor signals, controls movement of the arm, hand, as well as eye movements.[7]

Various studies in the 1990s found that different regions of the posterior parietal cortex in macaques represent different parts of space.

- The lateral intraparietal (LIP) contains a map of neurons (retinotopically-coded when the eyes are fixed[8]) representing the saliency of spatial locations, and attention to these spatial locations. It can be used by the oculomotor system for targeting eye movements, when appropriate.[9]

- The ventral intraparietal (VIP) area receives input from a number of senses (visual, somatosensory, auditory, and vestibular[10]). Neurons with tactile receptive fields represent space in a head-centered reference frame.[10] The cells with visual receptive fields also fire with head-centered reference frames[11] but possibly also with eye-centered coordinates[10]

- The medial intraparietal (MIP) area neurons encode the location of a reach target in nose-centered coordinates.[12]

- The anterior intraparietal (AIP) area contains neurons responsive to shape, size, and orientation of objects to be grasped[13] as well as for manipulation of hands themselves, both to viewed[13] and remembered stimuli.[14] The AIP has neurons that are responsible for grasping and manipulating objects through motor and visual inputs. The AIP and ventral premotor working together, are responsible for visuomotor transformations for actions of the hand.[7]

More recent FMRI studies have shown that humans have similar functional regions in and around the intraparietal sulcus and parietal-occipital junction.[15] The human "parietal eye fields" and "parietal reach region", equivalent to LIP and MIP in the monkey, also appear to be organized in gaze-centered coordinates so that their goal-related activity is "remapped" when the eyes move.[16] Both the left and right parietal systems play a determining role in self transcendence, the personality trait measuring predisposition to spirituality.[17]

6.3 Clinical significance

Features of parietal lobe lesions are as follows:

- **Unilateral parietal lobe**
    - Contralateral hemisensory loss
    - Astereognosis – inability to determine 3-D shape by touch.
    - Agraphaesthesia – inability to *read* numbers or letters drawn on hand, with eyes shut.
    - Contralateral homonymous Lower quadrantanopia
    - Asymmetry of optokinetic Nystagmus (OKN)
    - Sensory Seizures
    - Extinction phenomenon (contralateral)

- **Dominant hemisphere**
    - Dysphasia/Aphasia
    - Dyscalculia
    - Dyslexia – a general term for disorders that can involve difficulty in learning to read or interpret words, letters, and other symbols.
    - Apraxia – inability to perform complex movements in the presence of normal motor, sensory and cerebellar function.
    - Agnosia (tactile agnosia) – inability to recognize or discriminate.
    - Gerstmann syndrome – Characterized by acalculia, agraphia, finger anomia and difficulty in differentiation of right and left.

- **Non dominant hemisphere**
    - Spatial disorientation
    - Constructional apraxia
    - Dressing apraxia

- Anosognosia – a condition in which a person suffering disability seems to be unaware of the existence of his or her disability.

Damage to the right hemisphere of this lobe results in the loss of imagery, visualization of spatial relationships and neglect of left-side space and left side of the body. Even drawings may be neglected on the left side. Damage to the left hemisphere of this lobe will result in problems in mathematics, long reading, writing, and understanding symbols. The parietal association cortex enables individuals to read, write, and solve mathematical problems.The sensory inputs from the right side of the body go to the left side of the brain and vice versa.

The syndrome of hemispatial neglect is usually associated with large deficits of attention of the non-dominant hemisphere. Optic ataxia is associated with difficulties reaching toward objects in the visual field opposite to the side of the parietal damage. Some aspects of optic ataxia have been explained in terms of the functional organization described above.

Apraxia is a disorder of motor control which can be referred neither to "elemental" motor deficits nor to general cognitive impairment. The concept of apraxia was shaped by Hugo Liepmann about a hundred years ago.[18][19] Apraxia is predominantly a symptom of left brain damage, but some symptoms of apraxia can also occur after right brain damage.[20]

Amorphosynthesis is a loss of perception on one side of the body caused by a lesion in the parietal lobe. Usually, left-sided lesions cause agnosia, a full-body loss of perception, while right-sided lesions cause lack of recognition of the person's left side and extrapersonal space. The term amorphosynthesis was coined by D. Denny-Brown to describe patients he studied in the 1950s.[21]

Can also result in sensory impairment where one of the affected person's senses (sight, hearing, smell, touch, taste and spatial awareness) is no longer normal. [22]

Several studies have suggested that abnormal parietal function may be associated with schizophrenia. There is a possibility that grey matter abnormalities begin in parietal and occipital lobes, progress towards the frontal regions, causing schizophrenia structural and functional alterations.[23]

6.4 Additional images

- Lobes
- Drawing to illustrate the relations of the brain to the skull.
- Ventricles of brain and basal ganglia.Superior view. Horizontal section.Deep dissection

6.5 See also

- Lobes of the brain

6.6 References

[1] http://www.ruf.rice.edu/~{}lngbrain/cglidden/parietal.html

[2] Schacter, D. L., Gilbert, D. L. & Wegner, D. M. (2009). Psychology. (2nd ed.). New Work (NY): Worth Publishers.

[3] Blakemore & Frith (2005). *The Learning Brain*. Blackwell Publishing. ISBN 1-4051-2401-6

[4] Penfield, W., & Rasmussen, T. (1950). *The cerebral cortex of a man: A clinical study of localization of function*. New York: Macmillan.

[5] Mishkin M, Ungerleider LG. (1982) Contribution of striate inputs to the visuospatial functions of parieto-preoccipital cortex in monkeys. Behav Brain Res. 1982 Sep;6(1):57-77.

[6] Goodale MA, Milner AD. Separate visual pathways for perception and action. Trends Neurosci. 1992 Jan;15(1):20-5.

[7] Fogassi L, Luppino G. (2005).*Motor functions of the parietal lobe*. Current Opinion in Neurobiology, 15:626-631.

[8] Kusunoki M, Goldberg ME (March 2003). "The time course of perisaccadic receptive field shifts in the lateral intraparietal area of the monkey". *J. Neurophysiol.* **89** (3): 1519–27. doi:10.1152/jn.00519.2002. PMID 12612015.

[9] Goldberg ME, Bisley JW, Powell KD, Gottlieb J (2006). "Saccades, salience and attention: the role of the lateral intraparietal area in visual behavior". *Prog. Brain Res.* Progress in Brain Research. **155**: 157–75. doi:10.1016/S0079-6123(06)55010-1. ISBN 9780444519276. PMC 3615538. PMID 17027387.

[10] Avillac M, Deneve S, Olivier E, Pouget A, Duhamel JR (2005). "Reference frames for representing visual and tactile locations in parietal cortex". *Nat Neurosci.* **8** (7): 941–9. doi:10.1038/nn1480. PMID 15951810.

[11] Zhang T, Heuer HW, Britten KH (2004). "Parietal area VIP neuronal responses to heading stimuli are encoded in head-centered coordinates". *Neuron.* **42** (6): 993–1001. doi:10.1016/j.neuron.2004.06.008. PMID 15207243.

[12] Pesaran B, Nelson MJ, Andersen RA (2006). "Dorsal premotor neurons encode the relative position of the foot, eye, and goal during reach planning". *Neuron.* **51** (1): 125–34. doi:10.1016/j.neuron.2006.05.025. PMC 3066049. PMID 16815337.

[13] Murata A, Gallese V, Luppino G, Kaseda M, Sakata H (May 2000). "Selectivity for the shape, size, and orientation of objects for grasping in neurons of monkey parietal area AIP". *J. Neurophysiol.* **83** (5): 2580–601. PMID 10805659.

[14] Murata A, Gallese V, Kaseda M, Sakata H (May 1996). "Parietal neurons related to memory-guided hand manipulation". *J. Neurophysiol.* **75** (5): 2180–6. PMID 8734616.

[15] Culham JC, Valyear KF (2006). "Human parietal cortex in action". *Curr Opin Neurobiol.* **16** (2): 205–12. doi:10.1016/j.conb.2006.03.005. PMID 16563735.

[16] Medendorp WP, Goltz HC, Vilis T, Crawford JD. (2003) Gaze-centered updating of visual space in human parietal cortex. *J Neurosci.* 16;23(15):6209-14.

[17] Urgesi, Cosimo; S M Aglioti; M Skrap; F Fabbro (2010-02-11). "The Spiritual Brain: Selective Cortical Lesions Modulate Human Self-Transcendence". *Neuron.* **65** (3): 309–319. doi:10.1016/j.neuron.2010.01.026. Retrieved 2012-05-19.

[18] Goldenberg, George (2009). "Apraxia and the Parietal Lobes". *Neuropsychologia.* **47** (6): 1449–1459. doi:10.1016/j.neuropsychologia.2008.07.014. PMID 18692079. Retrieved 2012-05-19.

[19] Liepmann, 1900

[20] Khan AZ, Pisella L, Vighetto A, Cotton F, Luauté J, Boisson D, Salemme R, Crawford JD, Rossetti Y, et al. (2011). "Optic ataxia errors depend on remapped, not viewed, target location". *Nat Neurosci.* **8** (4): 418–20. doi:10.1038/nn1425. PMID 15768034.

[21] Denny-Brown, D., and Betty Q. Banker. "Amorphosynthesis from Left Parietal Lesion." A.M.A. Archives of Neurology and Psychiatry 71, no. 3 (March 1954): 302-13.

[22] http://alzheimers.about.com/library/blparietal.htm

[23] Murat Yildiz et al. "Parietal Lobes in Schizophrenia: Do They Matter?", Schizophrenia Research and Treatment Volume 2011 (2011)

Chapter 7

Temporal lobe

The **temporal lobe** is one of the four major lobes of the cerebral cortex in the brain of mammals. The temporal lobe is located beneath the lateral fissure on both cerebral hemispheres of the mammalian brain.[3]

The temporal lobe is involved in processing sensory input into derived meanings for the appropriate retention of visual memory, language comprehension, and emotion association.[4]:21

7.1 Structure

7.1.1 Medial temporal lobe

The medial temporal lobe consists of structures that are vital for declarative or long-term memory. Declarative (denotative) or explicit memory is conscious memory divided into semantic memory (facts) and episodic memory (events).[4]:194 Medial temporal lobe structures that are critical for long-term memory include the hippocampus, along with the surrounding hippocampal region consisting of the perirhinal, parahippocampal, and entorhinal neocortical regions.[4]:196 The hippocampus is critical for memory formation, and the surrounding medial temporal cortex is currently theorized to be critical for memory storage.[4]:21 The prefrontal and visual cortices are also involved in explicit memory.[4]:21

Research has shown that lesions in the hippocampus of monkeys results in limited impairment of function, whereas extensive lesions that include the hippocampus and the medial temporal cortex result in severe impairment.[5]

7.2 Function

7.2.1 Visual memories

The temporal lobe communicates with the hippocampus and plays a key role in the formation of explicit long-term memory modulated by the amygdala.[4]:349

7.2.2 Processing sensory input

Auditory Adjacent areas in the superior, posterior, and lateral parts of the temporal lobes are involved in high-level auditory processing. The temporal lobe is involved in primary auditory perception, such as hearing, and holds the primary auditory cortex.[6] The primary auditory cortex receives sensory information from the ears and secondary areas process the information into meaningful units such as speech and words.[6] The superior temporal gyrus includes an area (within the lateral fissure) where auditory signals from the cochlea first reach the cerebral cortex and are processed by the primary auditory cortex in the left temporal lobe.

Visual The areas associated with vision in the temporal lobe interpret the meaning of visual stimuli and establish object recognition. The ventral part of the temporal cortices appear to be involved in high-level visual processing of complex stimuli such as faces (fusiform gyrus) and scenes (parahippocampal gyrus). Anterior parts of this ventral stream for visual processing are involved in object perception and recognition.[6]

7.2.3 Language recognition

The left temporal lobe holds the primary auditory cortex, which is important for the processing of semantics in both speech and vision in humans. Wernicke's area, which spans the region between temporal and parietal lobes, plays a key role (in tandem with Broca's area in the frontal lobe) in speech comprehension.[7] The functions of the left temporal lobe are not limited to low-level perception but extend to comprehension, naming, and verbal memory.

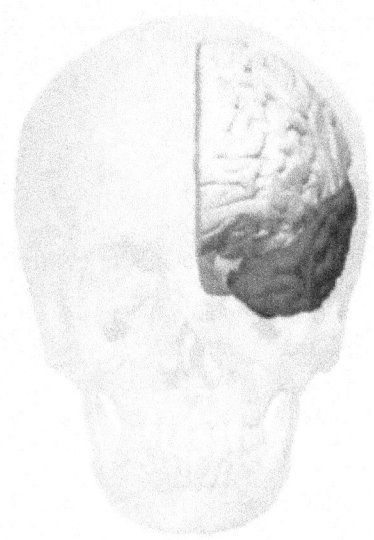

Animation of the human left temporal lobe

7.2.4 New memories

See also: Emotion and memory

The medial temporal lobes (near the sagittal plane) are thought to be involved in encoding declarative long term memory.[4]:194–199 The medial temporal lobes include the hippocampi, which are essential for memory storage, therefore damage to this area can result in impairment in new memory formation leading to permanent or temporary anterograde amnesia.[4]:194–199

7.3 Clinical significance

7.3.1 Unilateral temporal lesion

- Contralateral homonymous upper quadrantanopia (sector anopsia)
- Complex hallucinations (smell, sound, vision, memory)

7.3.2 Dominant hemisphere

- Receptive aphasia
  - Wernicke's aphasia
  - Anomic aphasia
- Dyslexia
- Impaired verbal memory
- Word agnosia, word deafness

7.3.3 Non-dominant hemisphere

- Impaired non-verbal memory
- Impaired musical skills
- Hiphopanasia

7.3.4 Bitemporal lesions (additional features)

- Deafness
- Apathy (affective indifference)
- Impaired learning and memory
- Amnesia, Korsakoff syndrome, Klüver–Bucy syndrome

7.3.5 Damage

Individuals who suffer from medial temporal lobe damage have a difficult time recalling visual stimuli. This neurotransmission deficit is due, not to lacking perception of visual stimuli but, to lacking perception of interpretation.[8] The most common symptom of inferior temporal lobe damage is visual agnosia, which involves impairment in the identification of familiar objects. Another less common type of inferior temporal lobe damage is prosopagnosia which is an impairment in the recognition of faces and distinction of unique individual facial features.[9]

Damage specifically to the anterior portion of the left temporal lobe can cause savant syndrome.[10]

7.3.6 Disorders

Pick's disease, also known as *frontotemporal amnesia*, is caused by atrophy of the frontotemporal lobe.[11] Emotional symptoms include mood changes, which the patient may be unaware of, including poor attention span and aggressive behavior towards themselves and/or others. Language symptoms include loss of speech, inability to read and/or write, loss of vocabulary and overall degeneration of motor ability.[12]

Temporal lobe epilepsy is a chronic neurological condition characterized by recurrent seizures; symptoms include a variety of sensory (visual, auditory, olfactory, and gustation) hallucinations, as well as an inability to process semantic and episodic memories.[13]

7.4 References

[1] Starr, Philip A.; Barbaro, Nicholas M.; Larson, Paul S. (30 November 2008). *Neurosurgical Operative Atlas: Functional Neurosurgery*. Thieme. pp. 16, 26. ISBN 9781588903990.

[2] Sekhar, Laligam N.; de Oliveira, Evandro (1999). *Cranial Microsurgery: Approaches and Techniques*. Thieme. p. 432. ISBN 9780865776982.

[3] "Temporal Lobe". *Langbrain*. Rice University. Retrieved 2 January 2011.

[4] Smith; Kosslyn (2007). *Cognitive Psychology: Mind and Brain*. New Jersey: Prentice Hall. pp. 21, 194–199, 349.

[5] Squire, LR; Stark, CE; Clark, RE (2004). "The medial temporal lobe" (PDF). *Annual Review of Neuroscience*. **27**: 279–306. doi:10.1146/annurev.neuro.27.070203.144130. PMID 15217334.

[6] Schacter, Daniel L.; Gilbert, Daniel T.; Wegner, Daniel M. (2010). *Psychology* (2nd ed.). New York: Worth Publishers. ISBN 9781429237192.

[7] Hickok, Gregory; Poeppel, David (May 2007). "The Cortical Organization of Speech Processing". *Nature Reviews Neuroscience*. **8** (5): 393–402. doi:10.1038/nrn2113. PMID 17431404. Retrieved 24 May 2014.

[8] Pertzov, Y., Miller, T. D., Gorgoraptis, N., Caine, D., Schott, J. M., Butler, C., & Husain, M. (2013). Binding deficits in memory following medial temporal lobe damage in patients with voltage-gated potassium channel complex antibody-associated limbic encephalitis. Brain: A Journal Of Neurology, 136(8), 2474-2485.

[9] Mizuno, T., & Takeda, K. (2009). [The symptomatology of frontal and temporal lobe damages]. Brain And Nerve = Shinkei Kenkyū No Shinpo, 61(11), 1209-1218.

[10] Treffert, D. A. (2009). "The savant syndrome: An extraordinary condition. A synopsis: Past, present, future". *Philosophical Transactions of the Royal Society B: Biological Sciences*. **364** (1522): 1351–7. doi:10.1098/rstb.2008.0326. PMC 2677584. PMID 19528017.

[11] Takeda, N.; Kishimoto, Y.; Yokota, O. (2012). "Pick's disease". *Advances in Experimental Medicine and Biology*. **724**: 300–316. doi:10.1007/978-1-4614-0653-2_23.

[12] Yokota, O.; Tsuchiya, K.; Arai, T.; Yagishita, S.; Matsubara, O.; Mochizuki, A.; Akiyama, H. (2009). "Clinicopathological characterization of Pick's disease versus frontotemporal lobar degeneration with ubiquitin/TDP-43-positive inclusions". *Acta Neuropathologica*. **117** (4): 429–444. doi:10.1007/s00401-009-0493-4.

[13] Lah, S., & Smith, M. (2013). Semantic and Episodic Memory in Children With Temporal Lobe Epilepsy: Do They Relate to Literacy Skills?. *Neuropsychology* doi:10.1037/neu0000029

7.5 External links

- The medial temporal lobe memory system
- H. M.'s Medial Temporal Lobe Lesion: Findings from Magnetic Resonance Imaging

Chapter 8

Occipital lobe

The **occipital lobe** is one of the four major lobes of the cerebral cortex in the brain of mammals. The occipital lobe is the visual processing center of the mammalian brain containing most of the anatomical region of the visual cortex.[1] The primary visual cortex is Brodmann area 17, commonly called V1 (visual one). Human V1 is located on the medial side of the occipital lobe within the calcarine sulcus; the full extent of V1 often continues onto the posterior pole of the occipital lobe. V1 is often also called striate cortex because it can be identified by a large stripe of myelin, the Stria of Gennari. Visually driven regions outside V1 are called extrastriate cortex. There are many extrastriate regions, and these are specialized for different visual tasks, such as visuospatial processing, color differentiation, and motion perception. The name derives from the overlying occipital bone, which is named from the Latin **ob**, *behind*, and **caput**, *the head*. Bilateral lesions of the occipital lobe can lead to cortical blindness (See Anton's syndrome).

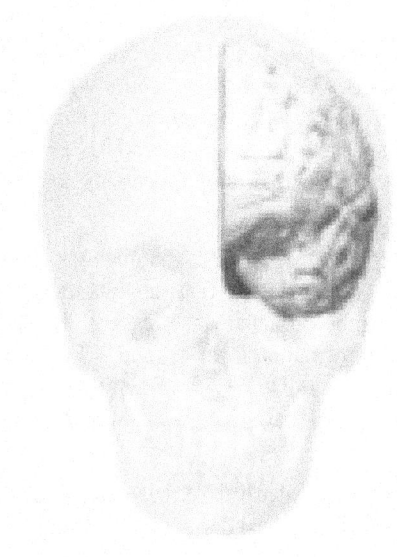

Animation. Occipital lobe (red) of left cerebral hemisphere.

8.1 Structure

The two occipital lobes are the smallest of four paired lobes in the human cerebral cortex. Located in the rearmost portion of the skull, the occipital lobes are part of the forebrain. None of the cortical lobes are defined by any internal structural features, but rather by the bones of the head bone that overlie them. Thus, the occipital lobe is defined as the part of the cerebral cortex that lies underneath the occipital bone. (See the human brain article for more information.)

The lobes rest on the tentorium cerebelli, a process of dura mater that separates the cerebrum from the cerebellum. They are structurally isolated in their respective cerebral hemispheres by the separation of the cerebral fissure. At the front edge of the occipital are several lateral occipital gyri, which are separated by lateral occipital sulcus.

The occipital aspects along the inside face of each hemisphere are divided by the calcarine sulcus. Above the medial, Y-shaped sulcus lies the cuneus, and the area below the sulcus is the lingual gyrus.

Damage to the primary visual areas of the occipital lobe can leave a person with partial or complete blindness.[2]

8.2 Function

The occipital lobe is divided into several functional visual areas. Each visual area contains a full map of the visual world. Although there are no anatomical markers distinguishing these areas (except for the prominent striations in the striate cortex), physiologists have used electrode recordings to divide the cortex into different functional regions.

The first functional area is the primary visual cortex. It contains a low-level description of the local orientation, spatial-frequency and color properties within small receptive fields. Primary visual cortex projects to the occipital areas of the ventral stream (visual area V2 and visual area V4), and the occipital areas of the dorsal stream—visual area V3, visual area MT (V5), and the dorsomedial area (DM).

The Ventral stream is known for the processing the "what" in vision, while the dorsal stream handles the "where/how." This is because the ventral stream provides important information for the identification of stimuli that are stored in memory. With this information in memory, the dorsal stream is able to focus on motor actions in response to the outside stimuli.

Although numerous studies have shown that the two systems are independent and structured separately from another, there is also evidence that both are essential for successful perception, especially as the stimuli takes on more complex forms. For example, a case study using fMRI was done on shape and location. The first procedure consisted of location tasks. The second procedure was in a lit-room where participants were shown stimuli on a screen for 600 ms. They found that the two pathways play a role in shape perception even though location processing continues to lie within the dorsal stream [3]

The dorsomedial (DM) is not as thoroughly studied. However, there is some evidence that suggests that this stream interacts with other visual areas. A case study on monkeys revealed that information from V1 and V2 areas make up half the inputs in the DM. The remaining inputs are from multiple sources that have to do with any sort of visual processing [4]

A significant functional aspect of the occipital lobe is that it contains the primary visual cortex.

Retinal sensors convey stimuli through the optic tracts to the lateral geniculate bodies, where optic radiations continue to the visual cortex. Each visual cortex receives raw sensory information from the outside half of the retina on the same side of the head and from the inside half of the retina on the other side of the head. The cuneus (Brodmann's area 17) receives visual information from the contralateral superior retina representing the inferior visual field. The lingula receives information from the contralateral inferior retina representing the superior visual field. The retinal inputs pass through a "way station" in the lateral geniculate nucleus of the thalamus before projecting to the cortex. Cells on the posterior aspect of the occipital lobes' gray matter are arranged as a spatial map of the retinal field. Functional neuroimaging reveals similar patterns of response in cortical tissue of the lobes when the retinal fields are exposed to a strong pattern.

8.3 Clinical significance

If one occipital lobe is damaged, the result can be homonymous hemianopsia vision loss from similarly positioned "field cuts" in each eye. Occipital lesions can cause visual hallucinations. Lesions in the parietal-temporal-occipital association area are associated with color agnosia, movement agnosia, and agraphia. Damage to the primary visual cortex, which is located on the surface of the posterior occipital lobe, can cause blindness due to the holes in the visual map on the surface of the visual cortex that resulted from the lesions.[5]

8.3.1 Epilepsy

Recent studies have shown that specific neurological findings have affected idiopathic occipital lobe epilepsies.[6] Occipital lobe seizures are triggered by a flash, or a visual image that contains multiple colors. These are called flicker stimulation (usually through TV) these seizures are referred to as photo-sensitivity seizures. Patients having experienced occipital seizures described their seizures as featuring bright colors, and severely blurring their vision (vomiting was also apparent in some patients). Occipital seizures are triggered mainly during the day, through television, video games or any flicker stimulatory system.[7] Occipital seizures originate from an epileptic focus confined within the occipital lobes. They may be spontaneous or triggered by external visual stimuli. Occipital lobe epilepsies are etiologically idiopathic, symptomatic, or cryptogenic.[8] Symptomatic occipital seizures can start at any age, as well as any stage after or during the course of the underlying causative disorder. Idiopathic occipital epilepsy usually starts in childhood.[8] Occipital epilepsies account for approximately 5% to 10% of all epilepsies.[8]

8.4 Additional images

- Base of brain.

- Drawing to illustrate the relations of the brain to the skull.

- Occipital lobe in blue

- Occipital lobe

- Occipital lobe

- Ventricles of brain and basal ganglia.Superior view. Horizontal section.Deep dissection

8.5 See also

This article uses anatomical terminology; for an overview, see Anatomical terminology.

- Lobes of the brain
- Regions of the human brain
- Sulcus Lunatus
- Visual evoked potential
- Vertical occipital fasciculus

8.6 References

[1] "SparkNotes: Brain Anatomy: Parietal and Occipital Lobes". Archived from the original on 2007-12-31. Retrieved 2008-02-27.

[2] Schacter, D. L., Gilbert, D. L. & Wegner, D. M. (2009). Psychology. (2nd ed.). New Work (NY): Worth Publishers.

[3] (Valyear, Culham, Sharif, Westwood, & Goodale, 2006).

[4] (Valyear et al., 2006).

[5] Carlson, Neil R. (2007). *Psychology : the science of behaviour*. New Jersey, USA: Pearson Education. p. 115. ISBN 978-0-205-64524-4.

[6] Chilosi, Anna Maria; Brovedani (November 2006). "Neuropsychological Findings in Idiopathic Occipital Lobe Epilepsies". *Epilepsia*. **47** (s2): 76–78. doi:10.1111/j.1528-1167.2006.00696.x. PMID 17105468.

[7] Destina Yalçin, A., Kaymaz, A., & Forta, H. (2000). Reflex occipital lobe epilepsy. Seizure, 9(6), 436-441.

[8] Adcock, Jane E; Panayiotopoulos, Chrysostomos P (31 October 2012). "Journal of Clinical Neurophysiology". *Occipital Lobe Seizures and Epilepsies*. **29** (5): 397–407. doi:10.1097/wnp.0b013e31826c98fe. PMID 23027097.

Chapter 9

Limbic lobe

The **limbic lobe** is an arc-shaped region of cortex on the medial surface of each cerebral hemisphere of the mammalian brain, consisting of parts of the frontal, parietal and temporal lobes. The term is ambiguous, with some authors including the paraterminal gyrus, the subcallosal area, the cingulate gyrus, the parahippocampal gyrus, the dentate gyrus, the hippocampus and the subiculum;[1] while the Terminologia Anatomica includes the cingulate sulcus, the cingulate gyrus, the isthmus of cingulate gyrus, the fasciolar gyrus, the parahippocampal gyrus, the parahippocampal sulcus, the dentate gyrus, the fimbrodentate sulcus, the fimbria of hippocampus, the collateral sulcus, and the rhinal sulcus, and omits the hippocampus.

9.1 History

Broca named the limbic lobe in 1878, identifying it with the cingulate and parahippocampal gyri, and associating it with the sense of smell - Treviranus having earlier noted that, between species, the size of the parahippocampal gyrus varies with the size of the olfactory nerve.[2] In 1937 Papez theorized that a circuit including the hippocampal formation and the cingulate gyrus constitutes the neural substrate of emotional behavior,[3] and Klüver and Bucy reported that, in monkeys, resection involving the hippocampal formation and the amygdaloid complex has a profound effect on emotional responses.[4][5] As a consequence of these publications, the idea that the entire limbic lobe is dedicated to olfaction receded, and a direct connection between emotion and the limbic lobe was established.[6]

9.2 Gallery

- Limbic lobe (shown in red) of left cerebral hemisphere.

- Limbic lobe (shown in orange) of left cerebral hemisphere.

- Limbic lobe (shown in purple) of right cerebral hemisphere.

- Limbic lobe (shown in purple) of right cerebral hemisphere.

9.3 References

[1] Fix, JD (2008). "Gross anatomy of the brain". *Neuroanatomy* (fourth ed.). Philadelphia: Lippincott Williams & Wilkins. p. 6. ISBN 0-7817-7245-1.

[2] Finger, S (2001). "Defining and controlling the circuits of emotion". *Origins of neuroscience: a history of explorations into brain function*. Oxford/NewYork: Oxford University Press. p. 286. ISBN 0-19-506503-4.

[3] Papez, JW (1937). "A proposed mechanism of emotion". *Archives of neurology and psychiatry*. **38**: 725–43. doi:10.1001/archneurpsyc.1937.02260220069003.

[4] Klüver, H; Bucy, PC (1937). ""Psychic blindness" and other symptoms following bilateral temporal lobectomy in Rhesus monkeys". *American Journal of Physiology*. **119**: 352–53.

[5] Klüver, H; Bucy, PC (1939). "Preliminary analysis of functions of the temporal lobes in monkeys". *Archives of Neurology and psychiatry*. **42**: 979–1000. doi:10.1001/archneurpsyc.1939.02270240017001.

[6] Nieuwenhuys, R; Voogd, J; van Huijzen, C (2008). "The greater limbic system". *The human central nervous system* (fourth ed.). Berlin/Heidelberg/New York: Springer-Verlag. p. 917. ISBN 3-540-13441-7.

9.4 External links

- Explanatory diagram

- http://ahsmail.uwaterloo.ca/kin356/ltm/limbic_lobe.html

Chapter 10

Insular cortex

In each hemisphere of the mammalian brain the **insular cortex** (often called **insula**, **insulary cortex** or **insular lobe**) is a portion of the cerebral cortex folded deep within the lateral sulcus (the fissure separating the temporal lobe from the parietal and frontal lobes).

The insulae are believed to be involved in consciousness and play a role in diverse functions usually linked to emotion or the regulation of the body's homeostasis. These functions include perception, motor control, self-awareness, cognitive functioning, and interpersonal experience. In relation to these, it is involved in psychopathology.

The insular cortex is divided into two parts: the larger anterior insula and the smaller posterior insula in which more than a dozen field areas have been identified. The cortical area overlying the insula toward the lateral surface of the brain is the operculum (meaning *lid*). The opercula are formed from parts of the enclosing frontal, temporal, and parietal lobes.

10.1 Structure

10.1.1 Connections

The anterior part of the insula is subdivided by shallow sulci into three or four **short gyri**.

The anterior insula receives a direct projection from the basal part of the ventral medial nucleus of the thalamus and a particularly large input from the central nucleus of the amygdala. In addition, the anterior insula itself projects to the amygdala.

One study on rhesus monkeys revealed widespread reciprocal connections between the insular cortex and almost all subnuclei of the amygdaloid complex. The posterior insula projects predominantly to the dorsal aspect of the lateral and to the central amygdaloid nuclei. In contrast, the anterior insula projects to the anterior amygdaloid area as well as the medial, the cortical, the accessory basal magnocellular, the medial basal, and the lateral amygdaloid nuclei.[1]

The posterior part of the insula is formed by a **long gyrus**.

The posterior insula connects reciprocally with the secondary somatosensory cortex and receives input from spinothalamically activated ventral posterior inferior thalamic nuclei. It has also been shown that this region receives inputs from the ventromedial nucleus (posterior part) of the thalamus that are highly specialized to convey homeostatic information such as pain, temperature, itch, local oxygen status, and sensual touch.[2]

A human neuroimaging study using diffusion tensor imaging revealed that the anterior insula is interconnected to regions in the temporal and occipital lobe, opercular and orbitofrontal cortex, triangular and opercular parts of the inferior frontal gyrus. The same study revealed differences in the anatomical connection patterns between the left and right hemisphere.[3]

The '**circular sulcus of insula**' (or **sulcus of Reil**[4])is a semi-circular sulcus or *fissure*[4] that separates the insula from the neighboring gyri of the operculum[5] in the front, above, and behind.[4]

10.1.2 Cytoarchitecture

The insular cortex has regions of variable cell structure or cytoarchitecture, changing from granular in the posterior portion to agranular in the anterior portion. The insula also receives differential cortical and thalamic input along its length. John Allman and his colleagues have shown that the anterior insular cortex contains a population of neurons, called spindle neurons. These are also called *von Economo neurons*, identified as characterising a distinctive subregion as the agranular frontal insula.[6]

10.1.3 Development

The insular cortex is considered a separate lobe of the telencephalon by some authorities.[7] Other sources see the insula as a part of the temporal lobe.[8] It is also sometimes grouped with limbic structures deep in the brain into a limbic lobe. As a paralimbic cortex, the insular cortex is considered to be a relatively old structure.

10.2 Function

10.2.1 Interoceptive awareness

There is evidence that, in addition to its more conserved functions, the insula may play a role in certain "higher" functions that operate only in humans and great apes. The spindle neurons found at a higher density in the right frontal insular cortex are also found in the anterior cingulate cortex, which is another region that has reached a high level of specialization in great apes. It has been speculated that these neurons are involved in cognitive-emotional processes that are specific to primates including great apes, such as empathy and self-aware emotional feelings. This is supported by functional imaging results showing that the structure and function of the right frontal insula are correlated with the ability to feel one's own heartbeat, or to empathize with the pain of others. It is thought that these functions are not distinct from the "lower" functions of the insula but rather arise as a consequence of the role of the insula in conveying homeostatic information to consciousness.[9][10] The right anterior insula aids interoceptive awareness of body states, such as the ability to time one's own heartbeat. Moreover, greater right anterior insular gray matter volume correlates with increased accuracy in this subjective sense of the inner body, and with negative emotional experience.[11] It is also involved in the control of blood pressure,[12] in particular during and after exercise,[12] and its activity varies with the amount of effort a person believes he/she is exerting.[13][14]

The insular cortex also is where the sensation of pain is judged as to its degree.[15] Further, the insula is where a person imagines pain when looking at images of painful events while thinking about their happening to one's own body.[16] Those with irritable bowel syndrome have abnormal processing of visceral pain in the insular cortex related to dysfunctional inhibition of pain within the brain.[17]

Another perception of the right anterior insula is the degree of nonpainful warmth[18] or nonpainful coldness[19] of a skin sensation. Other internal sensations processed by the insula include stomach or gastric distension.[20][21] A full bladder also activates the insular cortex.[22]

One brain imaging study suggests that the unpleasantness of subjectively perceived dyspnea is processed in the right human anterior insula and amygdala.[23]

The cerebral cortex processing vestibular sensations extends into the insula,[24] with small lesions in the anterior insular cortex being able to cause loss of balance and vertigo.[25]

Other noninteroceptive perceptions include passive listening to music,[26] laughter, and crying,[27] empathy and compassion,[28] and language.[29]

10.2.2 Motor control

In motor control, it contributes to hand-and-eye motor movement,[30][31] swallowing,[32] gastric motility,[33] and speech articulation.[34][35] It has been identified as a "central command" centre that ensures that heart rate and blood pressure increase at the onset of exercise.[36] Research upon conversation links it to the capacity for long and complex spoken sentences.[37] It is also involved in motor learning[38] and has been identified as playing a role in the motor recovery from stroke.[39]

10.2.3 Homeostasis

It plays a role in a variety of homeostatic functions related to basic survival needs, such as taste, visceral sensation, and autonomic control. The insula controls autonomic functions through the regulation of the sympathetic and parasympathetic systems.[40][41] It has a role in regulating the immune system.[42][43][44]

10.2.4 Self

It has been identified as playing a role in the experience of bodily self-awareness,[45][46] sense of agency,[47] and sense body ownership.[48]

10.2.5 Social emotions

The anterior insula processes a person's sense of disgust both to smells[49] and to the sight of contamination and mutilation[50] — even when just imagining the experience.[51] This associates with a mirror neuron-like link between external and internal experiences.

In social experience, it is involved in the processing of norm violations,[52] emotional processing,[53] empathy,[54] and orgasms.[55]

10.2.6 Emotions

The insular cortex, in particular its most anterior portion, is considered a limbic-related cortex. The insula has increasingly become the focus of attention for its role in body representation and subjective emotional experience. In particular, Antonio Damasio has proposed that this region plays a role in mapping visceral states that are associated with emotional experience, giving rise to conscious feelings. This is in essence a neurobiological formulation of the ideas of William James, who first proposed that subjective emotional experience (i.e., feelings) arise from our brain's interpretation of bodily states that are elicited by emotional events. This is an example of embodied cognition.

In terms of function, the insula is believed to process convergent information to produce an *emotionally relevant context for sensory experience*. To be specific, the anterior insula is related more to *olfactory, gustatory, viceroautonomic, and limbic function*, whereas the posterior insula is related more to *auditory-somesthetic-skeletomotor* function. Functional imaging experiments have revealed that the insula has an important role in pain experience and the experience of a number of basic emotions, including anger, fear, disgust, happiness, and sadness.[56]

The anterior insular cortex (AIC) is believed to be responsible for emotional feelings, including maternal and romantic love, anger, fear, sadness, happiness, sexual arousal, disgust, aversion, unfairness, inequity, indignation, uncertainty,[57] disbelief, social exclusion, trust, empathy, sculptural beauty, a 'state of union with God', and hallucinogenic state.[58]

Functional imaging studies have also implicated the insula in conscious desires, such as food craving and drug craving. What is common to all of these emotional states is that they each change the body in some way and are associated with highly salient subjective qualities. The insula is well-situated for the integration of information relating to bodily states into higher-order cognitive and emotional processes. The insula receives information from "homeostatic afferent" sensory pathways via the thalamus and sends output to a number of other limbic-related structures, such as the amygdala, the ventral striatum, and the orbitofrontal cortex, as well as to motor cortices.[59]

A study using magnetic resonance imaging found that the right anterior insula is significantly thicker in people that meditate.[60]

Another study using voxel-based morphometry and MRI on experienced Vipassana meditators was done to extend the findings of Lazar et al., which found increased grey matter concentrations in this and other areas of the brain in experienced meditators.[61]

10.2.7 Salience

Functional imaging research suggests the insula is involved in two types of salience. Interoceptive information processing that links interoception with emotional salience to generate a subjective representation of the body. This involves, first, the anterior insular cortex with the pregenual anterior cingulate cortex (Brodmann area 33) and the anterior and posterior mid-cingulate cortices, and, second, a general salience system concerned with environmental monitoring, response selection, and skeletomotor body orientation that involves all of the insular cortex and the mid-cingulate cortex.[62]

An alternative or perhaps complementary proposal is that the right anterior insular regulates the interaction between the salience of the selective attention created to achieve a task (the dorsal attention system) and the salience of arousal created to keep focused upon the relevant part of the environment (ventral attention system).[63] This regulation of salience might be particularly important during challenging tasks where attention might fatigue and so cause careless mistakes but if there is too much arousal it risks creating poor performance by turning into anxiety.[63]

10.3 Clinical significance

10.3.1 Progressive non-fluent aphasia

Progressive non-fluent aphasia is the deterioration of normal language function that causes individuals to lose the ability to communicate fluently while still being able to comprehend single words and intact other non-linguistic cognition. It is found in a variety of degenerative neurological conditions including Pick's disease, motor neuron disease, corticobasal degeneration, frontotemporal dementia, and Alzheimer's disease. It is associated with hypometabolism[64] and atrophy of the left anterior insular cortex.[65]

10.3.2 Addiction

A number of functional brain imaging studies have shown that the insular cortex is activated when drug abusers are exposed to environmental cues that trigger cravings. This has been shown for a variety of drugs, including cocaine, alcohol, opiates, and nicotine. Despite these findings, the insula has been ignored within the drug addiction literature, perhaps because it is not known to be a direct target of the mesotelencephalic dopamine system, which is central to current dopamine reward theories of addiction. Research published in 2007[66] has shown that cigarette smokers suffering damage to the insular cortex, from a stroke for instance, have their addiction to cigarettes practically eliminated. These individuals were found to be up to 136 times more likely to undergo a disruption of smoking addiction than smokers with damage in other areas. Disruption of addiction was evidenced by self-reported behavior changes such as quitting smoking less than one day after the brain injury, quitting smoking with great ease, not smoking again after quitting, and having no urge to resume smoking since quitting. The study was conducted on average eight years after the strokes, which opens up the possibility that recall bias could have affected the results.[67] More recent prospective studies, which overcome this limitation, have corroborated these findings[68][69] This suggests a significant role for the insular cortex in the neurological mechanisms underlying addiction to nicotine and other drugs, and would make this area of the brain a possible target for novel anti-addiction medication. In addition, this finding suggests that functions mediated by the insula, especially conscious feelings, may be particularly important for maintaining drug addiction, although this view is not represented in any modern research or reviews of the subject.[70]

A recent study in rats by Contreras et al.[71] corroborates these findings by showing that reversible inactivation of the insula disrupts amphetamine conditioned place preference, an animal model of cue-induced drug craving. In this study, insula inactivation also disrupted "malaise" responses to lithium chloride injection, suggesting that the representation of negative interoceptive states by the insula plays a role in addiction. However, in this same study, the conditioned place preference took place immediately after the injection of amphetamine, suggesting that it is the immediate, pleasurable interoceptive effects of amphetamine administration, rather than the delayed, aversive effects of amphetamine withdrawal that are represented within the insula.

A model proposed by Naqvi et al. (see above) is that the insula stores a representation of the pleasurable interoceptive effects of drug use (e.g., the airway sensory effects of nicotine, the cardiovascular effects of amphetamine), and that this representation is activated by exposure to cues that have previously been associated with drug use. A number of functional imaging studies have shown the insula to be activated during the administration of drugs of abuse. Several functional imaging studies have also shown that the insula is activated when drug users are exposed to drug cues, and that this activity is correlated with subjective urges. In the cue-exposure studies, insula activity is elicited when there is no actual change in the level of drug in the body. Therefore, rather than merely representing the interoceptive effects of drug use as it occurs, the insula may play a role in memory for the pleasurable interoceptive effects of past drug use, anticipation of these effects in the future, or both. Such a representation may give rise to conscious urges that feel as if they arise from within the body. This may make addicts feel as if their bodies need to use a drug, and may result in persons with lesions in the insula reporting that their bodies have forgotten the urge to use, according to this study.

10.3.3 Subjective certainty in ecstatic seizures

A common quality in mystical experiences is ineffability, a strong feeling of certainty which cannot be expressed in words. Fabienne Picard proposes a neurological explanation for this subjective certainty, based on clinical research of epilepsy. [72] [73] According to Picard, this feeling of certainty may be caused by a dysfunction of the anterior insula, a part of the brain which is involved in interoception, self-reflection, and in avoiding uncertainty about the internal representations of the world by "anticipation of resolution of uncertainty or risk". This avoidance of uncertainty functions through the comparison between predicted states and actual states, that is, "signaling that we do not understand, i.e., that there is ambiguity."[74] Picard notes that "the concept of insight is very close to that of certainty," and refers to Archimedes "Eureka!"[75][76] Picard hypothesizes that in ecstatic seizures the comparison between predicted states and actual states no longer functions, and that mismatches between predicted state and actual state are no longer processed, blocking "negative emotions and negative arousal arising from predictive unceertainty," which will be experienced as emotional confidence.[77] Picard concludes that "[t]his could lead to a spiritual intepretation in some individuals."[77]

10.3.4 Other clinical conditions

The insular cortex has been suggested to have a role in anxiety disorders,[78] and emotion dysregulation.[79]

10.4 History

The insula was first described by Johann Christian Reil while describing cranial and spinal nerves and plexi.[80] Henry Gray in Gray's Anatomy is responsible for it being known as the *Island of Reil*.[80]

10.5 Additional images

- The insula of the left side, exposed by removing the opercula.

- Coronal section of brain through intermediate mass of third ventricle.

- Coronal section through anterior cornua of lateral ventricles.

- Coronal section of brain through anterior commissure.

- Orbital surface of left frontal lobe.
- Section of brain showing upper surface of temporal lobe.
- Horizontal section of left cerebral hemisphere.
- Human brain frontal (coronal) section
- Human brain view on transverse temporal and insular gyri

10.6 See also

This article uses anatomical terminology; for an overview, see Anatomical terminology.

- List of regions in the human brain

10.7 References

[1] MUFSON, E; MESULAM, M; PANDYA, D (1 July 1981). "Insular interconnections with the amygdala in the rhesus monkey" (PDF). *Neuroscience*. **6** (7): 1231–1248. doi:10.1016/0306-4522(81)90184-6. PMID 6167896.

[2] Craig AD, Chen K, Bandy D, Reiman EM (2000). "Thermosensory activation of insular cortex". *Nat. Neurosci.* **3** (2): 184–90. doi:10.1038/72131. PMID 10649575.

[3] JAKAB, A; MOLNAR, P; BOGNER,P; BERES,M; BERENYI, E (1 Oct 2011). "Connectivity-based parcellation reveals interhemispheric differences in the insula". *Brain Topography*. **25** (3): 264–271. doi:10.1007/s10548-011-0205-y. PMID 22002490.

[4] Johannes Sobotta. "Sobotta's Atlas and Text-book of human anatomy 1909". *Sobotta's Atlas and Text-book of human anatomy 1909*. p. 145. Retrieved November 10, 2013.

[5] "Definition: 'Circular Sulcus Of Insula'". MediLexicon. Retrieved 2012-03-30.

[6] Bauernfeind A; et al. (April 2013). "A volumetric comparison of the insular cortex and its subregions in primates". *Human Evolution*. **64** (4): 263–279. doi:10.1016/j.hevol.2012.12.003 (inactive 2016-06-29).

[7] Brain, MSN Encarta. Archived 2009-10-31.

[8] Kolb, Bryan; Whishaw, Ian Q. (2003). *Fundamentals of human neuropsychology* (5th ed.). [New York]: Worth. ISBN 0-7167-5300-6.

[9] Benedetto De Martino; Dharshan Kumaran; Ben Seymour; Raymond J. Dolan (August 2006). "Frames, Biases, and Rational Decision-Making in the Human Brain". *Science*. **313** (6): 684–687. Bibcode:2006Sci...313..684D. doi:10.1126/science.1128356. PMC 2631940. PMID 16888142.

[10] Gui Xue; Zhonglin Lu; Irwin P. Levin d; Antoine Bechara (2010). "The impact of prior risk experiences on subsequent risky decision-making: The role of the insula". *NeuroImage*. **50** (2): 709–716. doi:10.1016/j.neuroimage.2009.12.097. PMC 2828040. PMID 20045470.

[11] Critchley HD, Wiens S, Rotshtein P, Ohman A, Dolan RJ (February 2004). "Neural systems supporting interoceptive awareness". *Nat. Neurosci.* **7** (2): 189–95. doi:10.1038/nn1176. PMID 14730305.

[12] Lamb K, Gallagher K, McColl R, Mathews D, Querry R, Williamson JW (April 2007). "Exercise-induced decrease in insular cortex rCBF during postexercise hypotension". *Med Sci Sports Exerc.* **39** (4): 672–9. doi:10.1249/mss.0b013e31802f04e0. PMID 17414805.

[13] Williamson JW, McColl R, Mathews D, Mitchell JH, Raven PB, Morgan WP (April 2001). "Hypnotic manipulation of effort sense during dynamic exercise: cardiovascular responses and brain activation". *J. Appl. Physiol.* **90** (4): 1392–9. PMID 11247939.

[14] Williamson JW, McColl R, Mathews D, Ginsburg M, Mitchell JH (September 1999). "Activation of the insular cortex is affected by the intensity of exercise". *J. Appl. Physiol.* **87** (3): 1213–9. PMID 10484598.

[15] Baliki MN, Geha PY, Apkarian AV (February 2009). "Parsing pain perception between nociceptive representation and magnitude estimation". *J. Neurophysiol.* **101** (2): 875–87. doi:10.1152/jn.91100.2008. PMC 3815214. PMID 19073802.

[16] Ogino Y, Nemoto H, Inui K, Saito S, Kakigi R, Goto F (May 2007). "Inner experience of pain: imagination of pain while viewing images showing painful events forms subjective pain representation in human brain". *Cereb. Cortex.* **17** (5): 1139–46. doi:10.1093/cercor/bhl023. PMID 16855007.

[17] Song GH, Venkatraman V, Ho KY, Chee MW, Yeoh KG, Wilder-Smith CH (December 2006). "Cortical effects of anticipation and endogenous modulation of visceral pain assessed by functional brain MRI in irritable bowel syndrome patients and healthy controls". *Pain.* **126** (1–3): 79–90. doi:10.1016/j.pain.2006.06.017. PMID 16846694.

[18] Olausson H, Charron J, Marchand S, Villemure C, Strigo IA, Bushnell MC (November 2005). "Feelings of warmth correlate with neural activity in right anterior insular cortex". *Neurosci. Lett.* **389** (1): 1–5. doi:10.1016/j.neulet.2005.06.065. PMID 16051437.

10.7. REFERENCES

[19] Craig AD, Chen K, Bandy D, Reiman EM (February 2000). "Thermosensory activation of insular cortex". *Nat. Neurosci.* **3** (2): 184–90. doi:10.1038/72131. PMID 10649575.

[20] Ladabaum U, Minoshima S, Hasler WL, Cross D, Chey WD, Owyang C (February 2001). "Gastric distention correlates with activation of multiple cortical and subcortical regions". *Gastroenterology.* **120** (2): 369–76. doi:10.1053/gast.2001.21201. PMID 11159877.

[21] Hamaguchi T, Kano M, Rikimaru H, et al. (June 2004). "Brain activity during distention of the descending colon in humans". *Neurogastroenterol. Motil.* **16** (3): 299–309. doi:10.1111/j.1365-2982.2004.00498.x. PMID 15198652.

[22] Matsuura S, Kakizaki H, Mitsui T, Shiga T, Tamaki N, Koyanagi T (November 2002). "Human brain region response to distention or cold stimulation of the bladder: a positron emission tomography study". *J. Urol.* **168** (5): 2035–9. doi:10.1097/01.ju.0000027600.26331.11 (inactive 2016-06-29). PMID 12394703.

[23] von Leupoldt, A.; Sommer, T.; Kegat, S.; Baumann, H. J.; Klose, H.; Dahme, B.; Buchel, C. (24 January 2008). "The Unpleasantness of Perceived Dyspnea Is Processed in the Anterior Insula and Amygdala" (PDF). *American Journal of Respiratory and Critical Care Medicine.* **177** (9): 1026–1032. doi:10.1164/rccm.200712-1821OC. PMID 18263796.

[24] Kikuchi M, Naito Y, Senda M, et al. (April 2009). "Cortical activation during optokinetic stimulation — an fMRI study". *Acta Otolaryngol.* **129** (4): 440–3. doi:10.1080/00016480802610226. PMID 19116795.

[25] Papathanasiou ES, Papacostas SS, Charalambous M, Eracleous E, Thodi C, Pantzaris M (2006). "Vertigo and imbalance caused by a small lesion in the anterior insula". *Electromyogr Clin Neurophysiol.* **46** (3): 185–92. PMID 16918202.

[26] Brown S, Martinez MJ, Parsons LM (September 2004). "Passive music listening spontaneously engages limbic and paralimbic systems". *NeuroReport.* **15** (13): 2033–7. doi:10.1097/00001756-200409150-00008. PMID 15486477.

[27] Sander K, Scheich H (October 2005). "Left auditory cortex and amygdala, but right insula dominance for human laughing and crying". *J Cogn Neurosci.* **17** (10): 1519–31. doi:10.1162/089892905774597227. PMID 16269094.

[28] http://ccare.stanford.edu/node/89

[29] Bamiou DE, Musiek FE, Luxon LM (May 2003). "The insula (Island of Reil) and its role in auditory processing. Literature review". *Brain Res. Brain Res. Rev.* **42** (2): 143–54. doi:10.1016/S0165-0173(03)00172-3. PMID 12738055.

[30] Anderson TJ, Jenkins IH, Brooks DJ, Hawken MB, Frackowiak RS, Kennard C (October 1994). "Cortical control of saccades and fixation in man. A PET study". *Brain.* **117** (Pt 5): 1073–84. doi:10.1093/brain/117.5.1073. PMID 7953589.

[31] Fink GR, Frackowiak RS, Pietrzyk U, Passingham RE (April 1997). "Multiple nonprimary motor areas in the human cortex". *J. Neurophysiol.* **77** (4): 2164–74. PMID 9114263.

[32] Sörös P, Inamoto Y, Martin RE (August 2009). "Functional brain imaging of swallowing: an activation likelihood estimation meta-analysis". *Hum Brain Mapp.* **30** (8): 2426–39. doi:10.1002/hbm.20680. PMID 19107749.

[33] Penfield W, Faulk ME (1955). "The insula; further observations on its function". *Brain.* **78** (4): 445–70. doi:10.1093/brain/78.4.445. PMID 13293263.

[34] Dronkers NF (November 1996). "A new brain region for coordinating speech articulation". *Nature.* **384** (6605): 159–61. Bibcode:1996Natur.384..159D. doi:10.1038/384159a0. PMID 8906789.

[35] Ackermann H, Riecker A (May 2004). "The contribution of the insula to motor aspects of speech production: a review and a hypothesis". *Brain Lang.* **89** (2): 320–8. doi:10.1016/S0093-934X(03)00347-X. PMID 15068914.

[36] Nowak M, Holm S, Biering-Sørensen F, Secher NH, Friberg L (June 2005). ""Central command" and insular activation during attempted foot lifting in paraplegic humans". *Hum Brain Mapp.* **25** (2): 259–65. doi:10.1002/hbm.20097. PMID 15849712.

[37] Borovsky A, Saygin AP, Bates E, Dronkers N (June 2007). "Lesion correlates of conversational speech production deficits". *Neuropsychologia.* **45** (11): 2525–33. doi:10.1016/j.neuropsychologia.2007.03.023. PMID 17499317.

[38] Mutschler I, Schulze-Bonhage A, Glauche V, Demandt E, Speck O, Ball T (2007). Fitch T, ed. "A rapid sound-action association effect in human insular cortex". *PLoS ONE.* **2** (2): e259. Bibcode:2007PLoSO...2..259M. doi:10.1371/journal.pone.0000259. PMC 1800344. PMID 17327919.

[39] Weiller C, Ramsay SC, Wise RJ, Friston KJ, Frackowiak RS (February 1993). "Individual patterns of functional reorganization in the human cerebral cortex after capsular infarction". *Annals of Neurology.* **33** (2): 181–9. doi:10.1002/ana.410330208. PMID 8434880.

[40] Oppenheimer SM, Gelb A, Girvin JP, Hachinski VC (September 1992). "Cardiovascular effects of human insular cortex stimulation". *Neurology.* **42** (9): 1727–32. doi:10.1212/wnl.42.9.1727. PMID 1513461.

[41] Critchley HD (December 2005). "Neural mechanisms of autonomic, affective, and cognitive integration". *J. Comp. Neurol.* **493** (1): 154–66. doi:10.1002/cne.20749. PMID 16254997.

[42] Pacheco-López G, Niemi MB, Kou W, Härting M, Fandrey J, Schedlowski M (March 2005). "Neural substrates for behaviorally conditioned immunosuppression in the rat". *J. Neurosci.* **25** (9): 2330–7. doi:10.1523/JNEUROSCI.4230-04.2005. PMID 15745959.

[43] Ramírez-Amaya V, Alvarez-Borda B, Ormsby CE, Martínez RD, Pérez-Montfort R, Bermúdez-Rattoni F (June 1996). "Insular cortex lesions impair the acquisition of conditioned immunosuppression". *Brain Behav. Immun.* **10** (2): 103–14. doi:10.1006/brbi.1996.0011. PMID 8811934.

[44] Ramírez-Amaya V, Bermúdez-Rattoni F (March 1999). "Conditioned enhancement of antibody production is disrupted by insular cortex and amygdala but not hippocampal lesions". *Brain Behav. Immun.* **13** (1): 46–60. doi:10.1006/brbi.1998.0547. PMID 10371677.

[45] Karnath HO, Baier B, Nägele T (August 2005). "Awareness of the functioning of one's own limbs mediated by the insular cortex?". *J. Neurosci.* **25** (31): 7134–8. doi:10.1523/JNEUROSCI.1590-05.2005. PMID 16079395.

[46] Craig AD (January 2009). "How do you feel—now? The anterior insula and human awareness". *Nature Reviews Neuroscience.* **10** (1): 59–70. doi:10.1038/nrn2555. PMID 19096369.

[47] Farrer C, Frith CD (March 2002). "Experiencing oneself vs another person as being the cause of an action: the neural correlates of the experience of agency". *NeuroImage.* **15** (3): 596–603. doi:10.1006/nimg.2001.1009. PMID 11848702.

[48] Tsakiris M, Hesse MD, Boy C, Haggard P, Fink GR (October 2007). "Neural signatures of body ownership: a sensory network for bodily self-consciousness". *Cereb. Cortex.* **17** (10): 2235–44. doi:10.1093/cercor/bhl131. PMID 17138596.

[49] Wicker B, Keysers C, Plailly J, Royet JP, Gallese V, Rizzolatti G (October 2003). "Both of us disgusted in My insula: the common neural basis of seeing and feeling disgust". *Neuron.* **40** (3): 655–64. doi:10.1016/S0896-6273(03)00679-2. PMID 14642287.

[50] Wright P, He G, Shapira NA, Goodman WK, Liu Y (October 2004). "Disgust and the insula: fMRI responses to pictures of mutilation and contamination". *NeuroReport.* **15** (15): 2347–51. doi:10.1097/00001756-200410250-00009. PMID 15640753.

[51] Jabbi M, Bastiaansen J, Keysers C (2008). Lauwereyns J, ed. "A common anterior insula representation of disgust observation, experience and imagination shows divergent functional connectivity pathways". *PLoS ONE.* **3** (8): e2939. Bibcode:2008PLoSO...3.2939J. doi:10.1371/journal.pone.0002939. PMC 2491556. PMID 18698355.

[52] Sanfey AG, Rilling JK, Aronson JA, Nystrom LE, Cohen JD (June 2003). "The neural basis of economic decision-making in the Ultimatum Game". *Science.* **300** (5626): 1755–8. Bibcode:2003Sci...300.1755S. doi:10.1126/science.1082976. PMID 12805551.

[53] Phan KL, Wager T, Taylor SF, Liberzon I (June 2002). "Functional neuroanatomy of emotion: a meta-analysis of emotion activation studies in PET and fMRI". *NeuroImage.* **16** (2): 331–48. doi:10.1006/nimg.2002.1087. PMID 12030820.

[54] Singer T (2006). "The neuronal basis and ontogeny of empathy and mind reading: review of literature and implications for future research". *Neurosci Biobehav Rev.* **30** (6): 855–63. doi:10.1016/j.neubiorev.2006.06.011. PMID 16904182.

[55] Ortigue S, Grafton ST, Bianchi-Demicheli F (August 2007). "Correlation between insula activation and self-reported quality of orgasm in women". *NeuroImage.* **37** (2): 551–60. doi:10.1016/j.neuroimage.2007.05.026. PMID 17601749.

[56] Wager, Tor (June 2002). "Functional Neuroanatomy of Emotion: A Meta-Analysis of Emotion Activation Studies in PET and fMRI". *NeuroImage.* **16** (2): 331–48. doi:10.1006/nimg.2002.1087. PMID 12030820.

[57] Vilares I, Howard JD, Fernandes HL, Gottfried JA, Kording KP (2012). "Differential Representations of Prior and Likelihood Uncertainty in the Human Brain". *Current Biology.* **22** (18): 1641–1648. doi:10.1016/j.cub.2012.07.010. PMC 3461114. PMID 22840519.

[58] Craig, A. D. (Bud) (2009). "How do you feel — now? The anterior insula and human awareness" (PDF). *Nature Reviews Neuroscience.* **10** (1): 59–70. doi:10.1038/nrn2555. PMID 19096369.

[59] Craig, A. D. (Bud) (2002). "A new view of pain as a homeostatic emotion" (PDF). *Trends in Neuroscience.* **26** (6): 303–307. doi:10.1016/s0166-2236(03)00123-1.

[60] Sara W. Lazar; Catherine E. Kerr; Rachel H. Wasserman; Jeremy R. Gray; Douglas N. Greve; Michael T. Treadway; Metta McGarvey; Brian T. Quinn; Jeffery A. Dusek; Herbert Benson; Scott L. Rauch; Christopher I. Moore; Bruce Fischl (2005). "Meditation experience is associated with increased cortical thickness". *NeuroReport.* **16** (17): 1893–7. doi:10.1097/01.wnr.0000186598.66243.19. PMC 1361002. PMID 16272874.

[61] http://scan.oxfordjournals.org/cgi/content/full/3/1/55

[62] Taylor KS, Seminowicz DA, Davis KD (September 2009). "Two systems of resting state connectivity between the insula and cingulate cortex". *Hum Brain Mapp.* **30** (9): 2731–45. doi:10.1002/hbm.20705. PMID 19072897.

[63] Eckert MA, Menon V, Walczak A, Ahlstrom J, Denslow S, Horwitz A, Dubno JR (2009). "At the heart of the ventral attention system: the right anterior insula". *Hum Brain*

Mapp. **30** (8): 2530–41. doi:10.1002/hbm.20688. PMC 2712290. PMID 19072895.

[64] Nestor PJ, Graham NL, Fryer TD, Williams GB, Patterson K, Hodges JR (November 2003). "Progressive non-fluent aphasia is associated with hypometabolism centred on the left anterior insula". *Brain.* **126** (Pt 11): 2406–18. doi:10.1093/brain/awg240. PMID 12902311.

[65] Gorno-Tempini ML, Dronkers NF, Rankin KP, et al. (March 2004). "Cognition and anatomy in three variants of primary progressive aphasia". *Annals of Neurology.* **55** (3): 335–46. doi:10.1002/ana.10825. PMC 2362399. PMID 14991811.

[66] Nasir H. Naqvi; David Rudrauf; Hanna Damasio; Antoine Bechara. (January 2007). "Damage to the Insula Disrupts Addiction to Cigarette Smoking" (abstract). *Science.* **315** (5811): 531–4. Bibcode:2007Sci...315..531N. doi:10.1126/science.1135926. PMC 3698854. PMID 17255515.

[67] Vorel SR, Bisaga A, McKhann G, Kleber HD (July 2007). "Insula damage and quitting smoking". *Science.* **317** (5836): 318–9; author reply 318–9. doi:10.1126/science.317.5836.318c. PMID 17641181.

[68] Suner-Soler, R. (2011). "Smoking Cessation 1 Year Post-stroke and Damage to the Insular Cortex". *Stroke.* **43** (1): 131–136. doi:10.1161/STROKEAHA.111.630004. PMID 22052507.

[69] Gaznick, N. (2013). "Basal Ganglia Plus Insula Damage Yields Stronger Disruption of Smoking Addiction Than Basal Ganglia Damage Alone". *Nicotine.* **16** (4): 445–453. doi:10.1093/ntr/ntt172.

[70] Hyman, Steven E. (2005-08-01). "Addiction: A Disease of Learning and Memory" (abstract). *Am J Psychiatry.* **162** (8): 1414–22. doi:10.1176/appi.ajp.162.8.1414. PMID 16055762.

[71] Marco Contreras; Francisco Ceric; Fernando Torrealba (January 2007). "Inactivation of the Interoceptive Insula Disrupts Drug Craving and Malaise Induced by Lithium" (abstract). *Science.* **318** (5850): 655–8. Bibcode:2007Sci...318..655C. doi:10.1126/science.1145590. PMID 17962567.

[72] Picard, Fabienne (2013), "State of belief, subjective certainty and bliss as a product of cortical dysfuntion", *Cortex*, **49** (9): 2494–2500, doi:10.1016/j.cortex.2013.01.006, PMID 23415878

[73] Gschwind, Markus; Picard, Fabienne (2016), "Ecstatic Epileptic Seizures: a glimpse into the multiple roles of the insula", *Frontiers in Behavioral Neuroscience*, **10**, doi:10.3389/fnbeh.2016.00021

[74] Picard 2013, p.2496-2498

[75] Picard 2013, p.2497-2498

[76] See also satori in Japanese Zen

[77] Picard 2013, p.2498

[78] Paulus MP, Stein MB (August 2006). "An insular view of anxiety". *Biol. Psychiatry.* **60** (4): 383–7. doi:10.1016/j.biopsych.2006.03.042. PMID 16780813.

[79] Thayer JF, Lane RD (December 2000). "A model of neurovisceral integration in emotion regulation and dysregulation". *J Affect Disord.* **61** (3): 201–16. doi:10.1016/S0165-0327(00)00338-4. PMID 11163422.

[80] Binder DK, Schaller K, Clusmann H (November 2007). "The seminal contributions of Johann-Christian Reil to anatomy, physiology, and psychiatry". *Neurosurgery.* **61** (5): 1091–6; discussion 1096. doi:10.1227/01.neu.0000303205.15489.23. PMID 18091285.

10.8 External links

- Insular cortex in the *Brede Database* at the Technical University of Denmark. Location and literature citations for the insula

- *synd/1212* at Who Named It?

- Anatomy diagram: 13048.000-1 at Roche Lexicon - illustrated navigator, Elsevier

- Stained brain slice images which include the "insular cortex" at the BrainMaps project

- Thomas P. Naidicha; et al. (1 February 2004). "The Insula: Anatomic Study and MR Imaging Display at 1.5 T". *American Journal of Neuroradiology.* **25** (2): 222–32. PMID 14970021.

- Kakigia R, Nakataa H, Inuia K, Hiroea N, et al. (October 2005). "Intracerebral pain processing in a Yoga Master who claims not to feel pain during meditation". *Eur J Pain.* **9** (5): 581–9. doi:10.1016/j.ejpain.2004.12.006. PMID 16139187. As for fMRI recording, there were remarkable changes in levels of activity in the...SII-insula (mainly the insula)

Chapter 11

Cerebrum

This article is about the larger superior region of the brain. For the smaller inferior region of the brain, see Cerebellum.

The **cerebrum**[help 1] is a large part of the brain containing the cerebral cortex (of the two cerebral hemispheres), as well as several subcortical structures, including the hippocampus, basal ganglia, and olfactory bulb. In humans, the cerebrum is the uppermost region of the central nervous system. The **telencephalon** is the embryonic structure from which the cerebrum develops prenatally. In mammals, the dorsal telencephalon, or pallium, develops into the cerebral cortex, and the ventral telencephalon, or subpallium, becomes the basal ganglia. The cerebrum is also divided into approximately symmetric left and right cerebral hemispheres.

With the assistance of the cerebellum, the cerebrum controls all voluntary actions in the body.

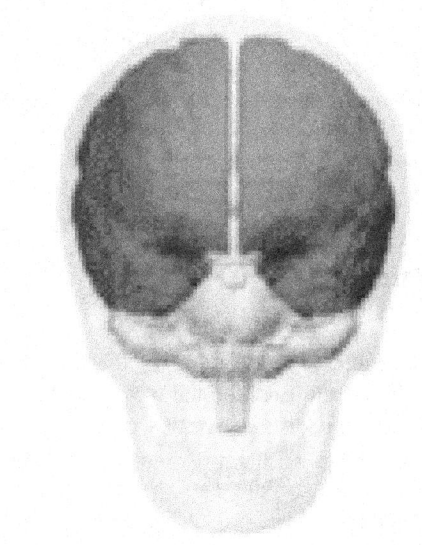

Location of the human cerebrum (red).

11.1 Structure

The cerebrum is the largest part of the brain. Depending upon the position of the animal it lies either in front or on top of the brainstem. In humans, the cerebrum is the largest and best-developed of the five major divisions of the brain. The cerebrum is the newest structure in the phylogenetic sense, and in mammals it is the largest and most developed, out of all known species.

The cerebrum is made up of the two cerebral hemispheres and their cortices, (the outer layers of grey matter), and the underlying regions of white matter.[1] Its subcortical structures include the hippocampus, basal ganglia and olfactory bulb. The cerebrum consists of two cerebral hemispheres, separated from each other by a deep fissure called the lateral sulcus.

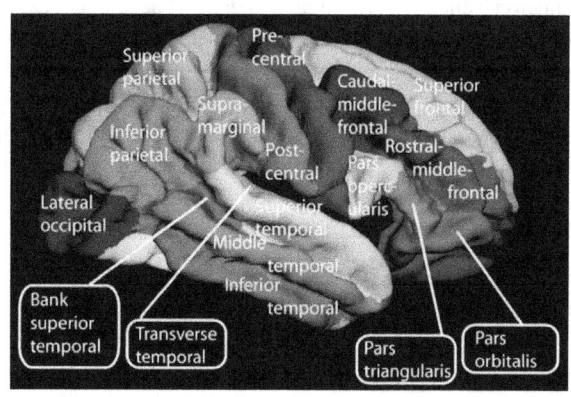

Surface of the cerebrum

11.1.1 Cerebral cortex

The cerebral cortex, the outer layer of gray matter of the cerebrum, is found only in mammals. In larger mammals, including humans, the surface of the cerebral cortex folds to create gyri (ridges) and sulci (furrows) which increase the surface area.[2]

The cerebral cortex is generally classified into four lobes: the frontal, parietal, temporal, and occipital lobes. The lobes are classified based on their overlying neurocranial bones.[3]

11.1.2 Cerebral hemispheres

The cerebrum is divided by the medial longitudinal fissure into two cerebral hemispheres, the right and the left. The right hemisphere controls and processes signals from the left side of the body, while the left hemisphere controls and processes signals from the right side of the body.[3] There is a strong but not complete bilateral symmetry between the hemispheres. The lateralization of brain function looks at the known and possible differences between the two.

11.2 Development

In the developing vertebrate embryo, the neural tube is subdivided into four unseparated sections which then develop further into distinct regions of the central nervous system; these are the prosencephalon (forebrain), the mesencephalon (midbrain) the rhombencephalon (hindbrain) and the spinal cord.[4] The prosencephalon develops further into the telencephalon and the diencephalon. The dorsal telencephalon gives rise to the pallium (cerebral cortex in mammals and reptiles) and the ventral telencephalon generates the basal ganglia. The diencephalon develops into the thalamus and hypothalamus, including the optic vesicles (future retina).[5] The dorsal telencephalon then forms two lateral telencephalic vesicles, separated by the midline, which develop into the left and right cerebral hemispheres. Birds and fish have a dorsal telencephalon, like all vertebrates, but it is generally unlayered and therefore not considered a cerebral cortex. Only a layered cytoarchitecture can be considered a cortex.

11.3 Functions

Note: As the cerebrum is a gross division with many subdivisions and sub-regions, it is important to state that this section lists the functions that the cerebrum *as a whole* serves. See main articles on cerebral cortex and basal ganglia for more information. The cerebrum is a major part of the brain, controlling emotions, hearing, vision, personality and much more. It controls all voluntary actions.

11.3.1 Movement

The cerebrum directs the conscious or volitional motor functions of the body. These functions originate within the primary motor cortex and other frontal lobe motor areas where actions are planned. Upper motor neurons in the primary motor cortex send their axons to the brainstem and spinal cord to synapse on the lower motor neurons, which innervate the muscles. Damage to motor areas of cortex can lead to certain types of motor neuron disease. This kind of damage results in loss of muscular power and precision rather than total paralysis.

It functions as the center of sensory perception, memory, thoughts and judgement; the cerebrum also functions as the center of voluntary motor activities

11.3.2 Sensory processing

The primary sensory areas of the cerebral cortex receive and process visual, auditory, somatosensory, gustatory, and olfactory information. Together with association cortical areas, these brain regions synthesize sensory information into our perceptions of the world.

11.3.3 Olfaction

Main article: Olfaction

The olfactory bulb, responsible for the sense of smell, takes up a large area of the cerebrum in most vertebrates. However, in humans, this part of the brain is much smaller and lies underneath the frontal lobe. The olfactory sensory system is unique since the neurons in the olfactory bulb send their axons directly to the olfactory cortex, rather than to the thalamus first. Damage to the olfactory bulb results in a loss of olfaction (the sense of smell).

11.3.4 Language and communication

Main article: Language

Speech and language are mainly attributed to the parts of the cerebral cortex. Motor portions of language are attributed to Broca's area within the frontal lobe. Speech comprehension is attributed to Wernicke's area, at the temporal-parietal lobe junction. These two regions are interconnected by a large white matter tract, the arcuate fasciculus. Damage to the Broca's area results in expressive aphasia (non-fluent aphasia) while damage to Wernicke's area results in receptive aphasia (also called fluent aphasia).

11.3.5 Learning and memory

Main article: Memory

Explicit or declarative (factual) memory formation is attributed to the hippocampus and associated regions of the medial temporal lobe. This association was originally described after a patient known as HM had both his left and right hippocampus surgically removed to treat chronic temporal lobe epilepsy. After surgery, HM had anterograde amnesia, or the inability to form new memories.

Implicit or procedural memory, such as complex motor behaviors, involves the basal ganglia.

Short-term or working memory involves association areas of the cortex, especially the dorsolateral prefrontal cortex, as well as the hippocampus.

11.4 Other animals

In the most primitive vertebrates, the hagfishes and lampreys, the cerebrum is a relatively simple structure receiving nerve impulses from the olfactory bulb. In cartilaginous and lobe-finned fishes and also in amphibians, a more complex structure is present, with the cerebrum being divided into three distinct regions. The lowermost (or ventral) region forms the basal nuclei, and contains fibres connecting the rest of the cerebrum to the thalamus. Above this, and forming the lateral part of the cerebrum, is the *paleopallium*, while the uppermost (or dorsal) part is referred to as the *archipallium*. The cerebrum remains largely devoted to olfactory sensation in these animals, in contrast to its much wider range of functions in amniotes.[6]

In ray-finned fishes the structure is somewhat different. The inner surfaces of the lateral and ventral regions of the cerebrum bulge up into the ventricles; these include both the basal nuclei and the various parts of the pallium and may be complex in structure, especially in teleosts. The dorsal surface of the cerebrum is membranous, and does not contain any nervous tissue.[6]

In the amniotes, the cerebrum becomes increasingly large and complex. In reptiles, the paleopallium is much larger than in amphibians and its growth has pushed the basal nuclei into the central regions of the cerebrum. As in the lower vertebrates, the grey matter is generally located beneath the white matter, but in some reptiles, it spreads out to the surface to form a primitive cortex, especially in the anterior part of the brain.[6]

In mammals, this development proceeds further, so that the cortex covers almost the whole of the cerebral hemispheres, especially in more developed species, such as the primates. The paleopallium is pushed to the ventral surface of the brain, where it becomes the olfactory lobes, while the archipallium becomes rolled over at the medial dorsal edge to form the hippocampus. In placental mammals, a corpus callosum also develops, further connecting the two hemispheres. The complex convolutions of the cerebral surface (see gyrus, gyrification) are also found only in higher mammals.[6] Although some large mammals (such as elephants) have particularly large cerebra, dolphins are the only species (other than humans) to have cerebra accounting for as much as 2 percent of their body weight.[7]

The cerebrum of birds are similarly enlarged to those of mammals, by comparison with reptiles. The increased size of bird brains was classically attributed to enlarged basal ganglia, with the other areas remaining primitive, but this view has been largely abandoned.[8] Birds appear to have undergone an alternate process of encephalization,[9] as they diverged from the other archosaurs, with few clear parallels to that experienced by mammals and their therapsid ancestors.

11.5 Additional Images

- Cerebrum. Lateral face.Deep dissection.
- Cerebrum. Medial face.Deep dissection.

11.6 See also

This article uses anatomical terminology; for an overview, see Anatomical terminology.

- List of regions in the human brain

11.7 Notes

[1] The word *cerebrum* is pronounced /ˈsɛrɨbrəm/ or /sɨˈriːbrəm/ (both are common). It comes from Latin, where it means "brain". In English it refers to the major part of the brain.

11.8 References

[1] Arnould-Taylor, William (1998). *A Textbook of Anatomy and Physiology*. Nelson Thornes. p. 52. Retrieved 27 January 2015.

[2] Angevine, J.; Cotman, C. (1981). *Principles of Neuroanatomy*. NY: Oxford University Press.

[3] Rosdahl, Caroline; Kowalski, Mary (2008). *Textbook of Basic Nursing* (9th ed.). Lippincott Williams & Wilkins. p. 189. Retrieved 28 January 2015.

[4] Gilbert, Scott F. (2014). *Developmental biology* (10th ed.). Sunderland, Mass.: Sinauer. ISBN 978-0-87893-978-7.

[5] Kandel, Eric R., ed. (2006). *Principles of neural science* (5th ed.). Appleton and Lange: McGraw-Hill. ISBN 978-0-07-139011-8.

[6] Romer, Alfred Sherwood; Parsons, Thomas S. (1977). *The Vertebrate Body*. Philadelphia, PA: Holt-Saunders International. pp. 536–543. ISBN 0-03-910284-X.

[7] T.L. Brink (2008). "Unit 4: The Nervous System.". *Psychology: A Student Friendly Approach.* (PDF). p. 62.

[8] Jarvis ED, Güntürkün O, Bruce L, et al. (2005). "Avian brains and a new understanding of vertebrate brain evolution.". *Nat. Rev. Neurosci.* **6** (2): 151–9. doi:10.1038/nrn1606. PMC 2507884. PMID 15685220.

[9] Emery NJ (2006-01-29). "Cognitive ornithology: the evolution of avian intelligence". *Philos. Trans. R. Soc. Lond., B, Biol. Sci.* **361** (1465): 23–43. doi:10.1098/rstb.2005.1736. PMC 1626540. PMID 16553307.

11.9 External links

- Cerebrum Medical Notes on rahulgladwin.com
- NIF Search - Cerebrum via the Neuroscience Information Framework

Chapter 12

White matter

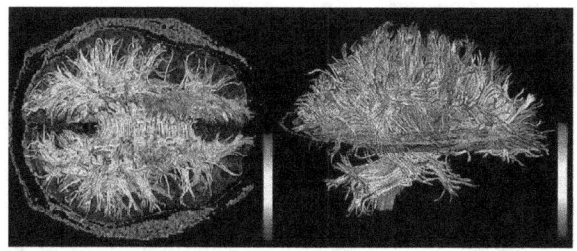

White matter structure of human brain (taken by MRI).

White matter, named for its relatively light appearance resulting from the lipid content of myelin, refers to axon tracts and commissures.

White matter tissue of the freshly cut brain appears pinkish white to the naked eye because myelin is composed largely of lipid tissue veined with capillaries. Its white color in prepared specimens is due to its usual preservation in formaldehyde.

White matter, long thought to be passive tissue, actively affects how the brain learns and functions. While grey matter is primarily associated with processing and cognition, white matter modulates the distribution of action potentials, acting as a relay and coordinating communication between different brain regions.[1]

12.1 Structure

White matter is composed of bundles of myelinated nerve cell projections (or axons), which connect various gray matter areas (the locations of nerve cell bodies) of the brain to each other, and carry nerve impulses between neurons. Myelin acts as an insulator, increasing the speed of transmission of all nerve signals.[2]

The total number of long range fibers within a cerebral hemisphere is 2% of the total number of cortico-cortical fibers (across cortical areas) and is roughly the same number as those that communicate between the two hemispheres in the brain's largest white tissue structure, the Corpus callosum.[3] Schüz and Braitenberg note "As a rough rule, the number of fibres of a certain range of lengths is inversely proportional to their length."[3]

The other main component of the brain is grey matter (actually pinkish tan due to blood capillaries), which is composed of neurons. The substantia nigra is a third colored component found in the brain that appears darker due to higher levels of melanin in dopaminergic neurons than its nearby areas. Note that white matter can sometimes appear darker than grey matter on a microscope slide because of the type of stain used. Cerebral- and spinal white matter do not contain dendrites, neural cell bodies, or shorter axons, which can only be found in grey matter.

White matter in nonelderly adults is 1.7–3.6% blood.[4]

12.1.1 Location

White matter forms the bulk of the deep parts of the brain and the superficial parts of the spinal cord. Aggregates of gray matter such as the basal ganglia (caudate nucleus, putamen, globus pallidus, subthalamic nucleus, nucleus accumbens) and brain stem nuclei (red nucleus, substantia nigra, cranial nerve nuclei) are spread within the cerebral white matter.

The cerebellum is structured in a similar manner as the cerebrum, with a superficial mantle of cerebellar cortex, deep cerebellar white matter (called the "arbor vitae") and aggregates of grey matter surrounded by deep cerebellar white matter (dentate nucleus, globose nucleus, emboliform nucleus, and fastigial nucleus). The fluid-filled cerebral ventricles (lateral ventricles, third ventricle, cerebral aqueduct, fourth ventricle) are also located deep within the cerebral white matter.

12.1.2 Myelinated axon length

Men have more white matter than females both in volume and in length of myelinated axons. At the age of 20, the total length of myelinated fibers in males is 176,000 km while that of a female is 149,000 km.[5] There is a decline in total length with age of about 10% each decade such that a man at 80 years of age has 97,200 km and a female 82,000 km.[5] Most of this reduction is due to the loss of thinner fibers.[5]

One study found that compared to women, men have approximately 6.5 times the amount of gray matter related to general intelligence; and compared to men, women have nearly 10 times the amount of white matter related to general intelligence. Gray matter represents information processing centers in the brain, and white matter represents the networking of – or connections between – these processing centers. [6]

12.2 Function

White matter is the tissue through which messages pass between different areas of gray matter within the central nervous system. The white matter is white because of the fatty substance (myelin) that surrounds the nerve fibers (axons). This myelin is found in almost all long nerve fibers, and acts as an electrical insulation. This is important because it allows the messages to pass quickly from place to place.

There are three different kinds of tracts, or bundles of axons, which connect one part of the brain to another and to the spinal cord, within the white matter:

1. Projection tract extend vertically between higher and lower brain and spinal cord centers, and carry information between the cerebrum and the rest of the body. The cortico spinal tracts, for example, carry motor signals from the cerebrum to the brainstem and spinal cord. Other projection tracts carry signals upward to the cerebral cortex. Superior to the brainstem, such tracts form a broad, dense sheet called the internal capsule between the thalamus and basal nuclei, then radiate in a diverging, fanlike array to specific areas of the cortex.

2. Commissural tracts cross from one cerebral hemisphere to the other through bridges called commissures. The great majority of commissural tracts pass through the large corpus callosum. A few tracts pass through the much smaller anterior and posterior commissures. Commissural tracts enable the left and right sides of the cerebrum to communicate with each other.

3. Association tracts connect different regions within the same hemisphere of the brain. Long association fibers connect different lobes of a hemisphere to each other whereas short association fibers connect different gyri within a single lobe. Among their roles, association tracts link perceptual and memory centers of the brain.[7]

The brain in general (and especially a child's brain) can adapt to white-matter damage by finding alternative routes that bypass the damaged white-matter areas, and can therefore maintain good connections between the various areas of gray matter.

Unlike gray matter, which peaks in development in a person's twenties, the white matter continues to develop, and peaks in middle age.[8] This claim has been disputed in recent years, however.

A 2009 paper by Jan Scholz and colleagues[9] used diffusion tensor imaging (DTI) to demonstrate changes in white matter volume as a result of learning a new motor task (e.g. juggling). The study is important as the first paper to correlate motor learning with white matter changes. Previously, many researchers had considered this type of learning to be exclusively mediated by dendrites, which are not present in white matter. The authors suggest that electrical activity in axons may regulate myelination in axons. Or, gross changes in the diameter or packing density of the axon might cause the change.[10] A more recent DTI study by Sampaio-Baptista and colleagues reported changes in white matter with motor learning along with increases in myelination.[11]

12.3 Clinical significance

Multiple sclerosis (MS) is one of the most common diseases which affect white matter. In MS lesions, the myelin shield around the axons has been destroyed by inflammation.

Alcohol use disorders are associated with decrease in white matter volume.[12] Animal studies suggest that alcohol may cause loss of white matter by damaging oligodendrocytes, the glial cell responsible for maintaining myelin.[13]

Changes in white matter known as amyloid plaques are associated with Alzheimer's disease and other neurodegenerative diseases. White matter injuries ("axonal shearing") may be reversible, while gray matter regeneration is less likely. Other changes that commonly occur with age include the development of leukoaraiosis, which is a rarefaction of the white matter that can be caused by a variety of conditions, including loss of myelin, axonal loss, and a breakdown of the blood–brain barrier.

The study of white matter has been advanced with the neuroimaging technique called diffusion tensor imaging where magnetic resonance imaging (MRI) brain scanners are used. As of 2007, more than 700 publications have been published on the subject.[14]

12.4 References

[1] Douglas Fields, R. (2008). "White Matter Matters". *Scientific American.* **298** (3): 54–61. Bibcode:2008SciAm.298c..54D. doi:10.1038/scientificamerican0308-54.

[2] Klein, S.B., & Thorne, B.M. Biological Psychology. Worth Publishers: New York. 2007.

[3] Schüz, Almut; Braitenberg, Valentino (2002). "The human cortical white matter: Quantitative aspects of cortico-cortical long-range connectivity". In Schüz, Almut; Braitenberg, Valentino. *Cortical Areas: Unity and Diversity, Conceptual Advances in Brain Research.* Taylor and Francis. pp. 377–86. ISBN 978-0-415-27723-5.

[4] Leenders, K. L.; Perani, D.; Lammertsma, A. A.; Heather, J. D.; Buckingham, P.; Jones, T.; Healy, M. J. R.; Gibbs, J. M.; Wise, R. J. S.; Hatazawa, J.; Herold, S.; Beaney, R. P.; Brooks, D. J.; Spinks, T.; Rhodes, C.; Frackowiak, R. S. J. (1990). "Cerebral Blood Flow, Blood Volume and Oxygen Utilization". *Brain.* **113**: 27–47. doi:10.1093/brain/113.1.27. PMID 2302536.

[5] Marner, Lisbeth; Nyengaard, Jens R.; Tang, Yong; Pakkenberg, Bente (2003). "Marked loss of myelinated nerve fibers in the human brain with age". *The Journal of Comparative Neurology.* **462** (2): 144–52. doi:10.1002/cne.10714. PMID 12794739.

[6] University Of California, Irvine. "Intelligence In Men And Women Is A Gray And White Matter." ScienceDaily. ScienceDaily, 22 January 2005. <www.sciencedaily.com/releases/2005/01/050121100142.htm>.

[7] Saladin, Kenneth (2012). *Anatomy & Physiology: The Unity of Form and Function.* New York: McGraw Hill. p. 531. ISBN 978-0-07-337825-1.

[8] Sowell, Elizabeth R.; Peterson, Bradley S.; Thompson, Paul M.; Welcome, Suzanne E.; Henkenius, Amy L.; Toga, Arthur W. (2003). "Mapping cortical change across the human life span". *Nature Neuroscience.* **6** (3): 309–15. doi:10.1038/nn1008. PMID 12548289.

[9] Scholz, Jan; Klein, Miriam C; Behrens, Timothy E J; Johansen-Berg, Heidi (2009). "Training induces changes in white-matter architecture". *Nature Neuroscience.* **12** (11): 1370–1. doi:10.1038/nn.2412. PMC 2770457. PMID 19820707.

[10] "White Matter Matters". Dolan DNA Learning Center. Archived from the original on 2009-11-12. Retrieved 2009-10-19.

[11] Sampaio-Baptista, C.; Khrapitchev, A. A.; Foxley, S.; Schlagheck, T.; Scholz, J.; Jbabdi, S.; Deluca, G. C.; Miller, K. L.; Taylor, A.; Thomas, N.; Kleim, J.; Sibson, N. R.; Bannerman, D.; Johansen-Berg, H. (2013). "Motor Skill Learning Induces Changes in White Matter Microstructure and Myelination". *Journal of Neuroscience.* **33** (50): 19499–503. doi:10.1523/JNEUROSCI.3048-13.2013. PMC 3858622. PMID 24336716.

[12] Monnig, Mollie A.; Tonigan, J. Scott; Yeo, Ronald A.; Thoma, Robert J.; McCrady, Barbara S. (2013). "White matter volume in alcohol use disorders: A meta-analysis". *Addiction Biology.* **18** (3): 581–92. doi:10.1111/j.1369-1600.2012.00441.x. PMC 3390447. PMID 22458455.

[13] Alfonso-Loeches, Silvia; Pascual, Maria; Gómez-Pinedo, Ulises; Pascual-Lucas, Maya; Renau-Piqueras, Jaime; Guerri, Consuelo (2012). "Toll-like receptor 4 participates in the myelin disruptions associated with chronic alcohol abuse". *Glia.* **60** (6): 948–64. doi:10.1002/glia.22327. PMID 22431236.

[14] Assaf, Yaniv; Pasternak, Ofer (2007). "Diffusion Tensor Imaging (DTI)-based White Matter Mapping in Brain Research: A Review". *Journal of Molecular Neuroscience.* **34** (1): 51–61. doi:10.1007/s12031-007-0029-0. PMID 18157658.

12.5 External links

- White Matter Atlas
- WebMD (2009). "white matter". *Webster's New World Medical Dictionary* (3rd ed.). Houghton Mifflin Harcourt. p. 456. ISBN 978-0-544-18897-6.

Chapter 13

Grey matter

For other uses, see Grey matter (disambiguation).

Grey matter (or **gray matter**) is a major component of the central nervous system, consisting of neuronal cell bodies, neuropil (dendrites and myelinated as well as unmyelinated axons), glial cells (astroglia and oligodendrocytes), synapses, and capillaries. Grey matter is distinguished from white matter, in that it contains numerous cell bodies and relatively few myelinated axons, while white matter contains relatively very few cell bodies and is composed chiefly of long-range myelinated axon tracts.[1] The colour difference arises mainly from the whiteness of myelin. In living tissue, grey matter actually has a very light grey colour with yellowish or pinkish hues, which come from capillary blood vessels and neuronal cell bodies.[2]

13.1 Structure

Grey matter refers to unmyelinated neurons and other cells of the central nervous system. It is present in the brain, brainstem and cerebellum, and present throughout the spinal cord.

Grey matter is distributed at the surface of the cerebral hemispheres (cerebral cortex) and of the cerebellum (cerebellar cortex), as well as in the depths of the cerebrum (thalamus; hypothalamus; subthalamus, basal ganglia – putamen, globus pallidus, nucleus accumbens; septal nuclei), cerebellar (deep cerebellar nuclei – dentate nucleus, globose nucleus, emboliform nucleus, fastigial nucleus), brainstem (substantia nigra, red nucleus, olivary nuclei, cranial nerve nuclei).

Grey matter in the spinal cord is known as the grey column which travels down the spinal cord distributed in three grey columns that are presented in an "H" shape. The forward-facing column is the anterior grey column, the rear-facing one is the posterior grey column and the interlinking one is the lateral grey column. The grey matter on the left and right side is connected by the gray commissure. The grey matter in the spinal cord consists of interneurons, as well as cell bodies.

- Diagram of a spinal vertebra. The grey matter is in the central part of the spinal cord.

- Cross-section of spinal cord with grey matter labelled.

Grey matter undergoes development and growth throughout childhood and adolescence.[3]

13.2 Function

Grey matter contains most of the brain's neuronal cell bodies. The grey matter includes regions of the brain involved in muscle control, and sensory perception such as seeing and hearing, memory, emotions, speech, decision making, and self-control.[4]

The grey matter in the spinal cord is split into three grey columns:

- The anterior grey column contains motor neurons. These synapse with interneurons and the axons of cells that have travelled down the pyramidal tract. These cells are responsible for the movement of muscles.

- The posterior grey column contains the points where sensory neurons synapse. These receives sensory information from the body, including fine touch, proprioception, and vibration. This information is sent from receptors of the skin, bones, and joints through sensory neurons whose cell bodies lie in the dorsal root ganglion. This information is then transmitted in axons up the spinal cord in spinal tracts, including the dorsal column-medial lemniscus tract and the spinothalamic tract.

- The lateral grey column is the third column of the spinal cord.

The grey matter of the spinal cord can be divided into different layers, called Rexed laminae. These describe, in general, the purpose of the cells within the grey matter of the spinal cord at a particular location.

- Interneurons present in the grey matter of the spinal cord
- Rexed laminae groups the grey matter in the spinal cord according to its function.

13.3 Clinical significance

See also: Neurobiological effects of physical exercise § Structural growth

13.3.1 Research

Volume and cognition in elderly people

Significant positive correlations have been found between grey matter volume in elderly persons and measures of semantic and short-term memory. No significant correlations with white matter volume were found. These results suggest that individual variability in specific cognitive functions that are relatively well preserved with aging is accounted for by the variability of grey matter volume in healthy elderly subjects.[5]

Volume associated with bipolar disorder

Some structural differences in grey matter may be associated with psychiatric disorders. There was no difference in whole-brain grey matter volume between patients with bipolar I disorder and healthy controls. Subjects with bipolar I disorder had smaller volumes in the left inferior parietal lobule, right superior temporal gyrus, right middle frontal gyrus, and left caudate. Only the volume of the right middle frontal gyrus was correlated with duration of illness and the number of episodes in patients.[6]

Volume associated with smoking

Older smokers lose grey matter and cognitive function at a greater rate than non-smokers. Chronic smokers who quit during the study lost fewer brain cells and retained better intellectual function than those who continued to smoke.[7]

Researchers discovered less grey matter in brains of men who view pornography in large amounts compared with those who do not. Yet a direct influence of pornographic material on shrinking size of grey matter could not have been made. Further studies are urged.[8][9]

Volume associated with poverty

Numerous reports have shown that children in lower-income families do worse on average on IQ and standardized tests compared to children from wealthier families. Later research has investigated the associations between poverty and neural development. This has shown an association between poverty and lower volume and surface area of grey matter in children growing up in poverty.[10] A U.S. study found that at birth there was no significant difference in total grey matter of newborn babies, but at the age of 2 a significant difference in grey matter was found, and by the age of 4 this difference was further pronounced.[11] Test scores have been linked to atypical structural development in the brain, with children living in poverty having a regional grey matter that was 3-4 percentage points below what was deemed the developmental-norm.[10] Similarly, a 2015 study conducted in the USA compared cortical grey matter of 8th graders in Massachusetts. Children from lower-income families not only scored lower in maths and English in standardised tests, but MRI showed that these children statistically had significantly lower grey matter volume and cortical thickness of the bilateral temporal and occipital lobes, compared to children from higher income families.[12] No significant differences were found in white matter volumes. Regional differences are not consistent from study to study. For instance, previous research examined prefrontal cortical thickness in healthy children – a region deemed essential to executive function, which is in turn associated with academic success, and has a long developmental trajectory, which may be susceptible to environmental factors. In a sub-sample of cases from a larger pool of 433 subjects, MRI scans showed the right anterior cingulate gyrus and the left frontal gyrus were significantly correlated with poverty.[13]

13.4 History

13.4.1 Etymology

In the current edition[14] of the official Latin nomenclature, *Terminologia Anatomica*, *substantia grisea* is used for English *grey matter*. The adjective *grisea* for *grey* is however not attested in classical Latin.[15] The adjective *grisea* is derived from the French word for grey, *gris*.[15] Alternative designations like *substantia cana*[16] and *substantia cinerea*[17] are being used alternatively. The adjective *cana*, attested in classical Latin,[18] can mean *grey*,[15] or *greyish white*.[19] The classical Latin *cinerea* means *ash-coloured*.[18]

13.5 Additional images

- Human brain right dissected lateral view
- Schematic representation of the chief ganglionic categories (I to V).

13.6 See also

- Grey matter heterotopia

13.7 References

[1] Purves, Dale, George J. Augustine, David Fitzpatrick, William C. Hall, Anthony-Samuel LaMantia, James O. McNamara, and Leonard E. White (2008). *Neuroscience. 4th ed.* Sinauer Associates. pp. 15–16. ISBN 978-0-87893-697-7.

[2] Kolb & Whishaw: Fundamentals of Human Neuropsychology (2003) page 49

[3] Sowell, Elizabeth; Thompson, Tessner, Toga (15 November 2001). "Mapping Continued Brain Growth and Gray Matter Density Reduction in Dorsal Frontal Cortex: Inverse Relationships during Postadolescent Brain Maturation". *The Journal of Neuroscience.* Cite uses deprecated parameter |coauthors= (help)

[4] Miller, A. K. H.; Alston, Corsellis (28 June 2008). "VARIATION WITH AGE IN THE VOLUMES OF GREY AND WHITE MATTER IN THE CEREBRAL HEMISPHERES OF MAN: MEASUREMENTS WITH AN IMAGE ANALYSER". *Neuropathology and Applied Neurobiology.* **6** (2): 119–132. doi:10.1111/j.1365-2990.1980.tb00283.x. PMID 7374914.

[5] Taki, Y; Kinomura, S; Sato, K; Goto, R; Wu, K; Kawashima, R; Fukuda, H (March 2011). "Correlation between gray/white matter volume and cognition in healthy elderly people.". *Brain and cognition.* **75** (2): 170–176. doi:10.1016/j.bandc.2010.11.008. PMID 21131121. (subscription required)

[6] Li, M; Cui, L; Deng, W; Ma, X; Huang, C; Jiang, L; Wang, Y; Collier, DA; Gong, Q; Li, T (February 28, 2011). "Voxel-based morphometric analysis on the volume of gray matter in bipolar I disorder". *Psychiatry Res.* **191** (2): 92–97. doi:10.1016/j.pscychresns.2010.09.006. PMID 21236649. (subscription required)

[7] "Smoking causes brain cell loss and cognitive decline". *The University of Western Australia.* 9 February 2011. Retrieved 2011-04-21.

[8] Tom Payne (30 May 2014). "Why watching too much porn could be bad for the brain". independent.co.uk.

[9] "Does porn affect the brain? Scientists urge more study", Retrieved December 4, 2015.

[10] Hair, N. L., Hanson, J. L., Wolfe, B. L., & Pollak, S. D. (2015). Association of child poverty, brain development, and academic achievement. JAMA pediatrics, 169(9), 822-829

[11] Hanson, J. L., Hair, N., Shen, D. G., Shi, F., Gilmore, J. H., Wolfe, B. L., & Pollak, S. D. (2013). Family poverty affects the rate of human infant brain growth.

[12] Mackey, A. P., Finn, A. S., Leonard, J. A., Jacoby-Senghor, D. S., West, M. R., Gabrieli, C. F., & Gabrieli, J. D. (2015). Neuroanatomical correlates of the income achievement gap. Psychological science, 0956797615572233.

[13] Lawson, G. M., Duda, J. T., Avants, B. B., Wu, J., & Farah, M. J. (2013). Associations between children's socioeconomic status and prefrontal cortical thickness. Developmental science, 16(5), 641-652.

[14] Federative Committee on Anatomical Terminology (FCAT) (1998). *Terminologia Anatomica.* Stuttgart: Thieme

[15] Triepel, H. (1910). *Die anatomischen Namen. Ihre Ableitung und Aussprache. Mit einem Anhang: Biographische Notizen.*(Dritte Auflage). Wiesbaden: Verlag J.F. Bergmann.

[16] Triepel, H. (1910). *Nomina Anatomica. Mit Unterstützung von Fachphilologen.* Wiesbaden: Verlag J.F. Bergmann.

[17] Schreger, C.H.Th.(1805). *Synonymia anatomica. Synonymik der anatomischen Nomenclatur.* Fürth: im Bureau für Literatur.

[18] Lewis, C.T. & Short, C. (1879). *A Latin dictionary founded on Andrews' edition of Freund's Latin dictionary.* Oxford: Clarendon Press.

[19] Stearn, W.T. (1983). *Botanical Latin. History, grammar, syntax, terminology and vocabulary.* (3rd edition). Newton Abbot London: David Charles.

13.8 External links

- Why Gray Matter is Gray

Chapter 14

Forebrain

In the anatomy of the brain of vertebrates, the **forebrain** or **prosencephalon** is the rostral-most (forward-most) portion of the brain. The forebrain, the midbrain (mesencephalon), and hindbrain (rhombencephalon) are the three primary portions of the brain during early development of the central nervous system. It controls body temperature, reproductive functions, eating, sleeping, and any display of emotions.

At the five-vesicle stage, the forebrain separates into the diencephalon (thalamus, hypothalamus, subthalamus, epithalamus, and pretectum) and the telencephalon which develops into the cerebrum. The cerebrum consists of the cerebral cortex, underlying white matter, and the basal ganglia.

By 5 weeks in utero, it is visible as a single portion toward the front of the fetus. At 8 weeks in utero, the forebrain splits into the left and right cerebral hemispheres.

When the embryonic forebrain fails to divide the brain into two lobes, it results in a condition known as holoprosencephaly.

14.1 See also

- List of regions in the human brain

14.2 External links

- NIF Search - Forebrain via the Neuroscience Information Framework

Chapter 15

Midbrain

The **midbrain** or **mesencephalon** (from the Greek **mesos**, *middle*, and **enkephalos**, *brain*[1]) is a portion of the central nervous system associated with vision, hearing, motor control, sleep/wake, arousal (alertness), and temperature regulation.[2]

15.1 Structure

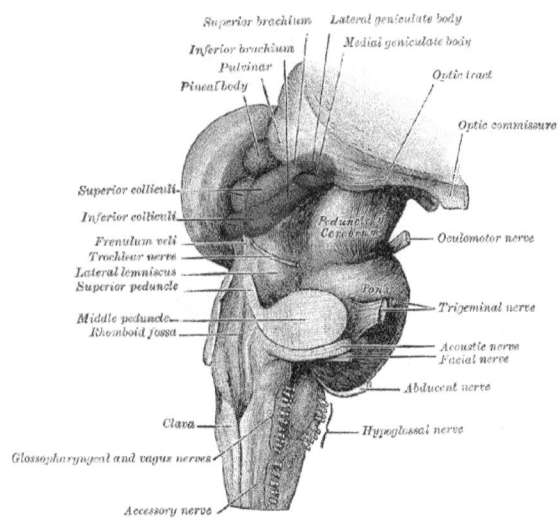

The midbrain comprises the tectum (or corpora quadrigemina), tegmentum, the cerebral aqueduct (or ventricular mesocoelia or "iter"), and the cerebral peduncles, as well as several nuclei and fasciculi. Caudally the midbrain adjoins the metencephalon (afterbrain) (pons and cerebellum); while rostrally it adjoins the diencephalon (thalamus, hypothalamus, etc.). The midbrain is located below the cerebral cortex, and above the hindbrain, placing it near the center of the brain.[3]

Specifically, the midbrain consists of:

- tectum
  - inferior colliculi
  - superior colliculi
- cerebral peduncle
  - midbrain tegmentum
  - crus cerebri
  - substantia nigra

- Brain Anatomy - Mid-fore-hindbrain.

15.1.1 Corpora quadrigemina

Main article: Corpora quadrigemina

The corpora quadrigemina ("quadruplet bodies") are four solid lobes on the dorsal side of the cerebral aqueduct, where the superior posterior pair are called the superior colliculi and the inferior posterior pair are called the inferior colliculi. The homologous structures are called *optic lobes* in some lower vertebrates (fishes, amphibians, and birds) where they integrate sensory information from the eyes and certain auditory reflexes.[4][5]

The four solid lobes help to decussate several fibres of the optic nerve. However, some fibers also show ipsilateral arrangement (i.e., they run parallel on the same side without decussating.)

The superior colliculus is involved with saccadic eye movements; while the inferior is a synapsing point for sound information. The trochlear nerve comes out of the posterior surface of the midbrain, below the inferior colliculus.

15.1.2 Cerebral peduncle

Main article: cerebral peduncles

The cerebral peduncles are paired structures, present on the ventral side of the cerebral aqueduct, and they further carry tegmentum on the dorsal side and cresta or pes on the ventral side, and both of them accommodate the corticospinal tract fibres, from the internal capsule (i.e., ascending + descending tracts = longitudinal tract.) the middle part of cerebral peduncles carry substantia nigra (literally "Black Matter"), which is a type of basal nucleus. It is the only part of the brain that carries melanin pigment.

Between the peduncles is the interpeduncular fossa, which is a cistern filled with cerebrospinal fluid. The oculomotor nerve comes out between the peduncles, and the trochlear nerve is visible wrapping around the outside of the peduncles. The oculomotor is responsible for pupil constriction (parasympathetic) and certain eye movements.[6]

15.1.3 Anatomical features of cross-sections through the midbrain

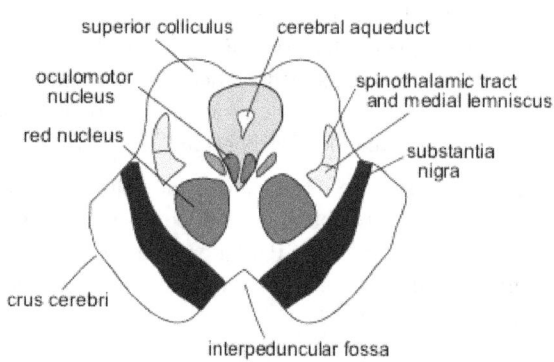

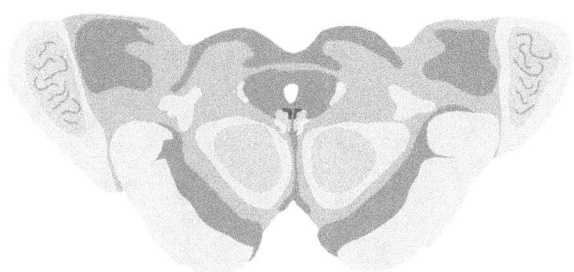

Cross-section of the midbrain at the level of the superior colliculus.

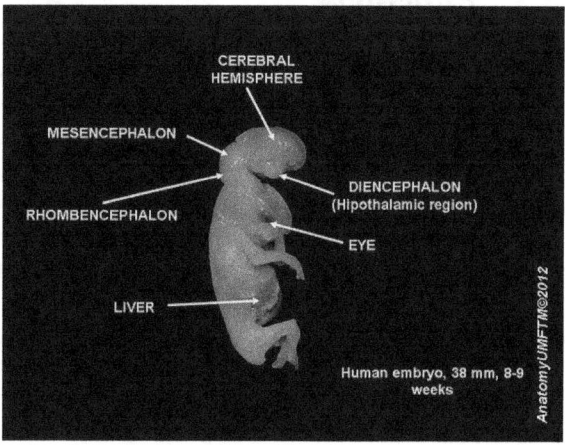

Mesencephalon of human embryo

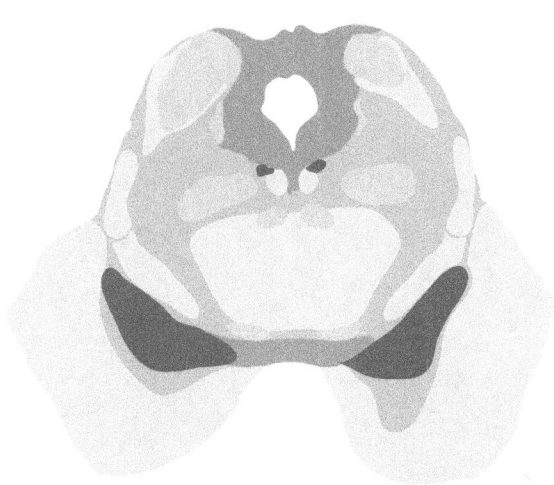

Cross-section of the midbrain at the level of the inferior colliculus.

The midbrain is usually sectioned at the level of the superior and inferior colliculi.

- A horizontal (transverse) cross-section at the level of the superior colliculus shows the red nucleus, the nuclei of the oculomotor nerve (and associated Edinger-Westphal nucleus), the cerebral peduncles or crus cerebri, and the substantia nigra.[7]

- A horizontal (transverse) cross-section at the level of the inferior colliculus still shows the substantia nigra. Also apparent are the trochlear nerve nucleus, and the decussation of the superior cerebellar peduncles.[8]

- Both sections will show the cerebral aqueduct, which connects the third and fourth ventricle and the periaqueductal gray.[9]

One mnemonic for remembering the structures of the midbrain involves visualizing the mesencephalic cross-section as an upside down bear face. The two red nuclei are the eyes of the bear and the cerebral crura are the ears. The tectum is the chin and the cerebral peduncles are the face and ears.

15.1.4 Development

During embryonic development, the midbrain arises from the second vesicle, also known as the mesencephalon, of the neural tube. Unlike the other two vesicles, the forebrain and hindbrain, the midbrain remains undivided for the remainder of neural development. It does not split into other brain areas. while the forebrain, for example, divides into the telencephalon and the diencephalon.[10]

Throughout embryonic development, the cells within the midbrain continually multiply and compress the still-forming cerebral aqueduct. Partial or total obstruction of the cerebral aqueduct during development can lead to congenital hydrocephalus.[11]

15.2 Function

The mesencephalon is considered part of the brainstem. Its substantia nigra is closely associated with motor system pathways of the basal ganglia. The human mesencephalon is archipallian in origin, meaning that its general architecture is shared with the most ancient of vertebrates. Dopamine produced in the substantia nigra and ventral tegmental area plays a role in excitation, motivation and habituation of species from humans to the most elementary animals such as insects. Laboratory house mice from lines that have been selectively bred for high voluntary wheel running have enlarged midbrains.[12] The midbrain helps to relay information for vision and hearing.

15.3 See also

- list of regions in the human brain

15.4 References

[1] Mosby's Medical, Nursing & Allied Health Dictionary, Fourth Edition, Mosby-Year Book 1994, p. 981

[2] Breedlove, Watson, & Rosenzweig. Biological Psychology, 6th Edition, 2010, pp. 45-46

[3] http://www.morris.umn.edu/~{}ratliffj/images/brain_slides/slide_5.htm

[4] Collins Dictionary of Biology, 3rd ed. © W. G. Hale, V. A. Saunders, J. P. Margham 2005

[5] Ferrier, David (1886). "Functions of the optic lobes or corpora quadrigemina". doi:10.1037/12789-005.

[6] Haines, Duane E. *Neuroanatomy: an atlas of structures, sections, and systems* (8th ed.). Philadelphia: Wolters Kluwer/ Lippincott Williams & Wilkins Health. p. 42. ISBN 978-1-60547-653-7.

[7] Martin. Neuroanatomy Text and Atlas, Second edition. 1996, pp. 522-525.

[8] Martin. Neuroanatomy Text and Atlas, Second edition. 1996, pp. 522-525.

[9] Martin. Neuroanatomy Text and Atlas, Second edition. 1996, pp. 522-525.

[10] Martin. Neuroanatomy Text and Atlas, Second Edition, 1996, pp. 35-36.

[11] "Hydrocephalus Fact Sheet". National Institute of Neurological Disorders and Stroke. February 2008. Retrieved 2011-03-23.

[12] Kolb, E. M., E. L. Rezende, L. Holness, A. Radtke, S. K. Lee, A. Obenaus, and Garland T, Jr. 2013. Mice selectively bred for high voluntary wheel running have larger midbrains: support for the mosaic model of brain evolution. Journal of Experimental Biology 216:515-523.

Chapter 16

Hindbrain

The **hindbrain** or **rhombencephalon** is a developmental categorization of portions of the central nervous system in vertebrates. It includes the medulla, pons, and cerebellum. Together they support vital bodily processes.[1]

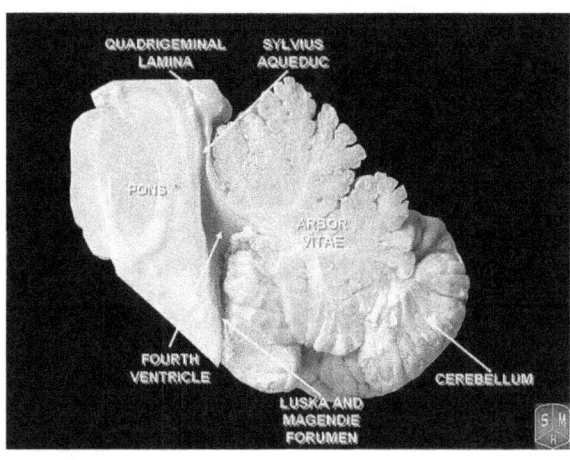

Rhombencephalon

The hindbrain can be subdivided in a variable number of transversal swellings called rhombomeres. In the human embryo eight rhombomeres can be distinguished, from caudal to rostral: Rh8-Rh1. Rostrally, the isthmus demarcates the boundary with the midbrain.

A rare disease of the rhombencephalon—"rhombencephalosynapsis"—is characterized by a missing vermis resulting in a fused cerebellum. Patients generally present with cerebellar ataxia.

The caudal rhombencephalon has been generally considered as the initiation site for neural tube closure.[2]

16.1 Myelencephalon

Rhombomeres Rh8-Rh4 form the myelencephalon.

The myelencephalon forms the medulla oblongata in the adult brain; it contains:

- a portion of the fourth ventricle,
- the glossopharyngeal nerve (CN IX),
- vagus nerve (CN X),
- accessory nerve (CN XI),
- hypoglossal nerve (CN XII),
- and a portion of the vestibulocochlear nerve (CN VIII).

16.2 Metencephalon

Rhombomeres Rh3-Rh1 form the metencephalon.

The metencephalon is composed of the pons and the cerebellum; it contains:

- a portion of the fourth ventricle,
- the trigeminal nerve (CN V),
- abducens nerve (CN VI),
- facial nerve (CN VII),
- and a portion of the vestibulocochlear nerve (CN VIII).

16.3 Evolution

The hindbrain is homologous to a part of the arthropod brain known as the sub-oesophageal ganglion, in terms of the genes that it expresses and its position in between the brain and the nerve cord.[3] On this basis, it has been suggested that the hindbrain first evolved in the Urbilaterian - the last common ancestor of chordates and arthropods - between 570 and 555 million years ago.[3][4]

16.4 Additional images

- Chicken embryo of thirty-three hours' incubation, viewed from the dorsal aspect. X 30.

- Human embryo between eighteen and twenty-one days.

- Hindbrain of human embryo

16.5 References

- Haycock DE (2011). *Being and Perceiving*. Manupod Press. p. 41. ISBN 978-0-9569621-0-2.

[1] "Brain atlas - Hindbrain". Lundbeck Institute - Brain explorer. Retrieved 2015-06-08.

[2] SpringerLink - Journal Article

[3] Ghysen A (2003). "The origin and evolution of the nervous system". *Int. J. Dev. Biol.* **47** (7–8): 555–62. PMID 14756331.

[4] Haycock, DE *Being and Perceiving*

16.6 External links

- NIF Search - Hindbrain via the Neuroscience Information Framework

Chapter 17

Ventricular system

The **ventricular system** is a set of four interconnected cavities (ventricles) in the brain, where the cerebrospinal fluid (CSF) is produced. Within each ventricle is a region of choroid plexus, a network of ependymal cells involved in the production of CSF. The ventricular system is continuous with the central canal of the spinal cord (from the fourth ventricle) allowing for the flow of CSF to circulate. All of the ventricular system and the central canal of the spinal cord is lined with ependyma, a specialised form of epithelium.

17.1 Structure

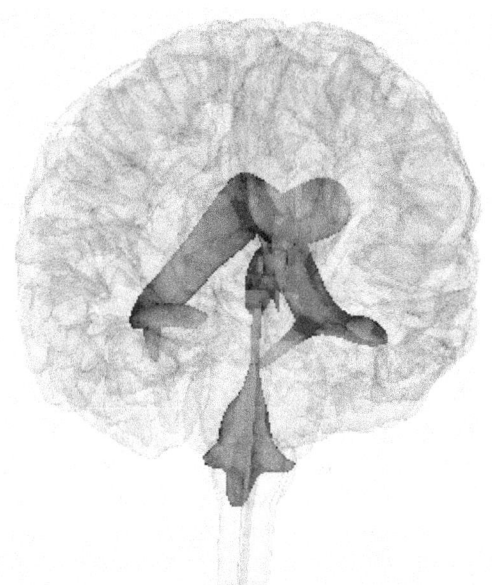

Rotating 3D rendering of the four ventricles.

The system comprises four ventricles:

- lateral ventricles right and left (one for each hemisphere)
- third ventricle
- fourth ventricle

There are several foramina, openings acting as channels, that connect the ventricles. The interventricular foramina (also called the foramina of Monro) connect the lateral ventricles to the third ventricle through which the cerebrospinal fluid can flow.

17.1.1 Ventricles

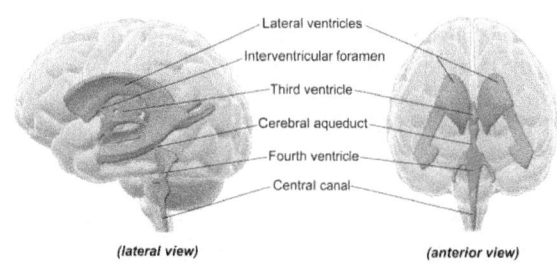

3D rendering of ventricles (lateral and anterior views).

The four cavities of the human brain are called ventricles.[1] The two largest are the lateral ventricles in the cerebrum; the third ventricle is in the diencephalon of the forebrain between the right and left thalamus; and the fourth ventricle is located at the back of the pons and upper half of the medulla oblongata of the hindbrain. The ventricles are concerned with the production and circulation of cerebrospinal fluid[2]

17.1.2 Development

The structures of the ventricular system are embryologically derived from the neural canal, the centre of the neural tube.

As the part of the primitive neural tube that will develop into the brainstem, the neural canal expands dorsally and

17.2 Function

laterally, creating the fourth ventricle, whereas the neural canal that does not expand and remains the same at the level of the midbrain superior to the fourth ventricle forms the cerebral aqueduct. The fourth ventricle narrows at the obex (in the caudal medulla), to become the central canal of the spinal cord.

In more detail, around the third week of development, the embryo is a three-layered disc. The embryo is covered on the dorsal surface by a layer of cells called ectoderm. In the middle of the dorsal surface of the embryo is a linear structure called the notochord. As the ectoderm proliferates, the notochord is dragged into the middle of the developing embryo. The notochord becomes a canal within the embryo known as the neural canal.[3]

As the brain develops, by the fourth week of embryological development several swellings have formed within the embryo around the canal, near where the head will develop. These swellings represent different components of the central nervous system, and are three in number: the prosencephalon, mesencephalon and rhombencephalon. These in turn divide into five sections. As these sections develop around the neural canal, the inner neural canal becomes known as *primitive* ventricles. These form the ventricular system of the brain:[3]

- The prosencephalon divides into the telencephalon, which forms the cortex of the developed brain, and the diencephalon. The ventricles contained within the telencephalon become the lateral ventricles, and the ventricles within the diencephalon become the third ventricle.

- The rhombencephalon divides into a metencephalon and myelencephalon. The ventricles contained within the rhombencephalon become the fourth ventricle, and the ventricles contained within the mesencephalon become the aqueduct of Sylvius.

Separating the anterior horns of the lateral ventricles is the septum pellucidum: a thin, triangular, vertical membrane which runs as a sheet from the corpus callosum down to the fornix. During the third month of fetal development, a space forms between two septal laminae, known as the cave of septum pellucidum (CSP), which is a marker for fetal neural maldevelopment. During the fifth month of development, the laminae start to close and this closure completes from about three to six months after birth. Fusion of the septal laminae is attributed to rapid development of the alvei of the hippocampus, amygdala, septal nuclei, fornix, corpus callosum and other midline structures. Lack of such limbic development interrupts this posterior-to-anterior fusion, resulting in the continuation of the CSP into adulthood.[4]

17.2.1 Flow of cerebrospinal fluid

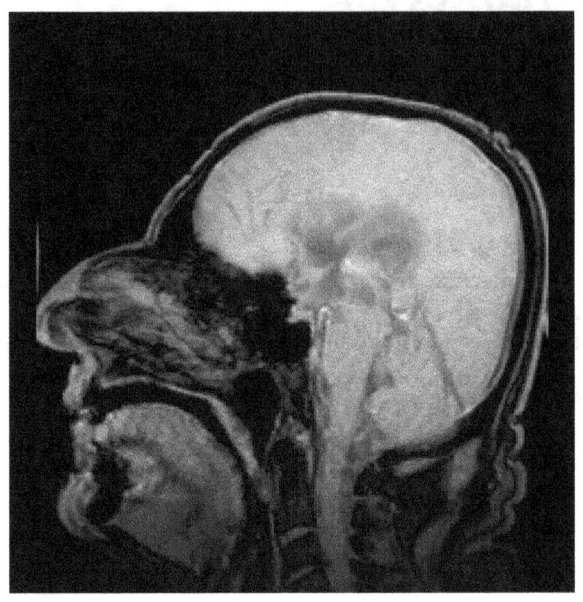

MRI showing flow of CSF

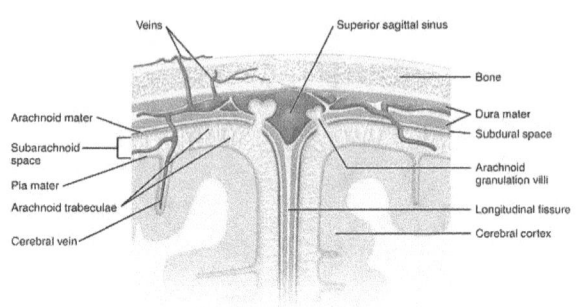

The cerebrospinal fluid passes out through arachnoid villi into the venous sinuses of the skull.

The ventricles are filled with cerebrospinal fluid (CSF) which bathes and cushions the brain and spinal cord within their bony confines. CSF is produced by modified ependymal cells of the choroid plexus found in all components of the ventricular system except for the cerebral aqueduct and the posterior and anterior horns of the lateral ventricles. CSF flows from the lateral ventricles via the foramina of Monro into the third ventricle, and then the fourth ventricle via the cerebral aqueduct in the brainstem. From the fourth ventricle it can pass into the central canal of the spinal cord or into the cisterns of the subarachnoid space via three small foramina: the central foramen of Magendie and the two lateral foramina of Luschka.

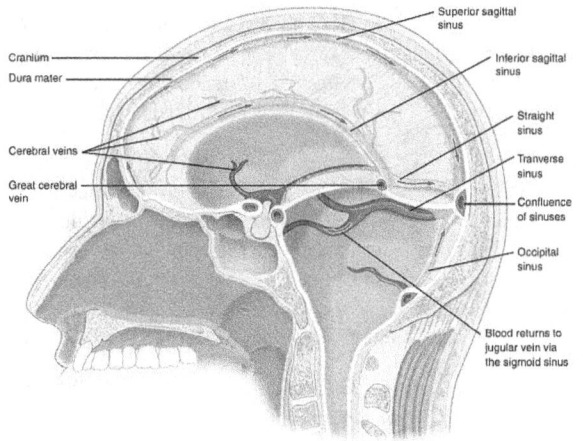

A schematic illustration of the venous sinuses surrounding the brain.

The fluid then flows around the superior sagittal sinus to be reabsorbed via the arachnoid villi (or granulation villi) into the venous sinuses, after which it passes through the jugular vein and major venous system. CSF within the spinal cord can flow all the way down to the lumbar cistern at the end of the cord around the cauda equina where lumbar punctures are performed.

The cerebral aqueduct between the third and fourth ventricles is very small, as are the foramina, which means that they can be easily blocked.

17.2.2 Protection of the brain

The brain and spinal cord are covered by the meninges, the three protective membranes of the tough dura mater, the arachnoid mater and the pia mater. The cerebrospinal fluid (CSF) within the skull and spine provides further protection and also buoyancy, and is found between the pia mater and the arachnoid mater.

The CSF that is produced in the ventricular system is also necessary for chemical stability, and the provision of nutrients needed by the brain. The CSF helps to protect the brain from jolts and knocks to the head and also provides buoyancy and support to the brain against gravity. (Since the brain and CSF are similar in density, the brain floats in neutral buoyancy, suspended in the CSF.) This allows the brain to grow in size and weight without resting on the floor of the cranium, which would destroy nervous tissue.[5][6]

17.3 Clinical significance

The narrowness of the cerebral aqueduct and foramina means that they can become blocked, for example, by blood following a haemorrhagic stroke. As cerebrospinal fluid is continually produced by the choroid plexus within the ventricles, a blockage of outflow leads to increasingly high pressure in the lateral ventricles. As a consequence, this commonly leads in turn to hydrocephalus. Medically one would call this post-haemorrhagic acquired hydrocephalus, but is often referred to colloquially by the layperson as "water on the brain". This is an extremely serious condition regardless of the cause of blockage. An endoscopic third ventriculostomy is a surgical procedure for the treatment of hydrocephalus in which an opening is created in the floor of the third ventricle using an endoscope placed within the ventricular system through a burr hole. This allows the cerebrospinal fluid to flow directly to the basal cisterns, thereby shortcutting any obstruction. A surgical procedure to make an entry hole to access any of the ventricles is called a ventriculostomy. This is done to drain accumulated cerebrospinal fluid either through a temporary catheter or a permanent shunt.

Other diseases of the ventricular system include inflammation of the membranes (meningitis) or of the ventricles (ventriculitis) caused by infection or the introduction of blood following trauma or haemorrhage (cerebral haemorrhage or subarachnoid haemorrhage).

During embryogenesis in the choroid plexus of the ventricles, choroid plexus cysts can form.

The scientific study of CT scans of the ventricles in the late 1970s gave new insight into the study of mental disorders. Researchers found that individuals with schizophrenia had (in terms of group averages) larger than usual ventricles. This became the first "evidence" that schizophrenia was biological in origin and led to a renewed interest in its study via the use of imaging techniques. Magnetic resonance imaging (MRI) has superseded the use of CT in research in the role of detecting ventricular abnormalities in psychiatric illness.

Whether the enlarged ventricles is a cause or a result of schizophrenia has not yet been established. Enlarged ventricles are also found in organic dementia and have been explained largely in terms of environmental factors.[7] They have also been found to be extremely diverse between individuals, such that the percentage difference in group averages in schizophrenia studies (+16%) has been described as "not a very profound difference in the context of normal variation" (ranging from 25% to 350% of the mean average).[8]

The cave of septum pellucidum has been loosely associated with schizophrenia,[9] post-traumatic stress disorder,[10] traumatic brain injury,[11] as well as with antisocial personality disorder.[4] CSP is one of the distinguishing features of individuals displaying symptoms of dementia pugilistica.[12]

17.4 Additional images

- Transverse dissection showing the ventricles of the brain.
- View of ventricles.
- Scheme showing relations of the ventricles to the surface of the brain.
- Drawing of a cast of the ventricular cavities, viewed from above.
- View of ventricles and choroid plexus

17.5 See also

This article uses anatomical terminology; for an overview, see Anatomical terminology.

- Blood–brain barrier
- Circumventricular organs

17.6 References

[1] *National Institutes of Health* (December 13, 2011). "Ventricles of the brain". nih.gov.

[2] International school of medicine and applied sciences kisumu library

[3] Schoenwolf, Gary C. (2009). ""Development of the Brain and Cranial Nerves"". *Larsen's human embryology* (4th ed.). Philadelphia: Churchill Livingstone/Elsevier. ISBN 9780443068119.

[4] Adrian Raine, Lydia Lee, Yaling Yang, Patrick Colletti (2010). "Neurodevelopmental marker for limbic maldevelopment in antisocial personality disorder and psychopathy". BJPsych. The British Journal of Psychiatry 197: 186–192. doi:10.1192/bjp.bp.110.078485.

[5] Klein, S.B., & Thorne, B.M. Biological Psychology. Worth Publishers: New York. 2007.

[6] Saladin, Kenneth S. Anatomy & Physiology. The Unit of Form and Function. 5th Edition. McGraw-Hill: New York. 2007

[7] Peper, Jiska S.; Brouwer, RM; Boomsma, DI; Kahn, RS; Hulshoff Pol, HE (2007). "Genetic influences on human brain structure: A review of brain imaging studies in twins". *Human Brain Mapping*. **28** (6): 464–73. doi:10.1002/hbm.20398. PMID 17415783.

[8] Allen JS, Damasio H, Grabowski TJ (August 2002). "Normal neuroanatomical variation in the human brain: an MRI-volumetric study". *American Journal of Physical Anthropology*. **118** (4): 341–58. doi:10.1002/ajpa.10092. PMID 12124914.

[9] Galarza M, Merlo A, Ingratta A, Albanese E, Albanese A (2004). "Cavum septum pellucidum and its increased prevalence in schizophrenia: a neuroembryological classification". *The Journal of neuropsychiatry and clinical neurosciences*. **16** (1): 41–6. doi:10.1176/appi.neuropsych.16.1.41. PMID 14990758.

[10] May F, Chen Q, Gilbertson M, Shenton M, Pitman R (2004). "Cavum septum pellucidum in monozygotic twins discordant for combat exposure: relationship to posttraumatic stress disorder". *Biol. Psychiatry*. **55** (6): 656–8. doi:10.1016/j.biopsych.2003.09.018. PMC 2794416. PMID 15013837.

[11] Zhang L, Ravdin L, Relkin N, Zimmerman R, Jordan B, Lathan W, Uluğ A (2003). "Increased diffusion in the brain of professional boxers: a preclinical sign of traumatic brain injury?". *AJNR. American journal of neuroradiology*. **24** (1): 52–7. PMID 12533327.

[12] Neuropathol Exp Neurol. 2009 Jul;68(7):709-35. doi: 10.1097/NEN.0b013e3181a9d503. Chronic traumatic encephalopathy in athletes: progressive tauopathy after repetitive head injury. McKee AC, Cantu RC, Nowinski CJ, Hedley-Whyte ET, Gavett BE, Budson AE, Santini VE, Lee HS, Kubilus CA, Stern RA.

17.7 External links

- Stained brain slice images which include the "Lateral%20Ventricle" at the BrainMaps project
- ventricular system and CSF (concise description, University of Washington)
- CSF at answers.com (brief description, excellent diagram of CSF flow)

Chapter 18

Medulla oblongata

The **medulla oblongata** (or medulla) is located in the hindbrain, anterior to the cerebellum. The medulla oblongata is a cone-shaped neuronal mass responsible for multiple autonomic (involuntary) functions ranging from vomiting to sneezing. The medulla contains the cardiac, respiratory, vomiting and vasomotor centers and therefore deals with the autonomic functions of breathing, heart rate and blood pressure.

The **bulb** is an archaic term for the medulla oblongata and in modern clinical usage the word **bulbar** (as in bulbar palsy) is retained for terms that relate to the medulla oblongata, particularly in reference to medical conditions. The word bulbar can refer to the nerves and tracts connected to the medulla, and also by association to those muscles innervated, such as those of the tongue, pharynx and larynx.

18.1 Anatomy

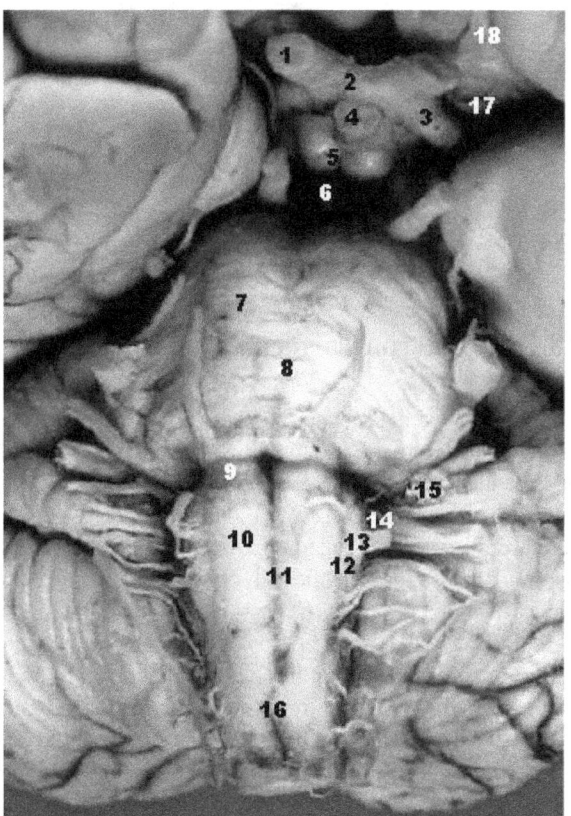

Medulla and parts (10-16) - (10) pyramid; (11) the anterior median fissure; (15) is the choroid plexus in the fourth ventricle; (13) olive and (7) the pons

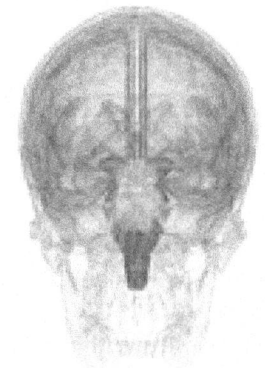

Medulla oblongata (animation)

The medulla can be thought of as being in two parts:

- an upper **open part** or superior part where the dorsal surface of the medulla is formed by the fourth ventricle.

- a lower **closed part** or inferior part where the fourth ventricle has narrowed at the obex in the caudal medulla, and surrounds part of the central canal.

18.1. ANATOMY

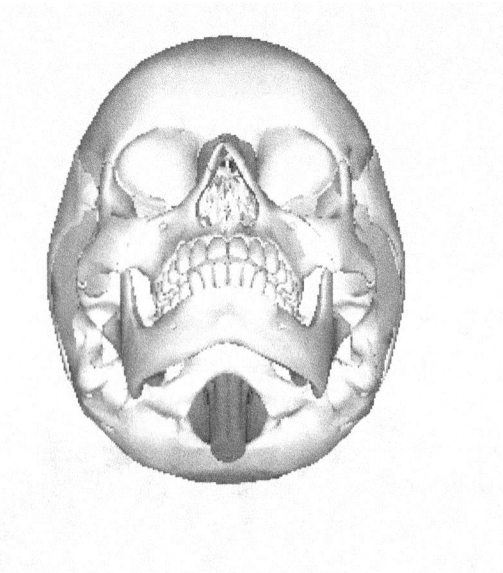

Medulla-animated as it protrudes from the foramen magnum of the skull-base, after which it gives rise to the spinal cord.

18.1.1 External surfaces

The anterior median fissure contains a fold of pia mater, and extends along the length of the medulla oblongata. It ends at the lower border of the pons in a small triangular area, termed the foramen cecum. On either side of this fissure are raised areas termed the medullary pyramids. The pyramids house the pyramidal tracts–the corticospinal and the corticobulbar tracts of the nervous system. At the caudal part of the medulla these tracts cross over in the decussation of the pyramids obscuring the fissure at this point. Some other fibers that originate from the anterior median fissure above the decussation of the pyramids and run laterally across the surface of the pons are known as the anterior external arcuate fibers.

The region between the anterolateral and posterolateral sulcus in the upper part of the medulla is marked by a pair of swellings known as olivary bodies (also called *olives*). They are caused by the largest nuclei of the olivary bodies, the inferior olivary nuclei.

The posterior part of the medulla between the posterior median sulcus and the posterolateral sulcus contains tracts that enter it from the posterior funiculus of the spinal cord. These are the gracile fasciculus, lying medially next to the midline, and the cuneate fasciculus, lying laterally. These fasciculi end in rounded elevations known as the gracile and the cuneate tubercles. They are caused by masses of gray matter known as the gracile nucleus and the cuneate nucleus. The soma (cell bodies) in these nuclei are the second-order neurons of the posterior column-medial lemniscus pathway, and their axons, called the internal arcuate fibers or fasciculi, decussate from one side of the medulla to the other to form the medial lemniscus.

Just above the tubercles, the posterior aspect of the medulla is occupied by a triangular fossa, which forms the lower part of the floor of the fourth ventricle. The fossa is bounded on either side by the inferior cerebellar peduncle, which connects the medulla to the cerebellum.

The lower part of the medulla, immediately lateral to the cuneate fasciculus, is marked by another longitudinal elevation known as the tuberculum cinereum. It is caused by an underlying collection of gray matter known as the spinal trigeminal nucleus. The gray matter of this nucleus is covered by a layer of nerve fibers that form the spinal tract of the trigeminal nerve.

The base of the medulla is defined by the commissural fibers, crossing over from the ipsilateral side in the spinal cord to the contralateral side in the brain stem; below this is the spinal cord.

18.1.2 Blood supply

Blood to the medulla is supplied by a number of arteries.

- Anterior spinal artery: This supplies the whole medial part of the medulla oblongata.

- Posterior inferior cerebellar artery: This is a major branch of the vertebral artery, and supplies the posterolateral part of the medulla, where the main sensory tracts run and synapse. It also supplies part of the cerebellum.

- Direct branches of the vertebral artery: The vertebral artery supplies an area between the other two main arteries, including the solitary nucleus and other sensory nuclei and fibers.

18.1.3 Development

The medulla oblongata forms in fetal development from the myelencephalon. The final differentiation of the medulla is seen at week 20 gestation.[1]

Neuroblasts from the alar plate of the neural tube at this level will produce the sensory nuclei of the medulla. The basal plate neuroblasts will give rise to the motor nuclei.

- Alar plate neuroblasts give rise to:
  - The solitary nucleus, which contains the general visceral afferent fibers for taste, as well as the special visceral afferent column.

- The spinal trigeminal nerve nuclei which contains the general somatic afferent column.
- The cochlear and vestibular nuclei, which contain the special somatic afferent column.
- The inferior olivary nucleus, which relays to the cerebellum.
- The dorsal column nuclei, which contain the gracile and cuneate nuclei.

- Basal plate neuroblasts give rise to:
  - The hypoglossal nucleus, which contains general somatic efferent fibers.
  - The nucleus ambiguus, which form the special visceral efferent.
  - The dorsal nucleus of vagus nerve and the inferior salivatory nucleus, both of which form the general visceral efferent fibers.

18.2 Function

The medulla oblongata connects the higher levels of the brain to the spinal cord, and is responsible for several functions of the autonomous nervous system which include:

- The control of ventilation via signals from the carotid and aortic bodies. Respiration is regulated by groups of chemoreceptors. These sensors detect changes in the acidity of the blood, thus if the blood is considered too acidic by the medulla oblongata electrical signals are sent to intercostal and phrenical muscle tissue increasing their contraction rate in order to reoxygenate the blood. The ventral respiratory group and the dorsal respiratory group are neurons involved in this regulation.
- Cardiovascular center – sympathetic, parasympathetic nervous system
- Vasomotor center – baroreceptors
- Reflex centers of vomiting, coughing, sneezing, and swallowing. These reflexes which include the pharyngeal reflex, the swallowing reflex (also known as the palatal reflex), and the masseter reflex can be termed, *bulbar* reflexes.[2]

18.3 Clinical significance

A blood vessel blockage (such as in a stroke) will injure the pyramidal tract, medial lemniscus, and the hypoglossal nucleus. This causes a syndrome called medial medullary syndrome.

Lateral medullary syndrome can be caused by the blockage of either the posterior inferior cerebellar artery or of the vertebral arteries.

18.4 Other animals

Both lampreys and hagfish possess a fully developed medulla oblongata.[3][4] Since these are both very similar to early agnathans, it has been suggested that the medulla evolved in these early fish, approximately 505 million years ago.[5] The status of the medulla as part of the primordial reptilian brain is confirmed by its disproportionate size in modern reptiles such as the crocodile, alligator, and monitor lizard.

18.5 Additional images

- Lobes
- Anteroinferior view of the medulla oblongata and pons.
- Section of the medulla oblongata through the lower part of the decussation of the pyramids
- Section of the medulla oblongata at the level of the decussation of the pyramids.
- Base of brain.
- Diagram showing the positions of the three principal subarachnoid cisternæ.
- Medulla oblongata
- Spinal cord. Brachial plexus. Cerebrum. Inferior view. Deep dissection
- Micrograph of the posterior portion of the *open part* of the medulla oblongata, showing the fourth ventricle (top of image) and the nuclei of CN XII (medial) and CN X (lateral). H&E-LFB stain.

18.6 References

This article incorporates text in the public domain from the 20th edition of Gray's Anatomy (1918)

[1] Carlson, Neil R. Foundations of Behavioral Neuroscience.63-65

[2] Hughes, T. (2003). "Neurology of swallowing and oral feeding disorders: Assessment and management". *Journal of Neurology, Neurosurgery & Psychiatry*. **74** (90003): 48iii. doi:10.1136/jnnp.74.suppl_3.iii48.

[3] Nishizawa H, Kishida R, Kadota T, Goris RC; Kishida, Reiji; Kadota, Tetsuo; Goris, Richard C. (1988). "Somatotopic organization of the primary sensory trigeminal neurons in the hagfish, Eptatretus burgeri". *J Comp Neurol*. **267** (2): 281–95. doi:10.1002/cne.902670210. PMID 3343402.

[4] Rovainen CM (1985). "Respiratory bursts at the midline of the rostral medulla of the lamprey". *J Comp Physiol A*. **157** (3): 303–9. doi:10.1007/BF00618120. PMID 3837091.

[5] Haycock, *Being and Perceiving*

- Haycock DE (2011). *Being and Perceiving*. Manupod Press. ISBN 978-0-9569621-0-2.

18.7 External links

- Stained brain slice images which include the "medulla" at the BrainMaps project

Chapter 19

Pons

For other uses, see Pons (disambiguation).

The **pons** is part of the brainstem, and in humans and other bipeds lies between the midbrain (above) and the medulla oblongata (below) and in front of the cerebellum.

The pons is also called the **pons Varolii** ("bridge of Varolius"), after the Italian anatomist and surgeon Costanzo Varolio (1543–75).[1] This region of the brainstem includes neural pathways or tracts that conduct signals from the brain down to the cerebellum and medulla, and tracts that carry the sensory signals up into the thalamus.[2]

The pons in humans measures about 2.5 centimetres (0.98 in) in length. Most of it appears as a broad anterior bulge rostral to the medulla. Posteriorly, it consists mainly of two pairs of thick stalks called cerebellar peduncles. They connect the cerebellum to the pons and midbrain.[2]

The pons contains nuclei that relay signals from the forebrain to the cerebellum, along with nuclei that deal primarily with sleep, respiration, swallowing, bladder control, hearing, equilibrium, taste, eye movement, facial expressions, facial sensation, and posture.[2]

Within the pons is the pneumotaxic center consisting of the subparabrachial and the medial parabrachial nuclei. This center regulates the change from inhalation to exhalation.[2]

The pons is implicated in sleep paralysis, and also plays a role in generating dreams.

19.1 Structure

The pons can be broadly divided into two parts: the basilar part of the pons, located ventrally, and the pontine tegmentum, located dorsally.

19.1.1 Development

During embryonic development, the metencephalon develops from the rhombencephalon and gives rise to two structures: the pons and the cerebellum.[2] The alar plate produces sensory neuroblasts, which will give rise to the solitary nucleus and its special visceral afferent (SVA) column; the cochlear and vestibular nuclei, which form the special somatic afferent (SSA) fibers of the vestibulocochlear nerve, the spinal and principal trigeminal nerve nuclei, which form the general somatic afferent column (GSA) of the trigeminal nerve, and the pontine nuclei which relays to the cerebellum.

Basal plate neuroblasts give rise to the abducens nucleus, which forms the general somatic efferent fibers (GSE); the facial and motor trigeminal nuclei, which form the special visceral efferent (SVE) column, and the superior salivatory nucleus, which forms the general visceral efferent fibers of the facial nerve.

19.1.2 Nucleus

A number of cranial nerve nuclei are present in the pons:

- mid-pons: the 'chief' or 'pontine' nucleus of the trigeminal nerve sensory nucleus (V)
- mid-pons: the motor nucleus for the trigeminal nerve (V)
- lower down in the pons: abducens nucleus (VI)
- lower down in the pons: facial nerve nucleus (VII)
- lower down in the pons: vestibulocochlear nuclei (vestibular nuclei and cochlear nuclei) (VIII)

19.2 Function

The functions of these four cranial nerves (V-VIII) include sensory roles in hearing, equilibrium, and taste, and in facial sensations such as touch and pain, as well as motor roles in eye movement, facial expressions, chewing, swallowing, and the secretion of saliva and tears.[2]

19.3 Clinical significance

- Central pontine myelinolysis is a demyelination disease that causes difficulty with sense of balance, walking, sense of touch, swallowing and speaking. In a clinical setting, it is often associated with transplant or rapid correction of blood sodium. Undiagnosed, it can lead to death or locked-in syndrome.

19.4 Other animals

19.4.1 Evolution

The pons first evolved as an offshoot of the medullary reticular formation.[3] Since lampreys possess a pons, it has been argued that it must have evolved as a region distinct from the medulla by the time the first agnathans appeared, 505 million years ago.[4]

19.5 Additional images

- Location and topography of Pons (animation)
- Axial section of the pons, at its upper part
- Hind- and mid-brains; posterolateral view
- Median sagittal section of brain
- Nuclei of the pons and brainstem
- Cerebrum. Deep dissection. Inferior dissection.

19.6 References

[1] Henry Gray (1862). *Anatomy, descriptive and surgical.* Blanchard and Lea. pp. 514–. Retrieved 10 November 2010.

[2] Saladin Kenneth S.(2007)

[3] Pritchard and Alloway *Medical Neuroscience*

[4] Butler and Hodos *Comparative vertebrate neuroanatomy: evolution and adaptation*

Saladin Kenneth S.(2007) Anatomy & physiology the unity of form and function. Dubuque, IA: McGraw-Hill

- Pritchard, TE & Alloway, D (1999). *Medical neuroscience*. Hayes Barton Press. ISBN 978-1-59377-200-0.
- Butler, AB & Hodos, W (2005). *Comparative vertebrate neuroanatomy: evolution and adaptation*. Wiley-Blackwell. ISBN 978-0-471-21005-4.

19.7 External links

- Diagram at UCC
- Stained brain slice images which include the "Pons" at the BrainMaps project

Chapter 20

Brainstem

In the anatomy of humans and of many other vertebrates, the **brainstem** (or **brain stem**) is the posterior part of the brain, adjoining and structurally continuous with the spinal cord. In humans it is usually described as including the medulla oblongata (myelencephalon), pons (part of metencephalon), and midbrain (mesencephalon).[1][2] Less frequently, parts of the diencephalon are included.

The brainstem provides the main motor and sensory innervation to the face and neck via the cranial nerves. Of the twelve pairs of cranial nerves, ten pairs come from the brainstem. Though small, this is an extremely important part of the brain as the nerve connections of the motor and sensory systems from the main part of the brain to the rest of the body pass through the brainstem. This includes the corticospinal tract (motor), the posterior column-medial lemniscus pathway (fine touch, vibration sensation, and proprioception), and the spinothalamic tract (pain, temperature, itch, and crude touch).

The brainstem also plays an important role in the regulation of cardiac and respiratory function. It also regulates the central nervous system, and is pivotal in maintaining consciousness and regulating the sleep cycle. The brainstem has many basic functions including heart rate, breathing, sleeping, and eating.

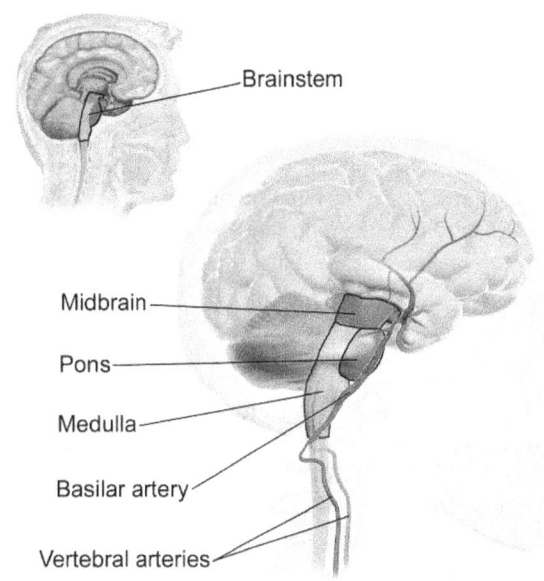

Structures of the brainstem

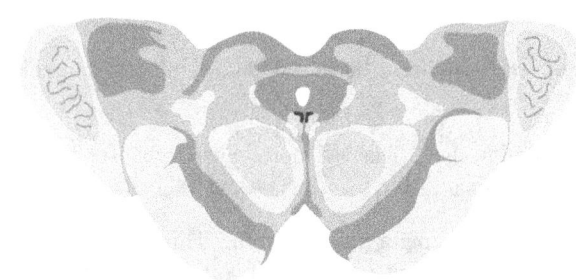

Cross-section of the midbrain at the level of the superior colliculus.

20.1 Structure

20.1.1 Midbrain

Main article: Midbrain

The midbrain is divided into three parts. The first is the tectum, (Latin:roof), which forms the ceiling. The tectum comprises the paired structure of the superior and inferior colliculi and is the dorsal covering of the cerebral aqueduct. The inferior colliculus, is the principal midbrain nucleus of the auditory pathway and receives input from several peripheral brainstem nuclei, as well as inputs from the auditory cortex. Its inferior brachium (arm-like process) reaches to the medial geniculate nucleus of the diencephalon. Superior to the inferior colliculus, the superior colliculus marks the rostral midbrain. It is involved in the special sense of vision and sends its superior brachium to the lateral geniculate body of the diencephalon. The second part is the tegmentum which forms the floor of the midbrain, and is ventral to the cerebral aqueduct. Sev-

20.1. STRUCTURE

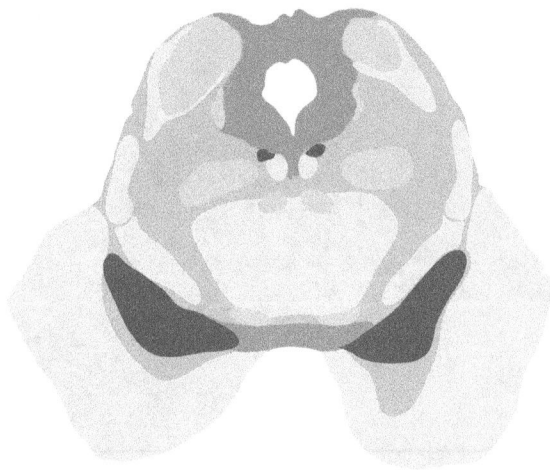

Cross-section of the midbrain at the level of the inferior colliculus.

eral nuclei, tracts, and the reticular formation are contained here. The third part, the ventral tegmentum is composed of paired cerebral peduncles. These transmit axons of upper motor neurons.

The midbrain consists of:

- Periaqueductal gray: The area of gray matter around the cerebral aqueduct, which contains various neurons involved in the pain desensitization pathway. Neurons synapse here and, when stimulated, cause activation of neurons in the nucleus raphe magnus, which then project down into the posterior grey column of the spinal cord and prevent pain sensation transmission.

- Oculomotor nerve nucleus: This is the third cranial nerve nucleus.

- Trochlear nerve nucleus: This is the fourth cranial nerve.

- Red nucleus: This is a motor nucleus that sends a descending tract to the lower motor neurons.

- Substantia nigra pars compacta: This is a concentration of neurons in the ventral portion of the midbrain that uses dopamine as its neurotransmitter and is involved in both motor function and emotion. Its dysfunction is implicated in Parkinson's disease.

- Reticular formation: This is a large area in the midbrain that is involved in various important functions of the midbrain. In particular, it contains lower motor neurons, is involved in the pain desensitization pathway, is involved in the arousal and consciousness systems, and contains the locus coeruleus, which is involved in intensive alertness modulation and in autonomic reflexes.

- Central tegmental tract: Directly anterior to the floor of the 4th ventricle, this is a pathway by which many tracts project up to the cortex and down to the spinal cord.

- Ventral tegmental area: A dopaminergic nucleus located close to the midline on the floor of the midbrain.

- Tail of the ventral tegmental area: A GABAergic nucleus located adjacent to the ventral tegmental area.

20.1.2 Pons

Main article: Pons

The pons lies between the medulla oblongata and the midbrain. It contains tracts that carry signals from the cerebrum to the medulla and to the cerebellum and also tracts that carry sensory signals to the thalamus. The pons is connected to the cerebellum by the cerebellar peduncles.

20.1.3 Medulla oblongata

Main article: Medulla oblongata

The medulla oblongata often just referred to as the medulla, is the lower half of the brainstem continuous with the spinal cord. Its upper part is continuous with the pons.[3] The medulla contains the cardiac, respiratory, vomiting and vasomotor centres dealing with heart rate, breathing and blood pressure.

20.1.4 Ventral view of medulla and pons

In the medial part of the medulla is the anterior median fissure. Moving laterally on each side are the medullary pyramids. The pyramids contain the fibers of the corticospinal tract (also called the pyramidal tract), or the upper motor neuronal axons as they head inferiorly to synapse on lower motor neuronal cell bodies within the anterior grey column of the spinal cord.

The anterolateral sulcus is lateral to the pyramids. Emerging from the anterolateral sulci are the CN XII (hypoglossal nerve) rootlets. Lateral to these rootlets and the anterolateral sulci are the olives. The olives are swellings in the medulla containing underlying inferior nucleary nuclei (containing various nuclei and afferent fibers). Lateral (and dorsal) to the olives are the rootlets for CN IX (glossopharyngeal), CN X (vagus) and CN XI (accessory nerve). The pyramids end at the pontine medulla junction, noted most obviously by the large basal pons. From this

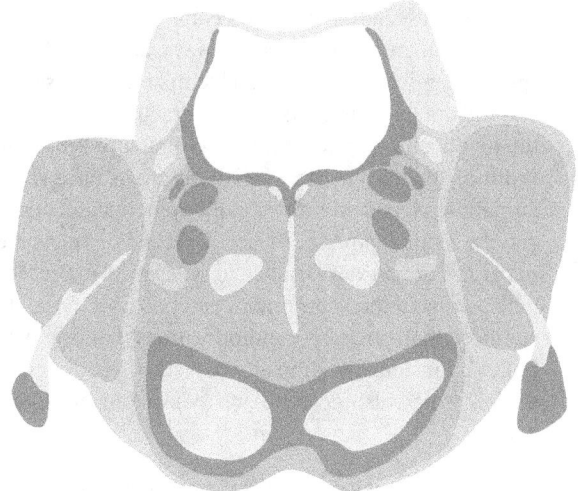
Cross-section of the middle pons (at the level of cranial nerve V).

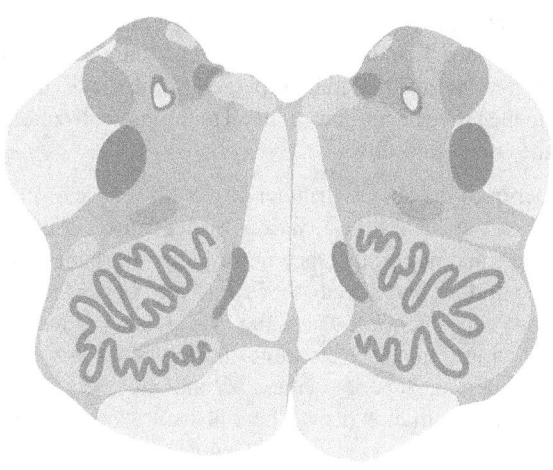
Cross-section of the middle medulla.

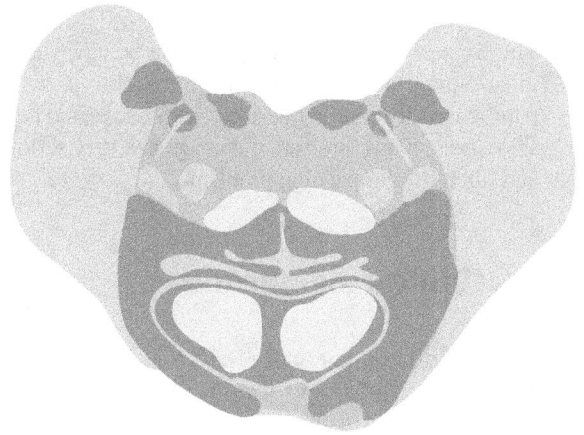
Cross-section of the inferior pons (at the level of the facial genu).

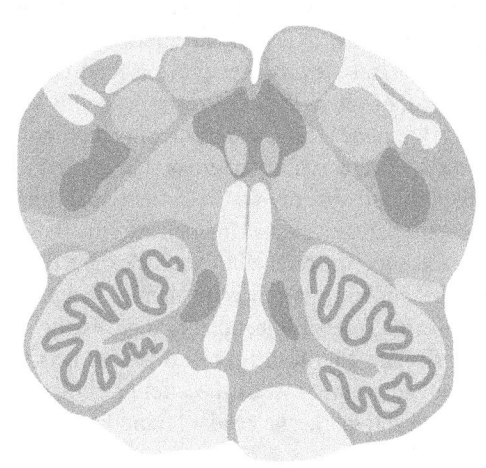
Cross-section of the inferior medulla.

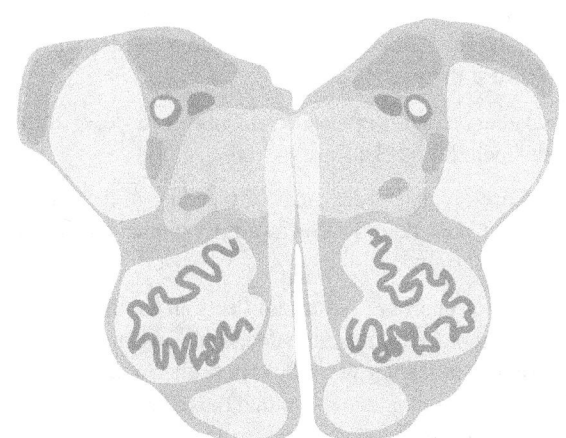
Cross-section of the rostral (superior) medulla.

junction, CN VI (abducens nerve), CN VII (facial nerve) and CN VIII (vestibulocochlear nerve) emerge. At the level of the midpons, CN V (the trigeminal nerve) emerges. Cranial nerve III (the occulomotor nerve) emerges ventrally from the midbrain, while the CN IV (the trochlear nerve) emerges out from the dorsal aspect of midbrain.

Between the two pyramids can be seen a decussation of fibres which marks the transition from the medulla to the spinal cord. The medulla is above the decussation and the spinal cord below.

20.1.5 Dorsal view of medulla and pons

The most medial part of the medulla is the posterior median sulcus. Moving laterally on each side is the fasciculus gracilis, and lateral to that is the fasciculus cuneatus. Superior to each of these, and directly inferior to the obex, are the gracile and cuneate tubercles, respectively. Underlying these are their respective nuclei. The obex marks the end of the 4th ventricle and the beginning of the central canal. The posterior intermediate sulci separates the fasciculi gracilis from the fasciculi cuneatus. Lateral to the fasciculi cuneatus is the lateral funiculus.

Superior to the obex is the floor of the 4th ventricle. In the floor of the 4th ventricle, various nuclei can be visualized by the small bumps that they make in the overlying tissue. In the midline and directly superior to the obex is the vagal trigone and superior to that it the hypoglossal trigone. Underlying each of these are motor nuclei for the respective cranial nerves. Superior to these trigones are fibers running laterally in both directions. These fibers are known collectively as the striae medullares. Continuing in a rostral direction, the large bumps are called the facial colliculi. Each facial colliculus, contrary to their names, do not contain the facial nerve nuclei. Instead, they have facial nerve axons traversing superficial to underlying abducens (CN VI) nuclei. Lateral to all these bumps previously discussed is an indented line, or sulcus that runs rostrally, and is known as the sulcus limitans. This separates the medial motor neurons from the lateral sensory neurons. Lateral to the sulcus limitans is the area of the vestibular system, which is involved in special sensation. Moving rostrally, the inferior, middle, and superior cerebellar peduncles are found connecting the midbrain to the cerebellum. Directly rostral to the superior cerebellar peduncle, there is the superior medullary velum and then the two trochlear nerves. This marks the end of the pons as the inferior colliculus is directly rostral and marks the caudal midbrain.

20.1.6 Development

The adult human brainstem emerges from two of the three primary vesicles formed of the neural tube. The mesencephalon is the second of the three primary vesicles, and does not further differentiate into a secondary vesicle. This will become the midbrain. The third primary vesicle, the rhombencephalon, will further differentiate into two secondary vesicles, the metencephalon and the myelencephalon. The metencephalon will become the cerebellum and the pons. The myelencephalon will become the medulla.

20.2 Function

There are three main functions of the brainstem:

1. The brainstem plays a role in conduction. That is, all information relayed from the body to the cerebrum and cerebellum and vice versa must traverse the brainstem. The ascending pathways coming from the body to the brain are the sensory pathways, and include the spinothalamic tract for pain and temperature sensation and the dorsal column, fasciculus gracilis, and cuneatus for touch, proprioception, and pressure sensation (both of the body). (The facial sensations have similar pathways, and will travel in the spinothalamic tract and the medial lemniscus also.) Descending tracts are upper motor neurons destined to synapse on lower motor neurons in the ventral horn and posterior horn. In addition, there are upper motor neurons that originate in the brainstem's vestibular, red, tectal, and reticular nuclei, which also descend and synapse in the spinal cord.

2. The cranial nerves III-XII emerge from the brainstem.[4] These cranial nerves supply the face, head, and viscera. (The first two pairs of cranial nerves arise from the cerebrum).

3. The brainstem has integrative functions being involved in cardiovascular system control, respiratory control, pain sensitivity control, alertness, awareness, and consciousness. Thus, brainstem damage is a very serious and often life-threatening problem.

20.3 Clinical significance

Diseases of the brainstem can result in abnormalities in the function of cranial nerves that may lead to visual disturbances, pupil abnormalities, changes in sensation, muscle weakness, hearing problems, vertigo, swallowing and speech difficulty, voice change, and co-ordination problems. Localizing neurological lesions in the brainstem may be very precise, although it relies on a clear understanding on the functions of brainstem anatomical structures and how to test them.

Brainstem stroke syndrome can cause a range of impairments including locked-in syndrome.

Duret haemorrhages are areas of bleeding in the midbrain and upper pons due to a downward traumatic displacement of the brainstem.[5]

Criteria for claiming brainstem death in the UK have developed in order to make the decision of when to stop ventilation of somebody who could not otherwise sustain life. These determining factors are that the patient is irreversibly unconscious and incapable of breathing unaided. All other possible causes must be ruled out that might otherwise indicate a temporary condition. The state of irreversible brain damage has to be unequivocal. There are brainstem reflexes that are checked for by two senior doctors so that imaging technology is unnecessary. The absence of the cough and gag reflexes, of the corneal reflex and the vestibulo-ocular reflex need to be established; the pupils of the eyes must be fixed and dilated; there must be an absence of motor response to stimulation and an absence of breathing marked by concentrations of carbon dioxide in the arterial blood. All of these tests must be repeated after a certain time before death can be declared.[6]

20.4 Additional images

- The midbrain, pons, and medulla oblongata are labelled on this coronal section of the human brain.
- Brainstem. Anterior face.Deep dissection
- Brainstem. Posterior face.Deep dissection

20.5 See also

This article uses anatomical terminology; for an overview, see Anatomical terminology.

- Cranial nerve nucleus
- Triune brain Reptilian brain

20.6 References

[1] "brainstem". *The Free Dictionary* (Medical Dictionary ed.). Farlex. Retrieved December 1, 2013.

[2] Brain Stem at the US National Library of Medicine Medical Subject Headings (MeSH)

[3] Dorland's (2012). *Dorland's Illustrated Medical Dictionary* (32nd ed.). Elsevier Saunders. p. 1121. ISBN 978-1-4160-6257-8.

[4] http://vanat.cvm.umn.edu/NeuroLectPDFs/CranialN-Lect.pdf

[5] Dorland's (2012). *Dowland's Illustrated Medical Dictionary* (32nd ed.). Elsevier. p. 842. ISBN 978-1-4160-6257-8.

[6] Black's Medical Dictionary 39th edition 1999

20.7 External links

- Comparative Neuroscience at Wikiversity
- http://www.meddean.luc.edu/lumen/Meded/Neuro/frames/nlBSsL/nl40fr.htm
- http://biology.about.com/library/organs/brain/blbrainstem.htm
- http://www.waiting.com/brainanatomy.html
- http://braininjuryhelp.com/video-tutorial/brain-injury-help-video-tutorial/
- http://www.martindalecenter.com/MedicalAnatomy_3_SAD.html
- NIF Search - Brainstem via the Neuroscience Information Framework

Chapter 21

Superior colliculus

The **superior colliculus**, (Latin, *upper hill*) is a paired structure of the mammalian midbrain. In other vertebrates the homologous structure is known as the **optic tectum** or simply **tectum**. The adjective form *tectal* is commonly used for mammals as well as other vertebrates.

The superior colliculus/optic tectum forms a major component of the midbrain. It is a layered structure, with a number of layers that varies by species. The layers can be grouped into the superficial layers (*stratum opticum* and above) and the deeper layers (the remaining layers). Neurons in the superficial layers receive direct input from the retina and respond almost exclusively to visual stimuli. Many neurons in the deeper layers also respond to other modalities, and some respond to stimuli in multiple modalities.[1] The deeper layers also contain a population of motor-related neurons, capable of activating eye movements as well as other responses.[2]

The general function of the tectal system is to direct behavioral responses toward specific points in egocentric ("body-centered") space. Each layer contains a topographic map of the surrounding world in retinotopic coordinates, and activation of neurons at a particular point in the map evokes a response directed toward the corresponding point in space. In primates, the superior colliculus has been studied mainly with respect to its role in directing eye movements. Visual input from the retina, or "command" input from the cerebral cortex, create a "bump" of activity in the tectal map, which, if strong enough, induces a saccadic eye movement. Even in primates, however, the superior colliculus is also involved in generating spatially directed head turns, arm-reaching movements,[3] and shifts in attention that do not involve any overt movements.[4] In other species, the tectum is involved in a wide range of responses, including whole-body turns in walking rats, swimming fishes, or flying birds; tongue-strikes toward prey in frogs; fang-strikes in snakes; etc.

In some vertebrates, including fish and birds, the tectum is one of the largest components of the brain. In mammals, and especially primates, the massive expansion of the cerebral cortex reduces the tectum ("superior colliculus") to a much smaller fraction of the whole brain. It remains nonetheless important in terms of function as the primary integrating center for eye movements.

Note on terminology: This article follows terminology established in the literature, using the term "superior colliculus" when discussing mammals and "optic tectum" when discussing either specific non-mammalian species or vertebrates in general.

21.1 Structure

The two superior colliculi sit below the thalamus and surround the pineal gland in the vertebrate midbrain. It comprises the dorsal aspect of the midbrain, posterior to the periaqueductal gray and immediately superior to the inferior colliculus. The inferior and superior colliculi are known collectively as the corpora quadrigemina (Latin, *quadruplet bodies*). The **brachium of superior colliculus** (or **superior brachium**) extends laterally from the superior colliculus, and, passing between the pulvinar and medial geniculate body, is partly continued into an eminence called the lateral geniculate body, and partly into the optic tract.

21.1.1 Neural circuit

The microstructure of the optic tectum / superior colliculus varies across species. As a general rule, there is always a clear distinction between superficial layers, which receive input primarily from the visual system and show primarily visual responses, and deeper layers, which receive many types of input and project to numerous motor-related brain areas. The distinction between these two zones is so clear and consistent that some anatomists have suggested that they should be considered separate brain structures.

In mammals, neuroanatomists conventionally identify seven layers[6] The top three layers are called *superficial*:

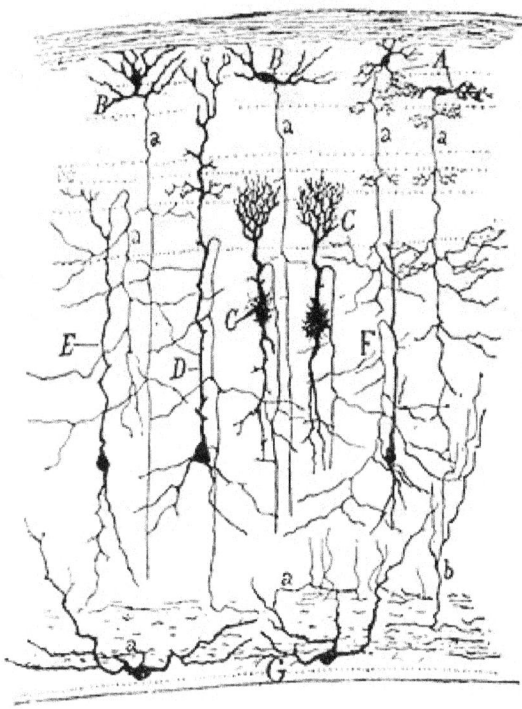

Drawing by Ramon y Cajal of several types of Golgi-stained neurons in the optic tectum of a sparrow.

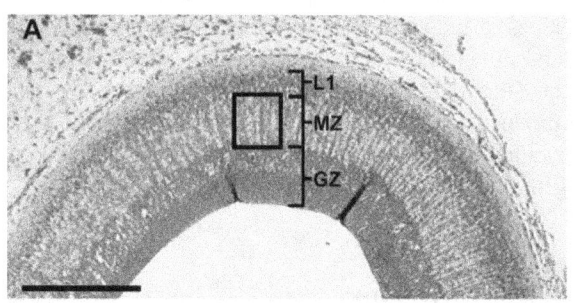

H&E stain of chicken optic tectum at E7 showing the generative zone (GZ), the migrating zone (MZ) and the first neuronal lamina (L1). Scale bar 200 μm. From Caltharp et al., 2007.[5]

- **Lamina I** or **SZ**, the *stratum zonale*, is a thin layer consisting of small myelinated axons together with marginal and horizontal cells.

- **Lamina II** or **SGS**, the *stratum griseum superficiale* ("superficial gray layer"), contains many neurons of various shapes and sizes.

- **Lamina III** or **SO**, the *stratum opticum* ("optic layer"), consists mainly of axons coming from the optic tract.

Next come two *intermediate layers*:

- **Lamina IV** or **SGI**, the *stratum griseum intermedium* ("intermediate gray layer"), is the thickest layer, and is filled with many neurons of many sizes. This layer is often as thick as all the other layers together. It is often subdivided into "upper" and "lower" parts.

- **Lamina V** or **SAI**, the *stratum album intermedium* ("intermediate white layer"), consists mainly of fibers from various sources.

Finally come the two *deep layers*:

- **Lamina VI** or **SGP**, the *stratum griseum profundum* ("deep gray layer"), consists of loosely packed neurons and myelinated fibers.

- **Lamina VII** or **SAP**, the *stratum album profundum* ("deep white layer"), lying directly above the periaqueductal gray, consists entirely of fibers.

The superficial layers receive input mainly from the retina, vision-related areas of the cerebral cortex, and two tectal-related structures called the pretectum and *parabigeminal nucleus*. The retinal input encompasses the entire superficial zone, and is bilateral, although the contralateral portion is more extensive. The cortical input comes most heavily from the primary visual cortex (area 17), the secondary visual cortex (areas 18 and 19), and the frontal eye fields. The parabigeminal nucleus plays a very important role in tectal function that is described below.

In contrast to the vision-dominated inputs to the superficial layers, the intermediate and deep layers receive inputs from a very diverse set of sensory and motor structures. Most areas of the cerebral cortex project to these layers, although the input from "association" areas tends to be heavier than the input from primary sensory or motor areas. However, the cortical areas involved, and the strength of their relative projections differs across species.[7] Another important input comes from the substantia nigra, pars reticulata, a component of the basal ganglia. This projection uses the inhibitory neurotransmitter GABA, and is thought to exert a "gating" effect on the superior colliculus. The intermediate and deep layers also receive input from the spinal trigeminal nucleus, which conveys somatosensory information from the face, as well as the hypothalamus, zona incerta, thalamus, and inferior colliculus.

In addition to their distinctive inputs, the superficial and deep zones of the superior colliculus also have distinctive outputs. One of the most important outputs goes to the pulvinar and lateral intermediate areas of the thalamus, which in turn project to areas of the cerebral cortex that are involved in controlling eye movements. There are also projections from the superficial zone to the pretectal nuclei, lateral geniculate nucleus of the thalamus, and the parabigeminal nucleus. The projections from the deeper layers are more extensive. There are two large descending pathways, traveling to the brainstem and spinal cord, and numerous ascending projections to a variety of sensory and motor centers, including several that are involved in generating eye movements.

21.1.2 Mosaic structure

On detailed examination the collicular layers are actually not smooth sheets, but divided into a honeycomb arrangement of discrete columns.[8] The clearest indication of columnar structure comes from the cholinergic inputs arising from the parabigeminal nucleus, whose terminals form evenly spaced clusters that extend from top to bottom of the tectum.[9] Several other neurochemical markers including calretinin, parvalbumin, GAP-43, and NMDA receptors, and connections with numerous other brain structures in the brainstem and diencephalon, also show a corresponding inhomogeneity.[10] The total number of columns has been estimated at around 100.[8] The functional significance of this columnar architecture is not clear, but it is interesting that recent evidence has implicated the cholinergic inputs as part of a recurrent circuit producing winner-take-all dynamics within the tectum, as described in more detail below.

All species that have been examined — including mammals and non-mammals — show compartmentalization, but there are some systematic differences in the details of the arrangement.[9] In species with a streak-type retina (mainly species with laterally placed eyes, such as rabbits and deer), the compartments cover the full extent of the SC. In species with a centrally placed fovea, however, the compartmentalization breaks down in the front (rostral) part of the SC. This portion of the SC contains many "fixation" neurons that fire continually while the eyes remain fixed in a constant position.

21.1.3 Related structures

The optic tectum is closely associated with an adjoining structure called *nucleus isthmii*, which has drawn great interest recently because of new evidence that it makes a very important contribution to tectal function. In mammals, where the term *superior colliculus* is generally used instead of *optic tectum*, this area is called the **parabigeminal nucleus**. Once again, this is simply a case of two different names being used for the same structure. The nucleus isthmii is divided into two parts, called *pars magnocellularis* (**Imc**; "the part with the large cells") and *pars parvocellularis* (**Ipc**; "the part with the small cells").

As illustrated in the adjoining diagram, connections between the three areas — tectum, Ipc, and Imc — are to-

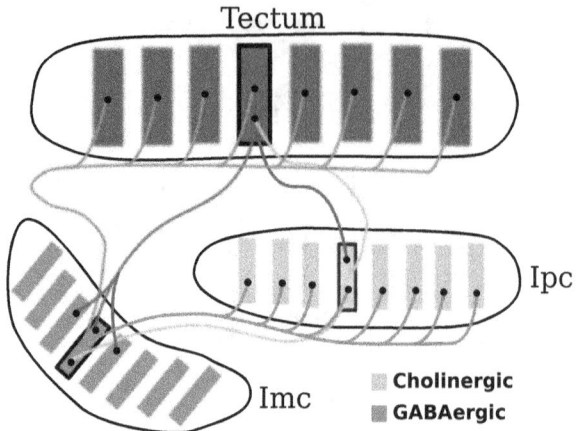

Schematic circuit diagram of topographic connections between the optic tectum and the two parts of nucleus isthmii.

pographic. Neurons in the superficial layers of the tectum project to corresponding points in Ipc and Imc. The projections to Ipc are tightly focused, while the projections to Imc are somewhat more diffuse. Ipc gives rise to tightly focused cholinergic projections both to Imc and the tectum. In the tectum, the cholinergic inputs from Ipc ramify to give rise to terminals that extend across an entire column, from top to bottom. Imc, in contrast, gives rise to GABAergic projections to Ipc and tectum that spread very broadly in the lateral dimensions, encompassing most of the retinotopic map. Thus, the tectum-Ipc-Imc circuit causes tectal activity to produce recurrent feedback that involves tightly focused excitation of a small column of neighboring tectal neurons, together with global inhibition of distant tectal neurons.

21.2 Function

The history of investigation of the optic tectum has been marked by several large shifts in opinion. Before about 1970, most studies involved non-mammals — fish, frogs, birds - that is, species in which the tectum is the dominant structure that receives input from the eyes. The general view then was that the tectum, in these species, is the main visual center in the non-mammalian brain, and, as a consequence, is involved in a wide variety of behaviors. From the 1970s to 1990s, however, neural recordings from mammals, mostly monkeys, focused primarily on the role of the superior colliculus in controlling eye movements. This line of investigation came to dominate the literature to such a degree that the majority opinion was that eye-movement control is the only important function in mammals, a view still reflected in many current textbooks.

In the late 1990s, however, experiments using animals whose heads were free to move showed clearly that the SC actually produces *gaze shifts*, usually composed of combined head and eye movements, rather than eye movements *per se*. This discovery reawakened interest in the full breadth of functions of the superior colliculus, and led to studies of multisensory integration in a variety of species and situations. Nevertheless, the role of the SC in controlling eye movements is understood in much greater depth than any other function.

Behavioral studies have shown that the SC is not needed for object recognition, but plays a critical role in the ability to direct behaviors toward specific objects, and can support this ability even in the absence of the cerebral cortex.[11] Thus, cats with major damage to the visual cortex cannot recognize objects, but may still be able to follow and orient toward moving stimuli, although more slowly than usual. If one half of the SC is removed, however, the cats will circle constantly toward the side of the lesion, and orient compulsively toward objects located there, but fail to orient at all toward objects located in the opposite hemifield. These deficits diminish over time but never disappear.

21.2.1 Eye movements

In primates, eye movements can be divided into several types: *fixation*, in which the eyes are directed toward a motionless object, with eye movements only to compensate for movements of the head; *smooth pursuit*, in which the eyes move steadily to track a moving object; *saccades*, in which the eyes move very rapidly from one location to another; and *vergence*, in which the eyes move simultaneously in opposite directions to obtain or maintain single binocular vision. The superior colliculus is involved in all of these, but its role in saccades has been studied most intensively.

Each of the two colliculi — one on each side of the brain — contains a two-dimensional map representing half of the visual field. The fovea — the region of maximum sensitivity — is represented at the front edge of the map, and the periphery at the back edge. Eye movements are evoked by activity in the deep layers of the SC. During fixation, neurons near the front edge — the foveal zone — are tonically active. During smooth pursuit, neurons a small distance from the front edge are activated, leading to small eye movements. For saccades, neurons are activated in a region that represents the point to which the saccade will be directed. Just prior to a saccade, activity rapidly builds up at the target location and decreases in other parts of the SC. The coding is rather broad, so that for any given saccade the activity profile forms a "hill" that encompasses a substantial fraction of the collicular map: The location of the peak of this "hill" represents the saccade target.

The SC encodes the target of a gaze shift, but it does not seem to specify the precise movements needed to get there.[12] The decomposition of a gaze shift into head and eye movements and the precise trajectory of the eye during a saccade depend on integration of collicular and non-collicular signals by downstream motor areas, in ways that are not yet well understood. Regardless of how the movement is evoked or performed, the SC encodes it in "retinotopic" coordinates: that is, a given SC activation pattern specifies a given offset from the current gaze direction, irrespective of the initial position of the eyes.[13]

There has been some controversy about whether the SC merely commands eye movements, and leaves the execution to other structures, or whether it actively participates in the performance of a saccade. In 1991, Munoz et al., on the basis of data they collected, argued that, during a saccade, the "hill" of activity in the SC moves gradually, to reflect the changing offset of the eye from the target location while the saccade is progressing.[14] At present, however, the predominant view is that, although the "hill" does shift slightly during a saccade, it does not shift in the steady and proportionate way that the "moving hill" hypothesis predicts.[15]

The output from the motor sector of the SC goes to a set of midbrain and brainstem nuclei, which transform the "place" code used by the SC into the "rate" code used by oculomotor neurons. Eye movements are generated by six muscles, arranged in three orthogonally-aligned pairs. Thus, at the level of the final common path, eye movements are encoded in essentially a Cartesian coordinate system.

Although the SC receives a strong input directly from the retina, in primates it is largely under the control of the cerebral cortex, which contains several areas that are involved in determining eye movements.[16] The frontal eye fields, a portion of the motor cortex, are involved in triggering intentional saccades, and an adjoining area, the supplementary eye fields, are involved in organizing groups of saccades into sequences. The parietal eye fields, farther back in the brain, are involved mainly in reflexive saccades, made in response to changes in the view.

The SC only receives visual inputs in its superficial layers, whereas the deeper layers of the colliculus receive also auditory and somatosensory inputs and are connected to many sensorimotor areas of the brain. The colliculus as a whole is thought to help orient the head and eyes toward something seen and heard.[4][17][18][19]

The superior colliculus also receives auditory information from the inferior colliculus. This auditory information is integrated with the visual information already present to produce the ventriloquist effect.

21.3 Other animals

21.3. OTHER ANIMALS

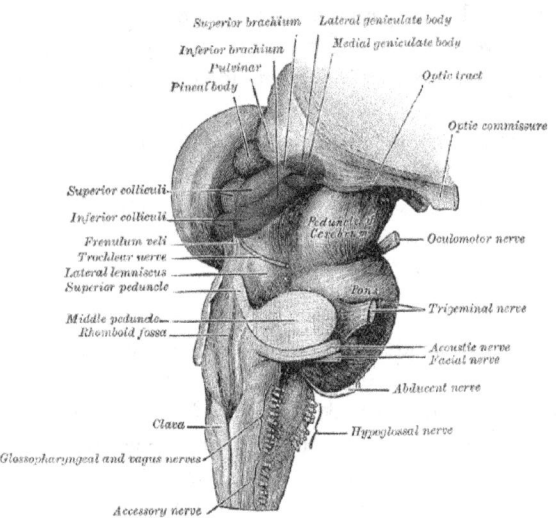

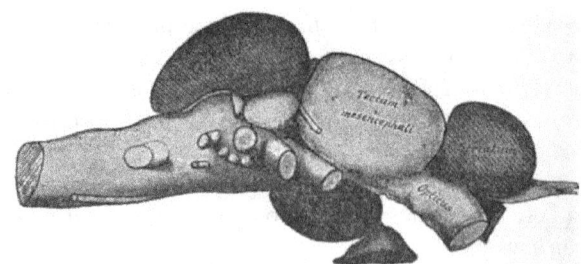

The brain of a cod, with the optic tectum highlighted

Hind- and mid-brains; postero-lateral view. Superior colliculus labeled in blue.

21.3.1 Primates

It is usually accepted that the primate superior colliculus is unique among mammals, in that it does not contain a complete map of the visual field seen by the contralateral eye. Instead, like the visual cortex and lateral geniculate nucleus, each colliculus represents only the contralateral half of the visual field, up to the midline, and excludes a representation of the ipsilateral half.[20] This functional characteristic is explained by the absence, in primates, of anatomical connections between the retinal ganglion cells in the temporal half of the retina and the contralateral superior colliculus. In other mammals, the retinal ganglion cells throughout the contralateral retina project to the contralateral colliculus. This distinction between primates and non-primates has been one of the key lines of evidence in support of the flying primates theory proposed by Australian neuroscientist Jack Pettigrew in 1986, after he discovered that flying foxes (megabats) resemble primates in terms of the pattern of anatomical connections between the retina and superior colliculus.[21]

21.3.2 Other vertebrates

The optic tectum is one of the fundamental components of the vertebrate brain, existing across the full range of species from hagfish to human.[22] (See the brain article for background.) Some aspects of the structure are very consistent, including a structure composed of a number of layers, with a dense input from the optic tracts to the superficial layers and another strong input conveying somatosensory input to deeper layers. Other aspects are highly variable, such as the total number of layers (from 3 in the African lungfish to 15 in the goldfish[23]), and the number of different types of cells (from 2 in the lungfish to 27 in the house sparrow[23]). In hagfish, lamprey, and shark it is a relatively small structure, but in teleost fish it is greatly expanded, in some cases becoming the largest structure in the brain. (See the adjoining drawing of a codfish brain.) In amphibians, reptiles, and especially birds it is also a very significant component, but in mammals it is dwarfed by the massive expansion of the cerebral cortex.[23]

In snakes that can detect infrared radiation, such as pythons and pit vipers, the initial neural input is through the trigeminal nerve instead of the optic tract. The rest of the processing is similar to that of the visual sense and, thus, involves the optic tectum.[24]

21.3.3 Lamprey

The lamprey has been extensively studied because it has a relatively simple brain that is thought in many respects to reflect the brain structure of early vertebrate ancestors. Beginning in the 1970s, Sten Grillner and his colleagues at the Karolinska Institute in Stockholm have used the lamprey as a model system to work out the fundamental principles of motor control in vertebrates, starting in the spinal cord and working upward into the brain.[25] In a series of studies, they found that neural circuits within the spinal cord are capable of generating the rhythmic motor patterns that underlie swimming, that these circuits are controlled by specific locomotor areas in the brainstem and midbrain, and that these areas in turn are controlled by higher brain structures including the basal ganglia and tectum. In a study of the lamprey tectum published in 2007,[26] they found that electrical stimulation could elicit eye movements, lateral bending movements, or swimming activity, and that the type, amplitude, and direction of movement varied as a function of the location within the tectum that was stimulated. These findings were interpreted as consistent with the idea that the tectum generates goal-directed locomotion in the lamprey as it does in other species.

21.3.4 Bats

Bats are not, in fact, blind, but they depend much more on echolocation than vision for navigation and prey capture. They obtain information about the surrounding world by emitting sonar chirps and then listening for the echoes. Their brains are highly specialized for this process, and some of these specializations appear in the superior colliculus.[27] In bats, the retinal projection occupies only a thin zone just beneath the surface, but there are extensive inputs from auditory areas, and outputs to motor areas capable of orienting the ears, head, or body. Echoes coming from different directions activate neurons at different locations in the collicular layers,[28] and activation of collicular neurons influences the chirps that the bats emit. Thus, there is a strong case that the superior colliculus performs the same sorts of functions for the auditory-guided behaviors of bats that it performs for the visual-guided behaviors of other species.

Bats are usually classified into two main groups: Microchiroptera (the most numerous, and commonly found throughout the world), and Megachiroptera (fruit bats, found in Asia, Africa and Australasia). With one exception, Megabats do not echolocate, and rely on a developed sense of vision to navigate. The visual receptive fields of neurons in the superior colliculus in these animals form a precise map of the retina, similar to that found in cats and primates.

21.4 See also

This article uses anatomical terminology; for an overview, see Anatomical terminology.

- inferior colliculus
- Lateral geniculate nucleus
- List of regions in the human brain

21.5 Additional images

- Superficial dissection of brain-stem. Lateral view.
- Dissection of brain-stem. Lateral view.
- Deep dissection of brain-stem. Lateral view.
- Deep dissection of brain-stem. Lateral view.
- Transverse section of mid-brain at level of superior colliculi.
-
- Superior colliculus
- Brainstem. Posterior view.

21.6 Notes

[1] Wallace et al., 1998

[2] Gandhi et al., 2011

[3] Lunenburger et al., 2001

[4] Kustov & Robinson, 1996

[5] Caltharp SA, Pira CU, Mishima N, Youngdale EN, McNeill DS, Liwnicz BH, Oberg KC (2007). "NOGO-A induction and localization during chick brain development indicate a role disparate from neurite outgrowth inhibition". *BMC Dev. Biol.* **7** (1): 32. doi:10.1186/1471-213X-7-32. PMC 1865376. PMID 17433109.

[6] Huerta & Harting, 1984

[7] Clemo HR, Stein BE (1984). "Topographic organization of somatosensory corticotectal influences in cat". *Journal of Neurophysiology.* **51** (5): 843–858. PMID 6726314.

[8] Chavalier & Mana, 2000

[9] Illing, 1996

[10] Mana & Chevalier, 2001

[11] Sprague, 1996

[12] Sparks & Gandhi, 2003

[13] Klier et al., 2001

[14] Munoz et al., 1991

[15] Soetedjo et al., 2002

[16] Pierrot-Deseilligny et al., 2003

[17] Klier et al., 2003

[18] Krauzlis et al., 2004

[19] Sparks, 1999

[20] Lane et al., 1973

[21] Pettigrew, 1986

[22] Maximino, 2008

[23] Northcutt, 2002

[24] Hartline et al., 1978

[25] Grillner, 2003

[26] Saitoh et al., 2007

[27] Ulanovsky & Moss, 2008

[28] Valentine & Moss, 1997

21.7 External links

•

• Stained brain slice images which include the "superior colliculus" at the BrainMaps project

21.8 References

- Chevalier, G; Mana S (2000). "Honeycomb-like structure of the intermediate layers of the rat superior colliculus, with additional observations in several other mammals: AChE patterning". *J Comp Neurol.* **419** (2): 137–53. doi:10.1002/(SICI)1096-9861(20000403)419:2<137::AID-CNE1>3.0.CO;2-6. PMID 10722995.

- Dean, P; Redgrave P; Westby GW (1989). "Event or emergency? Two response systems in the mammalian superior colliculus". *Trends Neurosci.* **12** (4): 137–47. doi:10.1016/0166-2236(89)90052-0. PMID 2470171.

- Gandhi, NJ; Katani HA (2011). "Motor Functions of the Superior Colliculus". *Ann Rev Neurosci.* **34**: 205–231. doi:10.1146/annurev-neuro-061010-113728. PMC 3641825. PMID 21456962.

- Grillner, S (2003). "The motor infrastructure: from ion channels to neuronal networks". *Nature Reviews Neuroscience.* **4** (7): 573–86. doi:10.1038/nrn1137. PMID 12838332.

- Hartline, PH; Kass L; Loop MS (1978). "Merging of modalities in the optic tectum: infrared and visual integration in rattlesnakes". *Science.* **199** (4334): 1225–9. doi:10.1126/science.628839. PMID 628839.

- Huerta, MF; Harting JK (1984). Vanegas H, ed. *Comparative Neurology of the Optic Tectum.* New York: Plenum Press. pp. 687–773. ISBN 978-0-306-41236-3.

- Illing, R-B (1996). "The mosaic architecture of the superior colliculus". *Prog Brain Res.* **112**: 17–34. doi:10.1016/S0079-6123(08)63318-X. PMID 8979818.

- King, AJ; Schnupp JWH; Carlile S; Smith AL; Thompson ID (1996). "The development of topographically-aligned maps of visual an auditory space in the superior colliculus". *Prog Brain Res.* **112**: 335–350. doi:10.1016/S0079-6123(08)63340-3. PMID 8979840.

- Klier, EM; Wang H; Crawford JD (2001). "The superior colliculus encodes gaze commands in retinal coordinates" (PDF). *Nat Neurosci.* **4** (6): 627–32. doi:10.1038/88450. PMID 11369944.

- Klier, E; Wang H; Crawford D (2003). "Three-dimensional eye-head coordination is implemented downstream from the superior colliculus". *J Neurophysiol.* **89** (5): 2839–53. doi:10.1152/jn.00763.2002. PMID 12740415.

- Krauzlis, R; Liston D; Carello C (2004). "Target selection and the superior colliculus: goals, choices and hypotheses". *Vision Res.* **44** (12): 1445–51. doi:10.1016/j.visres.2004.01.005. PMID 15066403.

- Kustov, A; Robinson D (1996). "Shared neural control of attentional shifts and eye movements". *Nature.* **384** (6604): 74–77. doi:10.1038/384074a0. PMID 8900281.

- Lane, RH; Allman JM; Kaas JH; Miezin FM (1973). "The visuotopic organization of the superior colliculus of the owl monkey (*Aotus trivirgatus*) and the bush baby (*Galago senegalensis*)". *Brain Res.* **60** (2): 335–49. doi:10.1016/0006-8993(73)90794-4. PMID 4202853.

- Lunenburger, L; Kleiser R; Stuphorn V; Miller LE; Hoffmann KP (2001). "A possible role of the superior colliculus in eye–hand coordination". *Prog Brain Res.* **134**: 109–25. doi:10.1016/S0079-6123(01)34009-8. PMID 11702538.

- Mana, S; Chevalier G (2001). "Honeycomb-like structure of the intermediate layers of the rat superior colliculus: afferent and efferent connections". *Neuroscience.* **103** (3): 673–93. doi:10.1016/S0306-4522(01)00026-4. PMID 11274787.

- Maximino, C; Soares, Daphne (2008). Soares, Daphne, ed. "Evolutionary changes in the complexity of the tectum of nontetrapods: a cladistic approach". *PLOS ONE.* **3** (10): e385. doi:10.1371/journal.pone.0003582. PMC 2571994. PMID 18974789.

- Munoz, DP; Pélisson D; Guitton D (1991). "Movement of activity on the superior colliculus motor map during gaze shifts" (PDF). *Science.* **251** (4999): 1358–60. doi:10.1126/science.2003221. PMID 2003221.

- Northcutt, RG (2002). "Understanding vertebrate brain evolution". *Integr Comp Biol.* **42** (4): 743–6. doi:10.1093/icb/42.4.743. PMID 21708771.

- Pettigrew, JD (1986). "Flying primates? Megabats have the advanced pathway from eye to midbrain". *Science.* **231** (4743): 1304–6. doi:10.1126/science.3945827. PMID 3945827.

- Pierrot-Deseilligny, C; Müri RM; Ploner CJ; Gaymard B; Rivaud-Péchoux S (2003). "Cortical control of ocular saccades in humans: a model for motricity". *Prog Brain Res.* **142**: 3–17. doi:10.1016/S0079-6123(03)42003-7. PMID 12693251.

- Saitoh, K; Ménard A; Grillner S (2007). "Tectal control of locomotion, steering, and eye movements in lamprey". *J Neurophysiol.* **97** (4): 3093–108. doi:10.1152/jn.00639.2006. PMID 17303814.

- Soetedjo, R; Kaneko CR; Fuchs AF (2002). "Evidence against a moving hill in the superior colliculus during saccadic eye movements in the monkey". *J Neurophysiol.* **87** (6): 2778–89. PMID 12037180.

- Sparks, DL (1999). "Conceptual issues related to the role of the superior colliculus in the control of gaze". *Current Opinion in Neurobiology.* **6** (6): 698–707. doi:10.1016/S0959-4388(99)00039-2. PMID 10607648.

- Sparks, DL; Gandhi NJ (2003). "Single-cell signals: an oculomotor perspective". *Prog Brain Res.* **142**: 35–53. doi:10.1016/S0079-6123(03)42005-0. PMID 12693253.

- Sprague, JM (1996). "Neural mechanisms of visual orienting responses". *Prog Brain Res.* **112**: 1–15. doi:10.1016/S0079-6123(08)63317-8. PMID 8979817.

- Stein, BE; Clamman HP (1981). "Control of pinna movements and sensorimotor register in cat superior colliculus". *Brain Behav Evol.* **19** (3-4): 180–192. doi:10.1159/000121641. PMID 7326575.

- Ulanovsky, N; Moss CF (2008). "What the bat's voice tells the bat's brain". *PNAS.* **105** (25): 8491–98. doi:10.1073/pnas.0703550105. PMC 2438418. PMID 18562301.

- Valentine, D; Moss CF (1997). "Spatially selective auditory responses in the superior colliculus of the echolocating bat". *J Neurosci.* **17** (5): 1720–33. PMID 9030631.

- Wallace, MT; Meredith MA; Stein BE (1998). "Multisensory integration in the superior colliculus of the alert cat". *J Neurophysiol.* **80** (2): 1006–10. PMID 9705489.

Chapter 22

Thalamus

The **thalamus** (from Greek θάλαμος, "chamber")[1] is a midline symmetrical structure of two halves, within the vertebrate brain, situated between the cerebral cortex and the midbrain. Some of its functions are the relaying of sensory and motor signals to the cerebral cortex,[2][3] and the regulation of consciousness, sleep, and alertness. The medial surface of the two halves constitute the upper lateral wall of the third ventricle. It is the main product of the embryonic diencephalon.

22.1 Anatomy

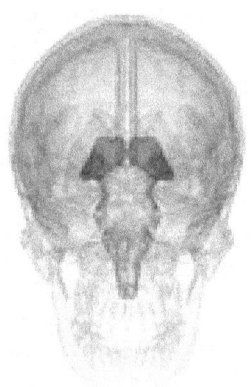

The thalamus in a 360° rotation

The thalamus is located in the forebrain which is superior to the midbrain, near the center of the brain, with nerve fibers projecting out to the cerebral cortex in all directions. The medial surface of the thalamus constitutes the upper part of the lateral wall of the third ventricle, and is connected to the corresponding surface of the opposite thalamus by a flattened gray band, the interthalamic adhesion.

22.1.1 Morphology

The two halves of the thalamus are prominent bulb-shaped masses, about 5.7 cm in length, located obliquely (about 30°) and symmetrically on each side of the third ventricle. Both parts of the thalamus, in the human, are about the size and shape of a walnut.[2] These are about three centimetres in length, at the widest part 2.5 centimetres across and about 2 centimetres in height (comparable to an unshelled walnut, with the nut-shell joining in the horizontal plane).

22.1.2 Blood supply

The thalamus derives its blood supply from a number of arteries: the polar artery (posterior communicating artery), paramedian thalamic-subthalamic arteries, inferolateral (thalamogeniculate) arteries, and posterior (medial and lateral) choroidal arteries.[4] These are all branches of the posterior cerebral artery.[5]

Some people have the artery of Percheron, which is a rare anatomic variation in which a single arterial trunk arises from the posterior cerebral artery to supply both parts of the thalamus.

22.1.3 Thalamic nuclei

See also: List of thalamic nuclei
The thalamus is part of a nuclear complex structured of four parts, the hypothalamus, epithalamus, prethalamus (formerly called ventral thalamus), and dorsal thalamus.[6]

Derivatives of the diencephalon also include the dorsally-located epithalamus (essentially the habenula and annexes) and the perithalamus (prethalamus) containing the zona incerta and the thalamic reticular nucleus. Due to their different ontogenetic origins, the epithalamus and the perithalamus are formally distinguished from the thalamus proper.

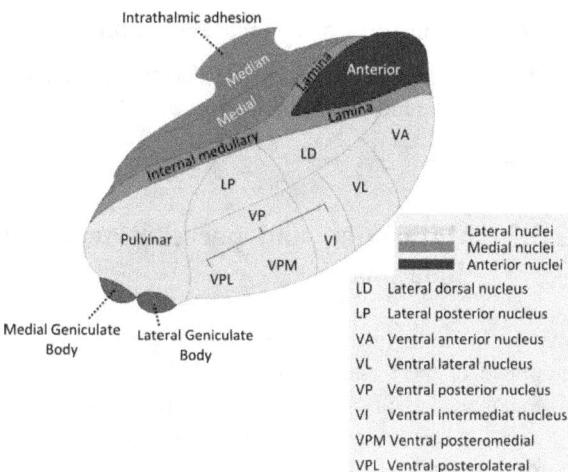

Nuclei of the thalamus

The thalamus comprises a system of lamellae (made up of myelinated fibers) separating different thalamic subparts. Other areas are defined by distinct clusters of neurons, such as the periventricular nucleus, the intralaminar elements, the "nucleus limitans", and others.[7] These latter structures, different in structure from the major part of the thalamus, have been grouped together into the *allothalamus* as opposed to the *isothalamus*.[8] This distinction simplifies the global description of the thalamus.

22.1.4 Connections

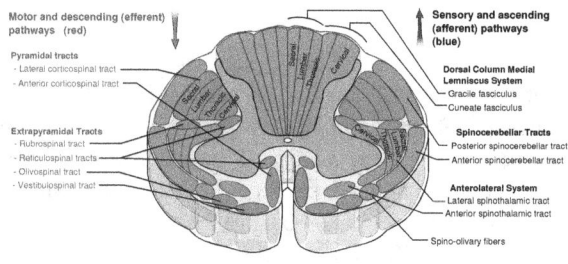

The thalamus is connected to the spinal cord via the spinothalamic tract.

The thalamus is manifoldly connected to the hippocampus via the mammillo-thalamic tract, this tract comprises the mammillary bodies and fornix.[9]

The thalamus is connected to the cerebral cortex via the thalamocortical radiations.[10]

The spinothalamic tract is a sensory pathway originating in the spinal cord. It transmits information to the thalamus about pain, temperature, itch and crude touch. There are two main parts: the lateral spinothalamic tract, which transmits pain and temperature, and the anterior (or ventral) spinothalamic tract, which transmits crude touch and pressure.

22.2 Function

The thalamus has multiple functions. It may be thought of as a kind of hub of information. It is generally believed to act as a relay between different subcortical areas and the cerebral cortex. In particular, every sensory system (with the exception of the olfactory system) includes a thalamic nucleus that receives sensory signals and sends them to the associated primary cortical area. For the visual system, for example, inputs from the retina are sent to the lateral geniculate nucleus of the thalamus, which in turn projects to the visual cortex in the occipital lobe. The thalamus is believed to both process sensory information as well as relay it—each of the primary sensory relay areas receives strong feedback connections from the cerebral cortex. Similarly the medial geniculate nucleus acts as a key auditory relay between the inferior colliculus of the midbrain and the primary auditory cortex, and the ventral posterior nucleus is a key somatosensory relay, which sends touch and proprioceptive information to the primary somatosensory cortex.

The thalamus also plays an important role in regulating states of sleep and wakefulness.[11] Thalamic nuclei have strong reciprocal connections with the cerebral cortex, forming thalamo-cortico-thalamic circuits that are believed to be involved with consciousness. The thalamus plays a major role in regulating arousal, the level of awareness, and activity. Damage to the thalamus can lead to permanent coma.

The role of the thalamus in the more anterior pallidal and nigral territories in the basal ganglia system disturbances is recognized but still poorly understood. The contribution of the thalamus to vestibular or to tectal functions is almost ignored. The thalamus has been thought of as a "relay" that simply forwards signals to the cerebral cortex. Newer research suggests that thalamic function is more selective.[12] Many different functions are linked to various regions of the thalamus. This is the case for many of the sensory systems (except for the olfactory system), such as the auditory, somatic, visceral, gustatory and visual systems where localized lesions provoke specific sensory deficits. A major role of the thalamus is support of motor and language systems, and much of the circuitry implicated for these systems is shared. The thalamus is *functionally connected* to the hippocampus[13] as part of the extended hippocampal system at the thalamic anterior nuclei[14] with respect to spatial memory and spatial sensory datum they are crucial for human episodic memory and rodent event memory.[15][16] There is support for the hypothesis that thalamic regions connection to particular parts of the mesio-temporal lobe provide differentiation of the functioning of recollective and familiarity memory.[9]

The neuronal information processes necessary for motor control were proposed as a network involving the thalamus as a subcortical motor centre.[17] Through investigations of the anatomy of the brains of primates[18] the nature of the interconnected tissues of the cerebellum to the multiple motor cortices suggested that the thalamus fulfills a key function in providing the specific channels from the basal ganglia and cerebellum to the cortical motor areas.[19][20] In an investigation of the saccade and antisaccade[21] motor response in three monkeys, the thalamic regions were found to be involved in the generation of antisaccade eye-movement.[22]

22.3 Development

The thalamic complex is composed of the perithalamus (or prethalamus, previously also known as ventral thalamus), the mid-diencephalic organiser (which forms later the zona limitans intrathalamica (ZLI)) and the thalamus (dorsal thalamus).[23][24] The development of the thalamus can be subdivide into three steps[25] The thalamus is the largest structure deriving from the embryonic diencephalon, the posterior part of the forebrain situated between the midbrain and the cerebrum.

22.3.1 Early brain development

After neurulation the anlage of the prethalamus and the thalamus is induced within the neural tube. Data from different vertebrate model organisms support a model in which the interaction between two transcription factors, Fez and Otx, are of decisive importance. Fez is expressed in the prethalamus, and functional experiments show that Fez is required for prethalamus formation.[26][27] Posteriorly, Otx1 and Otx2 abut the expression domain of Fez and are required for proper development of the thalamus.[28][29]

22.3.2 The formation of the mid-diencephalic organiser (MDO)

At the interface between the expression domains of Fez and Otx, the mid-diencephalic organizer (MDO, also called the ZLI organiser) is induced within the thalamic anlage. The MDO is the central signalling organizer in the thalamus. A lack of the organizer leads to the absence of the thalamus. The MDO matures from ventral to dorsal during development. Members of the SHH family and of the Wnt family are the main principal signals emitted by the MDO.

Besides its importance as signalling center, the organizer matures into the morphological structure of the zona limitans intrathalamica (ZLI).

22.3.3 Maturation and parcellation of the thalamus

After its induction, the MDO starts to orchestrate the development of the thalamic anlage by release of signalling molecules such as SHH.[30] In mice, the function of signaling at the MDO has not been addressed directly due to a complete absence of the diencephalon in SHH mutants.[31]

Studies in chicks have shown that SHH is both necessary and sufficient for thalamic gene induction.[32] In zebrafish, it was shown that the expression of two SHH genes, SHH-a and SHH-b (formerly described as twhh) mark the MDO territory, and that SHH signaling is sufficient for the molecular differentiation of both the prethalamus and the thalamus but is not required for their maintenance and SHH signaling from the MDO/alar plate is sufficient for the maturation of prethalamic and thalamic territory while ventral Shh signals are dispensable.[33]

The exposure to SHH leads to differentiation of thalamic neurons. SHH signaling from the MDO induces a posterior-to-anterior wave of expression the proneural gene Neurogenin1 in the major (caudal) part of the thalamus, and Ascl1 (formerly Mash1) in the remaining narrow stripe of rostral thalamic cells immediately adjacent to the MDO, and in the prethalamus.[34][35]

This zonation of proneural gene expression leads to the differentiation of glutamatergic relay neurons from the Neurogenin1+ precursors and of GABAergic inhibitory neurons from the Ascl1+ precursors. In fish, selection of these alternative neurotransmitter fates is controlled by the dynamic expression of Her6 the homolog of HES1. Expression of this hairy-like bHLH transcription factor, which represses Neurogenin but is required for Ascl1, is progressively lost from the caudal thalamus but maintained in the prethalamus and in the stripe of rostral thalamic cells. In addition, studies on chick and mice have shown that blocking the Shh pathway leads to absence of the rostral thalamus and substantial decrease of the caudal thalamus. The rostral thalamus will give rise to the reticular nucleus mainly whereby the caudal thalamus will form the relay thalamus and will be further subdivided in the thalamic nuclei.[25]

In humans, a common genetic variation in the promotor region of the serotonin transporter (the SERT-long and -short allele: 5-HTTLPR) has been shown to affect the development of several regions of the thalamus in adults. People who inherit two short alleles (SERT-ss) have more neurons and a larger volume in the pulvinar and possibly the limbic regions of the thalamus. Enlargement of the thalamus provides an anatomical basis for why people who inherit two SERT-ss alleles are more vulnerable to major depression, posttraumatic stress disorder, and suicide.[36]

22.4 Clinical significance

A cerebrovascular accident (stroke) can lead to the thalamic syndrome,[37] which involves a one-sided burning or aching sensation often accompanied by mood swings. Bilateral ischemia of the area supplied by the paramedian artery can cause serious problems including akinetic mutism, and be accompanied by oculomotor problems. A related concept is thalamocortical dysrhythmia. The occlusion of the artery of Percheron can lead to a bilateral thalamus infarction.

Korsakoff's syndrome stems from damage to the mammillary body, the mammillothalamic fasciculus or the thalamus.

Fatal familial insomnia is a hereditary prion disease in which degeneration of the thalamus occurs, causing the patient to gradually lose his ability to sleep and progressing to a state of total insomnia, which invariably leads to death. In contrast, damage to the thalamus can result in coma.

22.5 Additional images

Images are circa 1858.[38]

- The left optic nerve and the optic tracts.
- Coronal section of lateral and 3rd ventricles.
- Dissection showing the ventricles of the brain.
- Section of brain showing upper surface of temporal lobe.
- Horizontal section of right cerebral hemisphere.
- Mesal aspect of a brain sectioned in the median sagittal plane.
- Schematic representation of the chief ganglionic categories (I to V).
- Scheme showing the course of the fibers of the lemniscus; medial lemniscus in blue, lateral in red.
- Deep dissection of brain-stem. Lateral view.
- Deep dissection of brain-stem. Ventral view.
- Coronal section of brain immediately in front of pons.

- Coronal section of brain through intermediate mass of third ventricle.
- Human brain frontal (coronal) section
-
- Thalamus
- Ventricles of brain and basal ganglia. Superior view. Horizontal section. Deep dissection
- Ventricles of brain and basal ganglia. Superior view. Horizontal section. Deep dissection

22.6 See also

- List of regions in the human brain
- List of thalamic nuclei
- Neothalamus
- Primate basal ganglia system
- Thalamic stimulator
- Thalamotomy
- 5-HT7 receptor
- Nonmotor region of the ventral nuclear group of the thalamus

22.7 References

[1] Harper - index & University of Washington Faculty Web Server & Search engine search page + Perseus Project tufts.edu Retrieved 2012-02-09

[2] Sherman, S. (2006). "Thalamus". *Scholarpedia*. **1** (9): 1583. doi:10.4249/scholarpedia.1583.

[3] Sherman, S. Murray; Guillery, R. W. (2000). *Exploring the Thalamus*. Academic Press. ISBN 978-0-12-305460-9.

[4] Percheron, G. (1982). "The arterial supply of the thalamus". In Schaltenbrand; Walker, A. E. *Stereotaxy of the human brain*. Stuttgart: Thieme. pp. 218–32.

[5] Knipe, H Jones, J et al. Thalamus http://radiopaedia.org/articles/thalamus

[6] Herrero, María-Trinidad; Barcia, Carlos; Navarro, Juana (2002). "Functional anatomy of thalamus and basal ganglia". *Child's Nervous System*. **18** (8): 386–404. doi:10.1007/s00381-002-0604-1.

22.7. REFERENCES

[7] Jones Edward G. (2007) "The Thalamus" Cambridge Uni. Press

[8] Percheron, G. (2003). "Thalamus". In Paxinos, G.; May, J. *The human nervous system* (2nd ed.). Amsterdam: Elsevier. pp. 592–675.

[9] Carlesimo, GA; Lombardi, MG; Caltagirone, C (2011). "Vascular thalamic amnesia: A reappraisal". *Neuropsychologia.* **49** (5): 777–89. doi:10.1016/j.neuropsychologia.2011.01.026. PMID 21255590.

[10] *University of Washington* (1991). "Thalamocortical radiations". washington.edu.

[11] Steriade, Mircea; Llinás, Rodolfo R. (1988). "The Functional States of the Thalamus and the Associated Neuronal Interplay". *Physiological Reviews.* **68** (3): 649–742. PMID 2839857.

[12] Leonard, Abigail W. (August 17, 2006). "Your Brain Boots Up Like a Computer". *LiveScience.*

[13] Stein, Thor; Moritz, Chad; Quigley, Michelle; Cordes, Dietmar; Haughton, Victor; Meyerand, Elizabeth (2000). "Functional Connectivity in the Thalamus and Hippocampus Studied with Functional MR Imaging". *American Journal of Neuroradiology.* **21** (8): 1397–401. PMID 11003270.

[14] Aggleton, John P.; Brown, Malcolm W. (1999). "Episodic memory, amnesia, and the hippocampal–anterior thalamic axis". *Behavioral and Brain Sciences.* **22** (3): 425–44; discussion 444–89. doi:10.1017/S0140525X99002034. PMID 11301518.

[15] Aggleton, John P.; O'Mara, Shane M.; Vann, Seralynne D.; Wright, Nick F.; Tsanov, Marian; Erichsen, Jonathan T. (2010). "Hippocampal-anterior thalamic pathways for memory: Uncovering a network of direct and indirect actions". *European Journal of Neuroscience.* **31** (12): 2292–307. doi:10.1111/j.1460-9568.2010.07251.x. PMC 2936113. PMID 20550571.

[16] Burgess, Neil; Maguire, Eleanor A; O'Keefe, John (2002). "The Human Hippocampus and Spatial and Episodic Memory". *Neuron.* **35** (4): 625–41. doi:10.1016/S0896-6273(02)00830-9. PMID 12194864.

[17] Evarts, E V; Thach, W T (1969). "Motor Mechanisms of the CNS: Cerebrocerebellar Interrelations". *Annual Review of Physiology.* **31**: 451–98. doi:10.1146/annurev.ph.31.030169.002315. PMID 4885774.

[18] Orioli, PJ; Strick, PL (1989). "Cerebellar connections with the motor cortex and the arcuate premotor area: An analysis employing retrograde transneuronal transport of WGA-HRP". *The Journal of Comparative Neurology.* **288** (4): 612–26. doi:10.1002/cne.902880408. PMID 2478593.

[19] Asanuma C, Thach WT, Jones EG (May 1983). "Cytoarchitectonic delineation of the ventral lateral thalamic region in the monkey". *Brain Research.* **286** (3): 219–35. doi:10.1016/0165-0173(83)90014-0. PMID 6850357.

[20] Kurata, K (2005). "Activity properties and location of neurons in the motor thalamus that project to the cortical motor areas in monkeys". *Journal of Neurophysiology.* **94** (1): 550–66. doi:10.1152/jn.01034.2004. PMID 15703228.

[21] http://www.optomotorik.de/blicken/anti-rev.htm[]

[22] Kunimatsu, J; Tanaka, M (2010). "Roles of the primate motor thalamus in the generation of antisacades". *Journal of Neuroscience.* **30** (14): 5108–17. doi:10.1523/JNEUROSCI.0406-10.2010. PMID 20371831.

[23] Kuhlenbeck, Hartwig (1937). "The ontogenetic development of the diencephalic centers in a bird's brain (chick) and comparison with the reptilian and mammalian diencephalon". *The Journal of Comparative Neurology.* **66**: 23–75. doi:10.1002/cne.900660103.

[24] Shimamura, K; Hartigan, DJ; Martinez, S; Puelles, L; Rubenstein, JL (1995). "Longitudinal organization of the anterior neural plate and neural tube". *Development.* **121** (12): 3923–33. PMID 8575293.

[25] Scholpp, Steffen; Lumsden, Andrew (2010). "Building a bridal chamber: Development of the thalamus". *Trends in Neurosciences.* **33** (8): 373–80. doi:10.1016/j.tins.2010.05.003. PMC 2954313. PMID 20541814.

[26] Hirata, T.; Nakazawa, M; Muraoka, O; Nakayama, R; Suda, Y; Hibi, M (2006). "Zinc-finger genes Fez and Fez-like function in the establishment of diencephalon subdivisions". *Development.* **133** (20): 3993–4004. doi:10.1242/dev.02585. PMID 16971467.

[27] Jeong, J.-Y.; Einhorn, Z.; Mathur, P.; Chen, L.; Lee, S.; Kawakami, K.; Guo, S. (2007). "Patterning the zebrafish diencephalon by the conserved zinc-finger protein Fezl". *Development.* **134** (1): 127–36. doi:10.1242/dev.02705. PMID 17164418.

[28] Acampora, D; Avantaggiato, V; Tuorto, F; Simeone, A (1997). "Genetic control of brain morphogenesis through Otx gene dosage requirement". *Development.* **124** (18): 3639–50. PMID 9342056.

[29] Scholpp, S.; Foucher, I.; Staudt, N.; Peukert, D.; Lumsden, A.; Houart, C. (2007). "Otx1l, Otx2 and Irx1b establish and position the ZLI in the diencephalon". *Development.* **134** (17): 3167–76. doi:10.1242/dev.001461. PMID 17670791.

[30] Puelles, L; Rubenstein, JL (2003). "Forebrain gene expression domains and the evolving prosomeric model". *Trends in Neurosciences.* **26** (9): 469–76. doi:10.1016/S0166-2236(03)00234-0. PMID 12948657.

[31] Ishibashi, M; McMahon, AP (2002). "A sonic hedgehog-dependent signaling relay regulates growth of diencephalic and mesencephalic primordia in the early mouse embryo". *Development.* **129** (20): 4807–19. PMID 12361972.

[32] Kiecker, C; Lumsden, A (2004). "Hedgehog signaling from the ZLI regulates diencephalic regional identity". *Nature Neuroscience.* **7** (11): 1242–9. doi:10.1038/nn1338. PMID 15494730.

[33] Scholpp, S.; Wolf, O; Brand, M; Lumsden, A (2006). "Hedgehog signalling from the zona limitans intrathalamica orchestrates patterning of the zebrafish diencephalon". *Development.* **133** (5): 855–64. doi:10.1242/dev.02248. PMID 16452095.

[34] Scholpp, S.; Delogu, A.; Gilthorpe, J.; Peukert, D.; Schindler, S.; Lumsden, A. (2009). "Her6 regulates the neurogenetic gradient and neuronal identity in the thalamus". *Proceedings of the National Academy of Sciences.* **106** (47): 19895–900. doi:10.1073/pnas.0910894106. PMC 2775703. PMID 19903880.

[35] Vue, Tou Yia; Bluske, Krista; Alishahi, Amin; Yang, Lin Lin; Koyano-Nakagawa, Naoko; Novitch, Bennett; Nakagawa, Yasushi (2009). "Sonic Hedgehog Signaling Controls Thalamic Progenitor Identity and Nuclei Specification in Mice". *Journal of Neuroscience.* **29** (14): 4484–97. doi:10.1523/JNEUROSCI.0656-09.2009. PMC 2718849. PMID 19357274.

[36] Young, Keith A.; Holcomb, Leigh A.; Bonkale, Willy L.; Hicks, Paul B.; Yazdani, Umar; German, Dwight C. (2007). "5HTTLPR Polymorphism and Enlargement of the Pulvinar: Unlocking the Backdoor to the Limbic System". *Biological Psychiatry.* **61** (6): 813–8. doi:10.1016/j.biopsych.2006.08.047. PMID 17083920.

[37] Dejerine, J.; Roussy, G. (1906). "Le syndrome thalamique". *Revue Neurologique.* **14**: 521–32.

[38] Gray, H. & Carter, H. V. (1858), Anatomy Descriptive and Surgical, London: John W. Parker and Son, Retrieved (*16 October 2011*) [2012-02-10] →

22.8 External links

- Stained brain slice images which include the "thalamus" at the BrainMaps project

Chapter 23

Hypothalamus

The **hypothalamus** (from Greek ὑπό, "under" and θάλαμος, thalamus) is a portion of the brain that contains a number of small nuclei with a variety of functions. One of the most important functions of the hypothalamus is to link the nervous system to the endocrine system via the pituitary gland (hypophysis).

The hypothalamus is located below the thalamus and is part of the limbic system.[1] In the terminology of neuroanatomy, it forms the ventral part of the diencephalon. All vertebrate brains contain a hypothalamus. In humans, it is the size of an almond.

The hypothalamus is responsible for certain metabolic processes and other activities of the autonomic nervous system. It synthesizes and secretes certain neurohormones, called releasing hormones or hypothalamic hormones, and these in turn stimulate or inhibit the secretion of pituitary hormones. The hypothalamus controls body temperature, hunger, important aspects of parenting and attachment behaviors, thirst,[2] fatigue, sleep, and circadian rhythms.

23.1 Structure

The hypothalamus is a brain structure made up of distinct nuclei as well as less anatomically distinct areas. It is found in all vertebrate nervous systems. In mammals, magnocellular neurosecretory cells in the paraventricular nucleus and the supraoptic nucleus of the hypothalamus produce neurohypophysial hormones, oxytocin and vasopressin. These hormones are released into the blood in the posterior pituitary.[3] Much smaller parvocellular neurosecretory cells, neurons of the paraventricular nucleus, release corticotropin-releasing hormone and other hormones into the hypophyseal portal system, where these hormones diffuse to the anterior pituitary.

23.1.1 Nuclei

The hypothalamic nuclei include the following:[4][5][6]

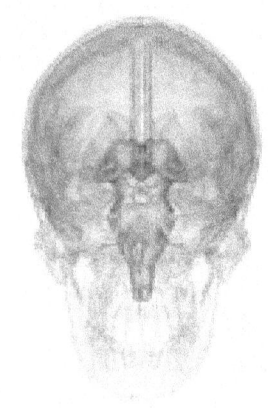

Human hypothalamus (shown in red)

See also: ventrolateral preoptic nucleus, periventricular nucleus.

- A cross section of the monkey hypothalamus displays 2 of the major hypothalamic nuclei on either side of the fluid-filled 3rd ventricle.

- Hypothalamic nuclei

- Hypothalamic nuclei on one side of the hypothalamus, shown in a 3-D computer reconstruction[1]

1. ^ Brain Research Bulletin 35:323-327, 1994

23.1.2 Neural connections

Further information: Lateral hypothalamus § Orexinergic projection system, and Tuberomammillary nucleus §

Histaminergic outputs

The hypothalamus is highly interconnected with other parts of the central nervous system, in particular the brainstem and its reticular formation. As part of the limbic system, it has connections to other limbic structures including the amygdala and septum, and is also connected with areas of the autonomous nervous system.

The hypothalamus receives many inputs from the brainstem, the most notable from the nucleus of the solitary tract, the locus coeruleus, and the ventrolateral medulla.

Most nerve fibres within the hypothalamus run in two ways (bidirectional).

- Projections to areas caudal to the hypothalamus go through the medial forebrain bundle, the mammillotegmental tract and the dorsal longitudinal fasciculus.

- Projections to areas rostral to the hypothalamus are carried by the mammillothalamic tract, the fornix and terminal stria.

- Projections to areas of the sympathetic motor system (lateral horn spinal segments T1-L2/L3) are carried by the hypothalamospinal tract and they activate the sympathetic motor pathway.

23.1.3 Sexual dimorphism

Several hypothalamic nuclei are sexually dimorphic; i.e., there are clear differences in both structure and function between males and females.

Some differences are apparent even in gross neuroanatomy: most notable is the sexually dimorphic nucleus within the preoptic area. However most of the differences are subtle changes in the connectivity and chemical sensitivity of particular sets of neurons.

The importance of these changes can be recognised by functional differences between males and females. For instance, males of most species prefer the odor and appearance of females over males, which is instrumental in stimulating male sexual behavior. If the sexually dimorphic nucleus is lesioned, this preference for females by males diminishes. Also, the pattern of secretion of growth hormone is sexually dimorphic, and this is one reason why in many species, adult males are much larger than females.

Responsiveness to ovarian steroids

Other striking functional dimorphisms are in the behavioral responses to ovarian steroids of the adult. Males and females respond to ovarian steroids in different ways, partly because the expression of estrogen-sensitive neurons in the hypothalamus is sexually dimorphic; i.e., estrogen receptors are expressed in different sets of neurons.

Estrogen and progesterone can influence gene expression in particular neurons or induce changes in cell membrane potential and kinase activation, leading to diverse non-genomic cellular functions. Estrogen and progesterone bind to their cognate nuclear hormone receptors, which translocate to the cell nucleus and interact with regions of DNA known as hormone response elements (HREs) or get tethered to another transcription factor's binding site. Estrogen receptor (ER) has been shown to transactivate other transcription factors in this manner, despite the absence of an estrogen response element (ERE) in the proximal promoter region of the gene. In general, ERs and progesterone receptors (PRs) are gene activators, with increased mRNA and subsequent protein synthesis following hormone exposure.

Male and female brains differ in the distribution of estrogen receptors, and this difference is an irreversible consequence of neonatal steroid exposure. Estrogen receptors (and progesterone receptors) are found mainly in neurons in the anterior and mediobasal hypothalamus, notably:

- the preoptic area (where LHRH neurons are located)

- the periventricular nucleus (where somatostatin neurons are located)

- the ventromedial hypothalamus (which is important for sexual behavior).

23.1.4 Development

In neonatal life, gonadal steroids influence the development of the neuroendocrine hypothalamus. For instance, they determine the ability of females to exhibit a normal reproductive cycle, and of males and females to display appropriate reproductive behaviors in adult life.

- If a *female rat* is injected once with testosterone in the first few days of postnatal life (during the "critical period" of sex-steroid influence), the hypothalamus is irreversibly masculinized; the adult rat will be incapable of generating an LH surge in response to estrogen (a characteristic of females), but will be capable of exhibiting *male* sexual behaviors (mounting a sexually receptive female).

- By contrast, a *male rat* castrated just after birth will be *feminized*, and the adult will show *female* sexual

23.2 Function

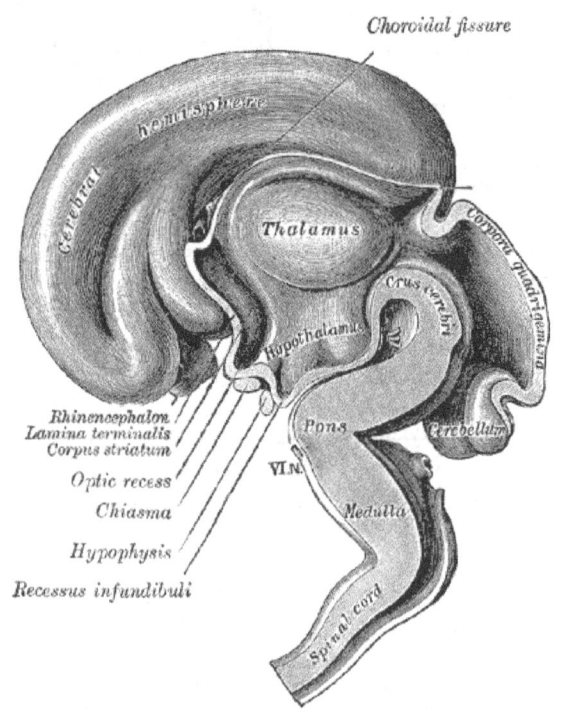

Median sagittal section of brain of human embryo of three months

behavior in response to estrogen (sexual receptivity, lordosis behavior).

In primates, the developmental influence of androgens is less clear, and the consequences are less understood. Within the brain, testosterone is aromatized to (estradiol), which is the principal active hormone for developmental influences. The human testis secretes high levels of testosterone from about week 8 of fetal life until 5–6 months after birth (a similar perinatal surge in testosterone is observed in many species), a process that appears to underlie the male phenotype. Estrogen from the maternal circulation is relatively ineffective, partly because of the high circulating levels of steroid-binding proteins in pregnancy.

Sex steroids are not the only important influences upon hypothalamic development; in particular, pre-pubertal stress in early life (of rats) determines the capacity of the adult hypothalamus to respond to an acute stressor.[9] Unlike gonadal steroid receptors, glucocorticoid receptors are very widespread throughout the brain; in the paraventricular nucleus, they mediate negative feedback control of CRF synthesis and secretion, but elsewhere their role is not well understood.

23.2 Function

23.2.1 Hormone release

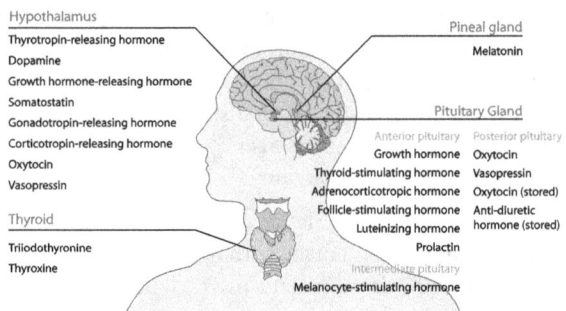

Endocrine glands in the human head and neck and their hormones

The hypothalamus has a central neuroendocrine function, most notably by its control of the anterior pituitary, which in turn regulates various endocrine glands and organs. Releasing hormones (also called releasing factors) are produced in hypothalamic nuclei then transported along axons to either the median eminence or the posterior pituitary, where they are stored and released as needed.[10]

Anterior pituitary

In the hypothalamic–adenohypophyseal axis, releasing hormones, also known as hypophysiotropic or hypothalamic hormones, are released from the median eminence, a prolongation of the hypothalamus, into the hypophyseal portal system, which carries them to the anterior pituitary where they exert their regulatory functions on the secretion of adenohypophyseal hormones.[11] These hypophysiotropic hormones are stimulated by parvocellular neurosecretory cells located in the periventricular area of the hypothalamus. After their release into the capillaries of the third ventricle, the hypophysiotropic hormones travel through what is known as the hypothalamo-pituitary portal circulation. Once they reach their destination in the anterior pituitary, these hormones bind to specific receptors located on the surface of pituitary cells. Depending on which cells are activated through this binding, the pituitary will either begin secreting or stop secreting hormones into the rest of the bloodstream. (Bear, Mark F. "Hypothalamic Control of the Anterior Pituitary." Neuroscience: Exploring the Brain. 4th ed. Philadelphia: Wolters Kluwer, 2016. 528. Print.)

Other hormones secreted from the median eminence include vasopressin, oxytocin, and neurotensin.[13][14][15][16]

Posterior pituitary

In the hypothalamic-neurohypophyseal axis, neurohypophysial hormones are released from the posterior pituitary, which is actually a prolongation of the hypothalamus, into the circulation.

It is also known that hypothalamic-pituitary-adrenal axis (HPA) hormones are related to certain skin diseases and skin homeostasis. There is evidence linking hyperactivity of HPA hormones to stress-related skin diseases and skin tumors.[17]

23.2.2 Stimulation

The hypothalamus coordinates many hormonal and behavioural circadian rhythms, complex patterns of neuroendocrine outputs, complex homeostatic mechanisms, and important behaviours. The hypothalamus must, therefore, respond to many different signals, some of which generated externally and some internally. Delta wave signalling arising either in the thalamus or in the cortex influences the secretion of releasing hormones; GHRH and prolactin are stimulated whilst TRH is inhibited.

The hypothalamus is responsive to:

- Light: daylength and photoperiod for regulating circadian and seasonal rhythms
- Olfactory stimuli, including pheromones
- Steroids, including gonadal steroids and corticosteroids
- Neurally transmitted information arising in particular from the heart, the stomach, and the reproductive tract
- Autonomic inputs
- Blood-borne stimuli, including leptin, ghrelin, angiotensin, insulin, pituitary hormones, cytokines, plasma concentrations of glucose and osmolarity etc.
- Stress
- Invading microorganisms by increasing body temperature, resetting the body's thermostat upward.

Olfactory stimuli

Olfactory stimuli are important for sexual reproduction and neuroendocrine function in many species. For instance if a pregnant mouse is exposed to the urine of a 'strange' male during a critical period after coitus then the pregnancy fails (the Bruce effect). Thus, during coitus, a female mouse forms a precise 'olfactory memory' of her partner that persists for several days. Pheromonal cues aid synchronisation of oestrus in many species; in women, synchronised menstruation may also arise from pheromonal cues, although the role of pheromones in humans is disputed.

Blood-borne stimuli

Peptide hormones have important influences upon the hypothalamus, and to do so they must pass through the blood–brain barrier. The hypothalamus is bounded in part by specialized brain regions that lack an effective blood–brain barrier; the capillary endothelium at these sites is fenestrated to allow free passage of even large proteins and other molecules. Some of these sites are the sites of neurosecretion - the neurohypophysis and the median eminence. However, others are sites at which the brain samples the composition of the blood. Two of these sites, the SFO (subfornical organ) and the OVLT (organum vasculosum of the lamina terminalis) are so-called circumventricular organs, where neurons are in intimate contact with both blood and CSF. These structures are densely vascularized, and contain osmoreceptive and sodium-receptive neurons that control drinking, vasopressin release, sodium excretion, and sodium appetite. They also contain neurons with receptors for angiotensin, atrial natriuretic factor, endothelin and relaxin, each of which important in the regulation of fluid and electrolyte balance. Neurons in the OVLT and SFO project to the supraoptic nucleus and paraventricular nucleus, and also to preoptic hypothalamic areas. The circumventricular organs may also be the site of action of interleukins to elicit both fever and ACTH secretion, via effects on paraventricular neurons.

It is not clear how all peptides that influence hypothalamic activity gain the necessary access. In the case of prolactin and leptin, there is evidence of active uptake at the choroid plexus from the blood into the cerebrospinal fluid (CSF). Some pituitary hormones have a negative feedback influence upon hypothalamic secretion; for example, growth hormone feeds back on the hypothalamus, but how it enters the brain is not clear. There is also evidence for central actions of prolactin.

Findings have suggested that thyroid hormone (T4) is taken up by the hypothalamic glial cells in the infundibular nucleus/ median eminence, and that it is here converted into T3 by the type 2 deiodinase (D2). Subsequent to this, T3 is transported into the thyrotropin-releasing hormone (TRH)-producing neurons in the paraventricular nucleus. Thyroid hormone receptors have been found in these neurons, indicating that they are indeed sensitive to T3 stimuli. In addition, these neurons expressed MCT8, a thyroid hormone transporter, supporting the theory that T3 is transported into them. T3 could then bind to the thyroid hormone receptor in these neurons and affect the production of thyrotropin-releasing hormone, thereby regulating thyroid hormone production.[18]

The hypothalamus functions as a type of thermostat for the body.[19] It sets a desired body temperature, and stimulates either heat production and retention to raise the blood temperature to a higher setting or sweating and vasodilation to cool the blood to a lower temperature. All fevers result from a raised setting in the hypothalamus; elevated body temperatures due to any other cause are classified as hyperthermia.[19] Rarely, direct damage to the hypothalamus, such as from a stroke, will cause a fever; this is sometimes called a *hypothalamic fever*. However, it is more common for such damage to cause abnormally low body temperatures.[19]

Steroids

The hypothalamus contains neurons that react strongly to steroids and glucocorticoids – (the steroid hormones of the adrenal gland, released in response to ACTH). It also contains specialized glucose-sensitive neurons (in the arcuate nucleus and ventromedial hypothalamus), which are important for appetite. The preoptic area contains thermosensitive neurons; these are important for TRH secretion.

Neural

Oxytocin secretion in response to suckling or vaginocervical stimulation is mediated by some of these pathways; vasopressin secretion in response to cardiovascular stimuli arising from chemoreceptors in the carotid body and aortic arch, and from low-pressure atrial volume receptors, is mediated by others. In the rat, stimulation of the vagina also causes prolactin secretion, and this results in pseudo-pregnancy following an infertile mating. In the rabbit, coitus elicits reflex ovulation. In the sheep, cervical stimulation in the presence of high levels of estrogen can induce maternal behavior in a virgin ewe. These effects are all mediated by the hypothalamus, and the information is carried mainly by spinal pathways that relay in the brainstem. Stimulation of the nipples stimulates release of oxytocin and prolactin and suppresses the release of LH and FSH.

Cardiovascular stimuli are carried by the vagus nerve. The vagus also conveys a variety of visceral information, including for instance signals arising from gastric distension or emptying, to suppress or promote feeding, by signalling the release of leptin or gastrin, respectively. Again this information reaches the hypothalamus via relays in the brainstem.

In addition hypothalamic function is responsive to—and regulated by—levels of all three classical monoamine neurotransmitters, noradrenaline, dopamine, and serotonin (5-hydroxytryptamine), in those tracts from which it receives innervation. For example, noradrenergic inputs arising from the locus coeruleus have important regulatory effects upon corticotropin-releasing hormone (CRH) levels.

23.2.3 Control of food intake

The extreme lateral part of the ventromedial nucleus of the hypothalamus is responsible for the control of food intake. Stimulation of this area causes increased food intake. Bilateral lesion of this area causes complete cessation of food intake. Medial parts of the nucleus have a controlling effect on the lateral part. Bilateral lesion of the medial part of the ventromedial nucleus causes hyperphagia and obesity of the animal. Further lesion of the lateral part of the ventromedial nucleus in the same animal produces complete cessation of food intake.

There are different hypotheses related to this regulation:[21]

1. Lipostatic hypothesis: This hypothesis holds that adipose tissue produces a humoral signal that is proportionate to the amount of fat and acts on the hypothalamus to decrease food intake and increase energy output. It has been evident that a hormone leptin acts on the hypothalamus to decrease food intake and increase energy output.

2. Gutpeptide hypothesis: gastrointestinal hormones like Grp, glucagons, CCK and others claimed to inhibit food intake. The food entering the gastrointestinal tract triggers the release of these hormones, which act on the brain to produce satiety. The brain contains both CCK-A and CCK-B receptors.

3. Glucostatic hypothesis: The activity of the satiety center in the ventromedial nuclei is probably governed by the glucose utilization in the neurons. It has been postulated that when their glucose utilization is low and consequently when the arteriovenous blood glucose difference across them is low, the activity across the neurons decrease. Under these conditions, the activity of the feeding center is unchecked and the individual feels hungry. Food intake is rapidly increased by intraventricular administration of 2-deoxyglucose therefore decreasing glucose utilization in cells.

4. Thermostatic hypothesis: According to this hypothesis, a decrease in body temperature below a given set-point stimulates appetite, whereas an increase above the set-point inhibits appetite.

23.2.4 Fear processing

The medial zone of hypothalamus is part of a circuitry that controls motivated behaviors, like defensive behaviors.[22] Analyses of Fos-labeling showed that a series of nuclei in the "behavioral control column" is important in regulating the expression of innate and conditioned defensive behaviors.[23]

Antipredatory defensive behavior

Exposure to a predator (such as a cat) elicits defensive behaviors in laboratory rodents, even when the animal has never been exposed to a cat.[24] In the hypothalamus, this exposure causes an increase in Fos-labeled cells in the anterior hypothalamic nucleus, the dorsomedial part of the ventromedial nucleus, and in the ventrolateral part of the premammillary nucleus (PMDvl).[25] The premammillary nucleus has an important role in expression of defensive behaviors towards a predator, since lesions in this nucleus abolish defensive behaviors, like freezing and flight.[25][26] The PMD does not modulate defensive behavior in other situations, as lesions of this nucleus had minimal effects on post-shock freezing scores.[26] The PMD has important connections to the dorsal periaqueductal gray, an important structure in fear expression.[27][28] In addition, animals display risk assessment behaviors to the environment previously associated with the cat. Fos-labeled cell analysis showed that the PMDvl is the most activated structure in the hypothalamus, and inactivation with muscimol prior to exposure to the context abolishes the defensive behavior.[25] Therefore, the hypothalamus, mainly the PMDvl, has an important role in expression of innate and conditioned defensive behaviors to a predator.

Social defeat

Likewise, the hypothalamus has a role in social defeat: Nuclei in medial zone are also mobilized during an encounter with an aggressive conspecific. The defeated animal has an increase in Fos levels in sexually dimorphic structures, such as the medial pre-optic nucleus, the ventrolateral part of ventromedial nucleus, and the ventral premammillary nucleus.[29] Such structures are important in other social behaviors, such as sexual and aggressive behaviors. Moreover, the premammillary nucleus also is mobilized, the dorsomedial part but not the ventrolateral part.[29] Lesions in this nucleus abolish passive defensive behavior, like freezing and the "on-the-back" posture.[29]

23.2.5 Sexual orientation

According to D. F. Swaab, writing in a July 2008 paper, "Neurobiological research related to sexual orientation in humans is only just gathering momentum, but the evidence already shows that humans have a vast array of brain differences, not only in relation to gender, but also in relation to sexual orientation."[30]

Swaab first reported on the relationship between sexual orientation in males and the hypothalamus's "clock", the suprachiasmatic nucleus (SCN). In 1990, Swaab and Hofman[31] reported that the suprachiasmatic nucleus in homosexual men was significantly larger than in heterosexual men. Then in 1995, Swaab et al.[32] linked brain development to sexual orientation by treating male rats both pre- and postnatally with ATD, an aromatase blocker in the brain. This produced an enlarged SCN and bisexual behavior in the adult male rats. In 1991, LeVay showed that part of the sexually dimorphic nucleus (SDN) known as the 3rd interstitial nucleus of the anterior hypothalamus (INAH 3), is nearly twice as large (in terms of volume) in heterosexual men than in homosexual men and heterosexual women. However, a study in 1992 has shown that the sexually dimorph nucleus of the preoptic area, which include the INAH3, are of similar size in homosexual males who died of AIDS to heterosexual males, and therefore larger than female. This clearly contradicts the hypothesis that homosexual males have a female hypothalamus. Furthermore, the SCN of homosexual males is extremely large (both the volume and the number of neurons are twice as many as in heterosexual males). These areas of the hypothalamus have not yet been explored in homosexual females nor bisexual males nor females. Although the functional implications of such findings still haven't been examined in detail, they cast serious doubt over the widely accepted Dörner hypothesis that homosexual males have a "female hypothalamus" and that the key mechanism of differentiating the "male brain from originally female brain" is the epigenetic influence of testosterone during prenatal development.[33][34]

In 2004 and 2006, two studies by Berglund, Lindström, and Savic[35][36] used positron emission tomography (PET) to observe how the hypothalamus responds to smelling common odors, the scent of testosterone found in male sweat, and the scent of estrogen found in female urine. These studies showed that the hypothalamus of heterosexual men and homosexual women both respond to estrogen. Also, the hypothalamus of homosexual men and heterosexual women both respond to testosterone. The hypothalamus of all four groups did not respond to the common odors, which produced a normal olfactory response in the brain.

23.3 See also

- Copeptin
- Hypothalamic-pituitary-adrenal axis (HPA axis)
- Hypothalamic–pituitary–gonadal axis (HPG axis)
- Hypothalamic–pituitary–thyroid axis (HPT axis)
- John Leonora
- Incertohypothalamic pathway
- Neuroendocrinology

- Neuroscience of sleep

23.4 Additional images

-
- Human brain left dissected midsagittal view
- Location of the hypothalamus

23.5 References

[1] Dr. Boeree, C. George. "The Emotional Nervous System". *The Limbic System*. Retrieved 2016-04-18.

[2] Definition of hypothalamus - NCI Dictionary of Cancer Terms

[3] Melmed, S; Polonsky, KS; Larsen, PR; Kronenberg, HM (2011). *Williams Textbook of Endocrinology* (12th ed.). Saunders. p. 107. ISBN 978-1437703245.

[4] Diagram of Nuclei (psycheducation.org)

[5] Diagram of Nuclei (universe-review.ca)

[6] Diagram of Nuclei (utdallas.edu)

[7] Unless else specified in table, then ref is: Guyton Twelfth Edition

[8] Malenka RC, Nestler EJ, Hyman SE (2009). "Chapter 6: Widely Projecting Systems: Monoamines, Acetylcholine, and Orexin". In Sydor A, Brown RY. *Molecular Neuropharmacology: A Foundation for Clinical Neuroscience* (2nd ed.). New York: McGraw-Hill Medical. pp. 175–176. ISBN 9780071481274. Within the brain, histamine is synthesized exclusively by neurons with their cell bodies in the tuberomammillary nucleus (TMN) that lies within the posterior hypothalamus. There are approximately 64000 histaminergic neurons per side in humans. These cells project throughout the brain and spinal cord. Areas that receive especially dense projections include the cerebral cortex, hippocampus, neostriatum, nucleus accumbens, amygdala, and hypothalamus. ... While the best characterized function of the histamine system in the brain is regulation of sleep and arousal, histamine is also involved in learning and memory ... It also appears that histamine is involved in the regulation of feeding and energy balance.

[9] Romeo, Russell D; Rudy Bellani; Ilia N. Karatsoreos; Nara Chhua; Mary Vernov; Cheryl D. Conrad; Bruce S. McEwen (2005). "Stress History and Pubertal Development Interact to Shape Hypothalamic-Pituitary-Adrenal Axis Plasticity". *Endocrinology*. The Endocrine Society. **147** (4): 1664–1674. doi:10.1210/en.2005-1432. PMID 16410296. Retrieved 3 November 2013.

[10] Bowen, R. "Overview of Hypothalamic and Pituitary Hormones". Retrieved 5 October 2014.

[11] Melmed S, Jameson JL (2005). "Disorders of the anterior pituitary and hypothalamus". In Kasper DL, Braunwald E, Fauci AS, et al. *Harrison's Principles of Internal Medicine* (16th ed.). New York, NY: McGraw-Hill. pp. 2076–97. ISBN 0-07-139140-1.

[12] Ben-Shlomo, Anat; Melmed, Shlomo (28 February 2010). "Pituitary somatostatin receptor signaling". *Trends in Endocrinology & Metabolism*. **21** (3): 123–133. doi:10.1016/j.tem.2009.12.003. PMC 2834886. PMID 20149677.

[13] Horn, A. M.; Robinson, I. C. A. F.; Fink, G. (1 February 1985). "Oxytocin and vasopressin in rat hypophysial portal blood: experimental studies in normal and Brattleboro rats". *Journal of Endocrinology*. **104** (2): 211–NP. doi:10.1677/joe.0.1040211. PMID 3968510.

[14] Date, Y; Mondal, MS; Matsukura, S; Ueta, Y; Yamashita, H; Kaiya, H; Kangawa, K; Nakazato, M (Mar 10, 2000). "Distribution of orexin/hypocretin in the rat median eminence and pituitary.". *Brain research. Molecular brain research*. **76** (1): 1–6. doi:10.1016/s0169-328x(99)00317-4. PMID 10719209.

[15] Watanobe, H; Takebe, K (April 1993). "In vivo release of neurotensin from the median eminence of ovariectomized estrogen-primed rats as estimated by push-pull perfusion: correlation with luteinizing hormone and prolactin surges.". *Neuroendocrinology*. **57** (4): 760–4. doi:10.1159/000126434. PMID 8367038.

[16] Spinazzi, R; Andreis, PG; Rossi, GP; Nussdorfer, GG (March 2006). "Orexins in the regulation of the hypothalamic-pituitary-adrenal axis.". *Pharmacological reviews*. **58** (1): 46–57. doi:10.1124/pr.58.1.4. PMID 16507882.

[17] Jung Eun Kim; Baik Kee Cho; Dae Ho Cho; Hyun Jeong Park (2013). "Expression of Hypothalamic-Pituitary-Adrenal Axis in Common Skin Diseases: Evidence of its Association with Stress-related Disease Activity". National Research Foundation of Korea. Retrieved 4 March 2014.

[18] Fliers, Eric; Unmehopa, Alkemade (7 June 2006). "Functional neuroanatomy of thyroid hormone feedback in the human hypothalamus and pituitary gland". *Molecular and Cellular Endocrinology*. **251** (1–2): 1–8. doi:10.1016/j.mce.2006.03.042. PMID 16707210.

[19] Fauci, Anthony; et al. (2008). *Harrison's Principles of Internal Medicine* (17 ed.). McGraw-Hill Professional. pp. 117–121. ISBN 978-0-07-146633-2.

[20] Malenka RC, Nestler EJ, Hyman SE (2009). "Chapter 10: Neural and Neuroendocrine Control of the Internal Milieu – Table 10:3". In Sydor A, Brown RY. *Molecular Neuropharmacology: A Foundation for Clinical Neuroscience* (2nd ed.). New York: McGraw-Hill Medical. p. 263. ISBN 9780071481274.

[21] Theologides A (1976). "Anorexia-producing intermediary metabolites". *Am J Clin Nutr.* **29** (5): 552–8. PMID 178168.

[22] Swanson, L.W. (2000). "Cerebral Hemisphere Regulation of Motivated Behavior". *Brain Research.* **886**: 113–164. doi:10.1016/S0006-8993(00)02905-X.

[23] Canteras, N.S. (2002). "The medial hypothalamic defensive system:Hodological organization and functional implications". *Pharmacology, Biochemistry & Behavior.* **71**: 481–491. doi:10.1016/S0091-3057(01)00685-2.

[24] Ribeiro-Barbosa, E.R.; et al. (2005). "An alternative experimental procedure for studying predator-related defensive responses.". *Neuroscience & Biobehavioral Reviews.* **29** (8): 1255–1263. doi:10.1016/j.neubiorev.2005.04.006.

[25] Cezário, A.F. (2008). "Hypothalamic sites responding to predator threats--the role of the dorsal premammillary nucleus in unconditioned and conditioned antipredatory defensive behavior.". *European Journal of Neuroscience.* **28** (5): 1003–1015. doi:10.1111/j.1460-9568.2008.06392.x.

[26] Blanchard, D.C. (2003). "Dorsal premammillary nucleus differentially modulates defensive behaviors induced by different threat stimuli in rats". *Neuroscience Letters.* **345** (3): 145–148. doi:10.1016/S0304-3940(03)00415-4.

[27] Canteras, N.S.; Swanson, L.W. (1992). "The dorsal premammillary nucleus: an unusual component of the mammillary body.". *PNAS.* **89** (21): 10089–10093. doi:10.1073/pnas.89.21.10089.

[28] Behbehani, M.M. (1995). "Functional characteristics of the midbrain periaqueductal gray.". *Progress in Neurobiology.* **46** (6): 575–605. doi:10.1016/0301-0082(95)00009-K.

[29] Motta, S.C.; et al. (2009). "Dissecting the brain's fear system reveals the hypothalamus is critical for responding in subordinate conspecific intruders." (PDF). *PNAS.* **106** (12): 4870–4875. doi:10.1073/pnas.0900939106.

[30] Swaab DF (2008). "Sexual orientation and its basis in brain structure and function". *PNAS.* **105** (30): 10273–10274. doi:10.1073/pnas.0805542105. PMC 2492513. PMID 18653758.

[31] Swaab DF, Hofman MA (1990). "An enlarged suprachiasmatic nucleus in homosexual men". *Brain Res.* **537** (1–2): 141–8. doi:10.1016/0006-8993(90)90350-K. PMID 2085769.

[32] Swaab DF, Slob AK, Houtsmuller EJ, Brand T, Zhou JN (1995). "Increased number of vasopressin neurons in the suprachiasmatic nucleus (SCN) of 'bisexual' adult male rats following perinatal treatment with the aromatase blocker ATD". *Developmental Brain Research.* **85** (2): 273–279. doi:10.1016/0165-3806(94)00218-O. PMID 7600674.

[33] http://www.hiim.unizg.hr/images/knjiga/CNS41.pdf - Judaš, M., Kostović, I., The Fundamentals of Neuroscience, ch. 41, Neurobiology of emotions and sexuality, p. 408 (in Croatian)

[34] "Gender and sexual orientation in relation to hypothalamic structures.". *Horm Res.* 38 Suppl 2: 51–61. 1992. doi:10.1159/000182597. PMID 1292983.

[35] Savic I, Berglund H, Lindström P (2005). "Brain response to putative pheromones in homosexual men". *PNAS.* **102** (20): 7356–7361. doi:10.1073/pnas.0407998102. PMC 1129091. PMID 15883379.

[36] Savic I, Berglund H, Lindström P (2006). "Brain response to putative pheromones in lesbian women". *PNAS.* **103** (21): 8269–8274. doi:10.1073/pnas.0600331103. PMC 1570103. PMID 16705035.

Bear, Mark F. "Hypothalamic Control of the Anterior Pituitary." Neuroscience: Exploring the Brain. 4th ed. Philadelphia: Wolters Kluwer, 2016. 528. Print.

23.6 Further reading

- de Vries, GJ, and Sodersten P (2009) Sex differences in the brain: the relation between structure and function. Hormones and Behavior 55:589-596.

23.7 External links

- Stained brain slice images which include the "Hypothalamus" at the BrainMaps project
- The Hypothalamus and Pituitary at endotexts.org
- NIF Search - Hypothalamus via the Neuroscience Information Framework
- Space-filling and cross-sectional diagrams of hypothalamic nuclei: right hypothalamus, anterior, tubular, posterior.

Chapter 24

Basal ganglia

The **basal ganglia** (or **basal nuclei**) comprise multiple subcortical nuclei, of varied origin, in the brains of vertebrates, which are situated at the base of the forebrain. Basal ganglia nuclei are strongly interconnected with the cerebral cortex, thalamus, and brainstem, as well as several other brain areas. The basal ganglia are associated with a variety of functions including: control of voluntary motor movements, procedural learning, routine behaviors or "habits" such as bruxism, eye movements, cognition[1] and emotion.[2]

The main components of the basal ganglia – as defined functionally – are the dorsal striatum (caudate nucleus and putamen), ventral striatum (nucleus accumbens and olfactory tubercle), globus pallidus, ventral pallidum, substantia nigra, and subthalamic nucleus.[3] It is important to note, however, that the dorsal striatum and globus pallidus may be considered anatomically distinct from the substantia nigra, nucleus accumbens, and subthalamic nucleus. Each of these components has a complex internal anatomical and neurochemical organization. The largest component, the striatum (dorsal and ventral), receives input from many brain areas beyond the basal ganglia, but only sends output to other components of the basal ganglia. The pallidum receives input from the striatum, and sends inhibitory output to a number of motor-related areas. The substantia nigra is the source of the striatal input of the neurotransmitter dopamine, which plays an important role in basal ganglia function. The subthalamic nucleus receives input mainly from the striatum and cerebral cortex, and projects to the globus pallidus.

Currently, popular theories implicate the basal ganglia primarily in action selection; that is, it helps determine the decision of which of several possible behaviors to execute at any given time. In more specific terms, the basal ganglia's primary function is likely to control and regulate activities of the motor and premotor cortical areas so that voluntary movements can be performed smoothly.[1][4] Experimental studies show that the basal ganglia exert an inhibitory influence on a number of motor systems, and that a release of this inhibition permits a motor system to become active. The "behavior switching" that takes place within the basal ganglia is influenced by signals from many parts of the brain, including the prefrontal cortex, which plays a key role in executive functions.[2][5]

The importance of these subcortical nuclei for normal brain function and behavior is emphasized by the numerous and diverse neurological conditions associated with basal ganglia dysfunction, which include: disorders of behavior control such as Tourette syndrome, hemiballismus, and obsessive–compulsive disorder; dystonia; addiction; and movement disorders, the most notable of which are Parkinson's disease, which involves degeneration of the dopamine-producing cells in the substantia nigra pars compacta, and Huntington's disease, which primarily involves damage to the striatum.[1][3] The basal ganglia have a limbic sector whose components are assigned distinct names: the nucleus accumbens, ventral pallidum, and ventral tegmental area (VTA). There is considerable evidence that this limbic part plays a central role in reward learning, particularly a pathway from the VTA to the nucleus accumbens that uses the neurotransmitter dopamine. A number of highly addictive drugs, including cocaine, amphetamine, and nicotine, are thought to work by increasing the efficacy of this dopamine signal. There is also evidence implicating overactivity of the VTA dopaminergic projection in schizophrenia.[6]

24.1 Structure

Main article: Anatomical subdivisions and connections of the basal ganglia

In terms of development, the human nervous system is often classified based on the original 3 primitive vesicles from which it develops: These primary vesicles form in the normal development of the neural tube of the human fetus and initially include prosencephalon, mesencephalon, and rhombencephalon, in rostral to caudal (from head to tail) orientation. Later in development of the nervous system each section itself turns into smaller components. During development, the cells that migrate tangentially to form the basal ganglia are directed by the lateral and medial ganglionic eminences.[7] The following table demonstrates this developmental classification and traces it to the anatomic structures found in the basal ganglia.[1][3][8] The structures relevant to the basal ganglia are shown in **bold**.

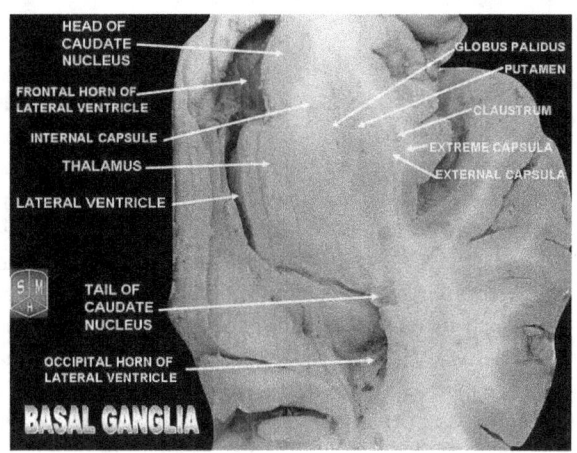

Basal ganglia

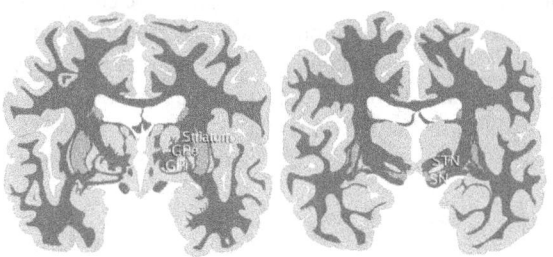

Coronal slices of human brain showing the basal ganglia. White matter is shown in dark gray, gray matter is shown in light gray.
Anterior: striatum, globus pallidus (GPe and GPi)
Posterior: subthalamic nucleus (STN), substantia nigra (SN)

The basal ganglia form a fundamental component of the cerebrum. In contrast to the cortical layer that lines the surface of the forebrain, the basal ganglia are a collection of distinct masses of gray matter lying deep in the brain not far from the junction of the thalamus. Like most parts of the brain, the basal ganglia consist of left and right sides that are virtual mirror images of each other.

In terms of anatomy, the basal ganglia are divided by anatomists into four distinct structures, depending on how superior or rostral they are (in other words depending on how close to the top of the head they are): Two of them, the striatum and the pallidum, are relatively large; the other two, the substantia nigra and the subthalamic nucleus, are smaller. In the illustration to the right, two coronal sections of the human brain show the location of the basal ganglia components. Of note, and not seen in this section, the subthalamic nucleus and substantia nigra lie farther back (posteriorly) in the brain than the striatum and pallidum.

24.1.1 Striatum

Main article: Striatum

The striatum is the largest component of the basal ganglia. The term "striatum" comes from the observation that this structure has a striped appearance when sliced in certain directions, arising from numerous large and small bundles of nerve fibers (white matter) that traverse it. Early anatomists, examining the human brain, perceived the striatum as two distinct masses of gray matter separated by a large tract of white matter called the internal capsule. They named these two masses the "caudate nucleus" and "putamen". More recent anatomists have concluded, on the basis of microscopic and neurochemical studies, that it is more appropriate to consider these masses as two separated parts of a single entity, the "striatum", in the same way that a city may be separated into two parts by a river. Numerous functional differences between the caudate and putamen have been identified, but these are taken to be consequences of the fact that each sector of the striatum is preferentially connected to specific parts of the cerebral cortex.

The internal organization of the striatum is extraordinarily complex. The great majority of neurons (about 96%) are of a type called "medium spiny neurons".[1] These are GABAergic cells (meaning that they inhibit their targets) with small cell bodies and dendrites densely covered with dendritic spines, which receive synaptic input primarily from the cortex and thalamus. Medium spiny neurons can be divided into subtypes in a number of ways, on the basis of neurochemistry and connectivity. The next most numerous type (around 2%) are a class of large cholinergic interneurons with smooth dendrites. There are also several other types of interneurons making up smaller fractions of the neural population.

Numerous studies have shown that the connections between cortex and striatum are, in general, topographic; that is, each part of the cortex sends stronger input to some parts of the striatum than to others. The nature of the topography has been difficult to understand, however—perhaps in part because the striatum is organized in three dimensions, whereas the cortex, as a layered structure, is organized in two. This dimensional discrepancy entails a great deal of distortion and discontinuity in mapping one structure to the other. It is interesting to note that the same topography applies to the striatal connections to the thalamus.[9]

24.1.2 Pallidum

The *pallidum* consists of a large structure called the globus pallidus ("pale globe") together with a smaller ventral extension called the ventral pallidum. The globus pallidus appears as a single neural mass, but can be divided into two functionally distinct parts, called the internal (or medial) and external (lateral) segments, abbreviated GPi and GPe.[1] Both segments contain primarily GABAergic neurons, which therefore have inhibitory effects on their targets. The two segments participate in distinct neural circuits. The external segment, or GPe, receives input mainly from the striatum, and projects to the subthalamic nucleus. The internal segment, or GPi, receives signals from the striatum via two pathways, called "direct" and "indirect". Pallidal neurons operate using a disinhibition principle. These neurons fire at steady high rates in the absence of input, and signals from the striatum cause them to pause or reduce their rate of firing. Because pallidal neurons themselves have inhibitory effects on their targets, the net effect of striatal input to the pallidum is a reduction of the tonic inhibition exerted by pallidal cells on their targets (disinhibition) with an increased rate of firing in the targets.

24.1.3 Substantia nigra

Main article: Substantia nigra

The substantia nigra is a mesencephalic gray matter por-

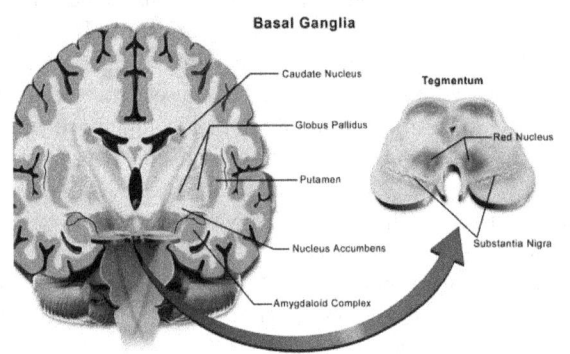

Location of the substantia nigra within the basal ganglia

tion of the basal ganglia that is divided into SNr (reticulata) and SNc (compacta). SNr often works in unison with GPi, and the SNr-GPi complex inhibits the thalamus. Substantia nigra pars compacta (SNc) however, produces the neurotransmitter dopamine, which is very significant in maintaining balance in the striatal pathway. The circuit portion below explains the role and circuit connections of each of the components of the basal ganglia.

24.1.4 Subthalamic nucleus

Main article: Subthalamic nucleus

The subthalamic nucleus (STN) is a diencephalic gray matter portion of the basal ganglia, and the only portion of the ganglia that produces an excitatory neurotransmitter, glutamate. The role of the subthalamic nucleus is to stimulate the SNr-GPi complex and it is part of the indirect pathway. The subthalamic nucleus receives inhibitory input from the external part of the globus pallidus and sends excitatory input to the GPi.

24.1.5 Circuit connections

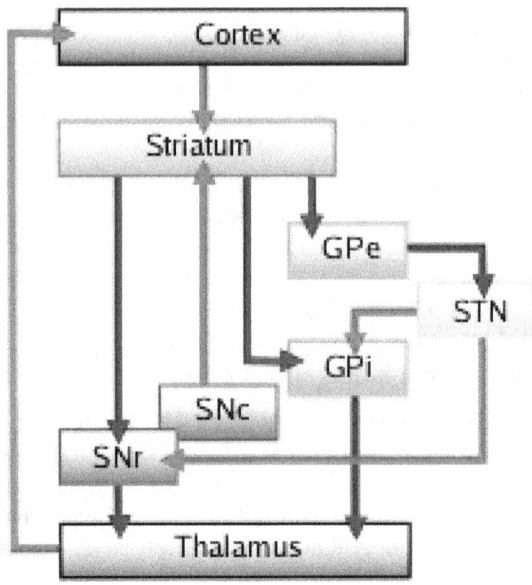

Connectivity diagram showing excitatory glutamatergic pathways as red, inhibitory GABAergic pathways as blue, and modulatory dopaminergic pathways as magenta. (Abbreviations: GPe: globus pallidus external; GPi: globus pallidus internal; STN: subthalamic nucleus; SNc: substantia nigra compacta; SNr: substantia nigra reticulata)

In order to understand the circuitry of the basal ganglia, one has to first understand the important participants in this circuit. Parts of the basal ganglia are in direct communication with the thalamus and the cortex. The cortex, thalamus, and the basal ganglia are, therefore, the three main participants in a circuit created by the basal ganglia.

At the top of the hierarchy lies the cerebral cortex. The cortex has many different areas with different functions. One such cortical area is called the primary motor cortex (along the pre-central gyrus). Specialized neurons from the primary motor cortex extend their axons all the way to the striatum portion of the basal ganglia. These cortical neurons release the neurotransmitter glutamate, which is excitatory in nature. Once excited by glutamate, the cells in the striatum project in two different directions giving rise to two major pathways: the "*direct*" and the "*indirect*" pathways:

In the direct pathway, cortical cells project excitatory inputs to the striatum, which in turn projects inhibitory neurons onto the cells of the SNr-GPi complex. The SNr-GPi complex projects directly onto the thalamus through the inhibitory ansa lenticularis pathway. The striatal inhibition of the SNr-GPi complex coupled with SNr-GPi inhibition of the thalamus therefore results in a net **reduction** of inhibition of the thalamus via the striatum. The thalamus projects excitatory glutamatergic neurons to the cortex itself. The direct pathway, therefore, results in the excitation of the motor cortex by the thalamus. Once stimulated, the cortex projects its own excitatory outputs to the brain stem and ultimately muscle fibers via the lateral corticospinal tract. The following diagram depicts the direct pathway:

Cortex (stimulates) → **Striatum** (inhibits) → **"SNr-GPi" complex** (less inhibition of thalamus) → **Thalamus** (stimulates) → **Cortex** (stimulates) → **Muscles, etc.** → (hyperkinetic state)

The indirect pathway also starts from neurons in the striatum. Once stimulated by the cortex, striatal neurons in the indirect pathway project inhibitory axons onto the cells of the globus pallidus externa (GPe), which tonically inhibits the subthalamic nucleus (STN). This inhibition (by the striatum) of the inhibitory projections of the GPe, results in the net reduction of inhibition of the STN. The STN, in turn, projects excitatory inputs to the SNr-GPi complex (which inhibits the thalamus). The end-result is inhibition of the thalamus and, therefore, decreased stimulation of the motor cortex by the thalamus and reduced muscle activity. The direct and indirect pathways are therefore antagonist in their functions. Following is a diagram of the indirect pathway:

Cortex (stimulates) → **Striatum** (inhibits) → **GPe** (less inhibition of STN) → **STN** (stimulates) → **"SNr-GPi" complex** (inhibits) → **Thalamus** (is stimulating less) → **Cortex** (is stimulating less) → **Muscles, etc.** → (hypokinetic state)

The antagonistic functions of the direct and indirect pathways are modulated by the substantia nigra pars compacta (SNc), which produces dopamine. In the presence of dopamine, D1-receptors in the basal ganglia stimulate the GABAergic neurons, favoring the direct pathway, and thus

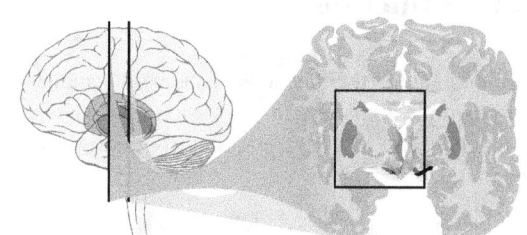

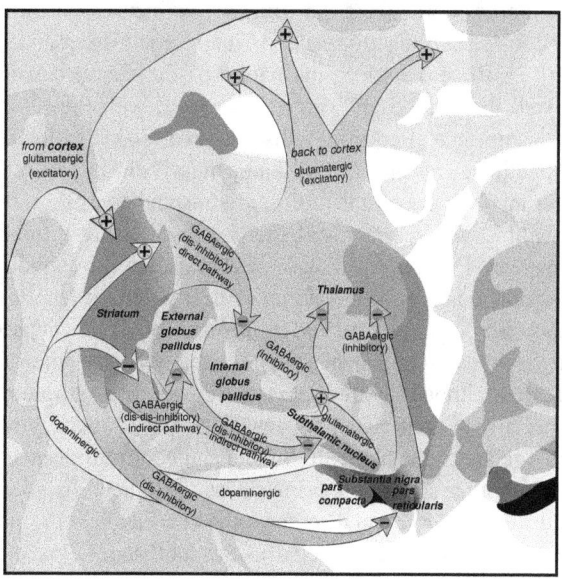

Main circuits of the basal ganglia. This diagram shows 2 coronal slices that have been superimposed to include the involved basal ganglia structures. The + and – signs at the point of the arrows indicate whether the pathway is excitatory or inhibitory, respectively, in effect. Green arrows refer to excitatory glutamatergic pathways, red arrows refer to inhibitory GABAergic pathways and turquoise arrows refer to dopaminergic pathways that are excitatory on the direct pathway and inhibitory on the indirect pathway.

increasing movement. The GABAergic neurons of the indirect pathway are stimulated by excitatory neurotransmitters acetylcholine and glutamate. This sets off the indirect pathway that ultimately results in inhibition of upper motor neurons, and less movement. In the presence of dopamine, D2-receptors in the basal ganglia inhibit these GABAergic neurons, which reduces the indirect pathways inhibitory effect. Dopamine therefore increases the excitatory effect of the direct pathway (causing movement) and reduces the inhibitory effect of the indirect pathway (preventing full inhibition of movement). Through these mechanisms the body is able to maintain balance between excitation and inhibition of motion. Lack of balance in this delicate system leads to pathologies such as Parkinson's disease. Parkinson's disease involves the loss of dopamine which means the direct pathway is less able to function (so no movement is initiated) and the indirect pathway is in overdrive (causing too much inhibition of movement).

24.2 Function

Information about the functions of the basal ganglia comes from anatomical studies, from physiology studies carried out mainly in rats and monkeys, and from the study of diseases that damage them.

The greatest source of insight into the functions of the basal ganglia has come from the study of two neurological disorders, Parkinson's disease and Huntington's disease. For both of these disorders, the nature of the neural damage is well understood and can be correlated with the resulting symptoms. Parkinson's disease involves major loss of dopaminergic cells in the substantia nigra; Huntington's disease involves massive loss of medium spiny neurons in the striatum. The symptoms of the two diseases are virtually opposite: Parkinson's disease is characterized by gradual loss of the ability to initiate movement, whereas Huntington's disease is characterized by an inability to prevent parts of the body from moving unintentionally. It is noteworthy that, although both diseases have cognitive symptoms, especially in their advanced stages, the most salient symptoms relate to the ability to initiate and control movement. Thus, both are classified primarily as movement disorders. A different movement disorder, called hemiballismus, may result from damage restricted to the subthalamic nucleus. Hemiballismus is characterized by violent and uncontrollable flinging movements of the arms and legs.

24.2.1 Eye movements

One of the most intensively studied functions of the basal ganglia (BG) is their role in controlling eye movements.[10] Eye movement is influenced by an extensive network of brain regions that converge on a midbrain area called the superior colliculus (SC). The SC is a layered structure whose layers form two-dimensional retinotopic maps of visual space. A "bump" of neural activity in the deep layers of the SC drives an eye movement directed toward the corresponding point in space.

The SC receives a strong inhibitory projection from the BG, originating in the substantia nigra *pars reticulata* (SNr).[10] Neurons in the SNr usually fire continuously at high rates, but at the onset of an eye movement they "pause", thereby releasing the SC from inhibition. Eye movements of all types are associated with "pausing" in the SNr; however, individual SNr neurons may be more strongly associated with some types of movements than others. Neurons in some parts of the caudate nucleus also show activity related to eye movements. Since the great majority of caudate cells fire at very low rates, this activity almost always shows up as an increase in firing rate. Thus, eye movements begin with activation in the caudate nucleus, which inhibits the SNr via the direct GABAergic projections, which in turn disinhibits the SC.

24.2.2 Role in motivation

Although the role of the basal ganglia in motor control is clear, there are also many indications that it is involved in the control of behavior in a more fundamental way, at the level of motivation. In Parkinson's disease, the ability to execute the components of movement is not greatly affected, but motivational factors such as hunger fail to cause movements to be initiated or switched at the proper times. The immobility of Parkinsonian patients has sometimes been described as a "paralysis of the will".[11] These patients have occasionally been observed to show a phenomenon called *kinesia paradoxica*, in which a person who is otherwise immobile responds to an emergency in a coordinated and energetic way, then lapses back into immobility once the emergency has passed.

The role in motivation of the "limbic" part of the basal ganglia—the nucleus accumbens (NA), ventral pallidum, and ventral tegmental area (VTA)—is particularly well established. Thousands of experimental studies combine to demonstrate that the dopaminergic projection from the VTA to the NA plays a central role in the brain's reward system. Animals with stimulating electrodes implanted along this pathway will bar-press very energetically if each press is followed by a brief pulse of electric current. Numerous things that people find rewarding, including addictive drugs, good-tasting food, and sex, have been shown to elicit activation of the VTA dopamine system. Damage to the NA or VTA can produce a state of profound torpor.

Although it is not universally accepted, some theorists have proposed a distinction between "appetitive" behaviors, which are initiated by the basal ganglia, and "consummatory" behaviors, which are not. For example, an animal with severe basal ganglia damage will not move toward food even if it is placed a few inches away, but, if the food is placed directly in the mouth, the animal will chew it and swallow it.

24.2.3 Neurotransmitters

In most regions of the brain, the predominant classes of neurons use glutamate as neurotransmitter and have excitatory effects on their targets. In the basal ganglia, however, the great majority of neurons use GABA as neurotransmitter and have inhibitory effects on their targets. The inputs from the cortex and thalamus to the striatum and STN are glutamatergic, but the outputs from the striatum, pallidum, and substantia nigra *pars reticulata* all use GABA. Thus, following the initial excitation of the striatum, the internal dynamics of the basal ganglia are dominated by inhibition and disinhibition.

Other neurotransmitters have important modulatory effects. The most intensively studied is dopamine, which is used by the projection from the substantia nigra *pars compacta* to the dorsal striatum, and also in the analogous projection from the ventral tegmental area to the ventral striatum (nucleus accumbens). Acetylcholine also plays an important role, being used both by several external inputs to the striatum, and by a group of striatal interneurons. Although cholinergic cells make up only a small fraction of the total population, the striatum has one of the highest acetylcholine concentrations of any brain structure.

24.3 Clinical significance

The following is a list of disorders that have been linked to the basal ganglia:

Main article: Basal ganglia disease

- Addiction
- Athetosis
- Athymhormic syndrome (PAP syndrome)
- Attention-deficit hyperactivity disorder (ADHD)
- Blepharospasm
- Bruxism
- Cerebral palsy: basal ganglia damage during second and third trimester of pregnancy
- Chorea
- Dystonia
- Fahr's disease
- Foreign accent syndrome (FAS)
- Huntington's disease
- Kernicterus
- Lesch–Nyhan syndrome
- Major Depressive Disorder [12]
- Obsessive-compulsive disorder[13][14]
- Other anxiety disorders [14]
- PANDAS

- Parkinson's disease
- Spasmodic dysphonia
- Stuttering[15]
- Sydenham's chorea
- Tardive dyskinesia, caused by chronic antipsychotic treatment
- Tourette's disorder
- Wilson's disease

24.4 History

The acceptance that the basal ganglia system constitutes one major cerebral system took time to arise. The first anatomical identification of distinct subcortical structures was published by Thomas Willis in 1664.[16] For many years, the term corpus striatum[17] was used to describe a large group of subcortical elements, some of which were later discovered to be functionally unrelated.[18] For many years, the putamen and the caudate nucleus were not associated with each other. Instead, the putamen was associated with the pallidum in what was called the nucleus lenticularis or nucleus lentiformis.

A thorough reconsideration by Cécile and Oskar Vogt (1941) simplified the description of the basal ganglia by proposing the term striatum to describe the group of structures consisting of the caudate nucleus, the putamen, and the mass linking them ventrally, the nucleus accumbens. The striatum was named on the basis of the striated (striped) appearance created by radiating dense bundles of striato-pallido-nigral axons, described by anatomist Samuel Alexander Kinnier Wilson (1912) as "pencil-like".

The anatomical link of the striatum with its primary targets, the pallidum and the substantia nigra, was discovered later. The name *globus pallidus* was attributed by Déjerine to Burdach (1822). For this, the Vogts proposed the simpler "pallidum". The term "locus niger" was introduced by Félix Vicq-d'Azyr as *tache noire* in (1786), though that structure has since become known as the substantia nigra, due to contributions by Von Sömmering in 1788. The structural similarity between the substantia nigra and globus pallidus was noted by Mirto in 1896. Together, the two are known as the pallidonigral ensemble, which represents the core of the basal ganglia. Altogether, the main structures of the basal ganglia are linked to each other by the striato-pallido-nigral bundle, which passes through the pallidum, crosses the internal capsule as the "comb bundle of Edinger", then finally reaches the substantia nigra.

Additional structures that later became associated with the basal ganglia are the "body of Luys" (1865) (nucleus of Luys on the figure) or subthalamic nucleus, whose lesion was known to produce movement disorders. More recently, other areas such as the central complex (centre médian-parafascicular) and the pedunculopontine complex have been thought to be regulators of the basal ganglia.

Near the beginning of the 20th century, the basal ganglia system was first associated with motor functions, as lesions of these areas would often result in disordered movement in humans (chorea, athetosis, Parkinson's disease).

24.4.1 Terminology

The nomenclature of the basal ganglia system and its components has always been problematic. Early anatomists, seeing the macroscopic anatomical structure but knowing nothing of the cellular architecture or neurochemistry, grouped together components that are now believed to have distinct functions (such as the internal and external segments of the globus pallidus), and gave distinct names to components that are now thought to be functionally parts of a single structure (such as the caudate nucleus and putamen).

The term "basal" comes from the fact that most of its elements are located in the basal part of the forebrain. The term ganglia is a misnomer: In modern usage, neural clusters are called "ganglia" only in the peripheral nervous system; in the central nervous system they are called "nuclei". For this reason, the basal ganglia are also occasionally known as the "basal nuclei".[19] Terminologia anatomica (1998), the international authority for anatomical naming, retained "nuclei basales", but this is not commonly used.

The International Basal Ganglia Society (IBAGS) informally considers the basal ganglia to be made up of the striatum, the pallidum (with two nuclei), the substantia nigra (with its two distinct parts), and the subthalamic nucleus. Percheron *et al.* in 1991 and Parent and Parent in 2005 included the central region (centre median-parafascicular) of the thalamus as part of the basal ganglia,[20][21] while Mena-Segovia *et al.* in 2004 included the pedunculopontine complex as well.[22]

Also, the names given to the various nuclei of the basal ganglia are different in different species. In particular, the internal segment of the globus pallidus in primates is called the entopeduncular nucleus in rodents. The "striatum" and "external segment of the globus pallidus" in primates are called the "paleostriatum augmentatum" and "paleostriatum primitivum," respectively, in birds.

24.5 In other animals

See also: Primate basal ganglia system

The basal ganglia form one of the basic components of the forebrain, and can be recognized in all species of vertebrates.[23] Even in the lamprey (generally considered one of the most primitive of vertebrates), striatal, pallidal, and nigral elements can be identified on the basis of anatomy and histochemistry.[24]

A clear emergent issue in comparative anatomy of the basal ganglia is the development of this system through phylogeny as a convergent cortically re-entrant loop in conjunction with the development and expansion of the cortical mantle. There is controversy, however, regarding the extent to which convergent selective processing occurs versus segregated parallel processing within re-entrant closed loops of the basal ganglia. Regardless, the transformation of the basal ganglia into a cortically re-entrant system in mammalian evolution occurs through a re-direction of pallidal (or "paleostriatum primitivum") output from midbrain targets such as the superior colliculus, as occurs in sauropsid brain, to specific regions of the ventral thalamus and from there back to specified regions of the cerebral cortex that form a subset of those cortical regions projecting into the striatum. The abrupt rostral re-direction of the pathway from the internal segment of the globus pallidus into the ventral thalamus—via the path of the ansa lenticularis— could be viewed as a footprint of this evolutionary transformation of basal ganglia outflow and targeted influence.

24.6 See also

This article uses anatomical terminology; for an overview, see Anatomical terminology.

- Nathaniel A. Buchwald
- Alexander Cools

24.7 References

[1] Stocco, Andrea; Lebiere, Christian; Anderson, John R. (2010). "Conditional Routing of Information to the Cortex: A Model of the Basal Ganglia's Role in Cognitive Coordination". *Psychological Review*. **117** (2): 541–74. doi:10.1037/a0019077. PMC 3064519. PMID 20438237.

[2] Weyhenmeyer, James A.; Gallman, Eve. A. (2007). *Rapid Review of Neuroscience*. Mosby Elsevier. p. 102. ISBN 0-323-02261-8.

[3] Fix, James D. (2008). "Basal Ganglia and the Striatal Motor System". *Neuroanatomy (Board Review Series)* (4th ed.). Baltimore: Wulters Kluwer & Lippincott Wiliams & Wilkins. pp. 274–281. ISBN 0-7817-7245-1.

[4] Chakravarthy, V. S.; Joseph, Denny; Bapi, Raju S. (2010). "What do the basal ganglia do? A modeling perspective". *Biological Cybernetics*. **103** (3): 237–53. doi:10.1007/s00422-010-0401-y. PMID 20644953.

[5] Cameron IG, Watanabe M, Pari G, Munoz DP (June 2010). "Executive impairment in Parkinson's disease: response automaticity and task switching". *Neuropsychologia*. Neuropsychologia. **48** (7): 1948–57. doi:10.1016/j.neuropsychologia.2010.03.015. PMID 20303998.

[6] Inta, D.; Meyer-Lindenberg, A.; Gass, P. (2010). "Alterations in Postnatal Neurogenesis and Dopamine Dysregulation in Schizophrenia: A Hypothesis". *Schizophrenia Bulletin*. **37** (4): 674–80. doi:10.1093/schbul/sbq134. PMC 3122276. PMID 21097511.

[7] Marín & Rubenstein. (2001). A Long, Remarkable Journey: Tangential Migration in the Telencephalon. *Nature Reviews Neuroscience, 2*.

[8] Regina Bailey. "Divisions of the Brain". about.com. Archived from the original on 2 December 2010. Retrieved 2010-11-30.

[9] Kamishina, H; Yurcisin, G; Corwin, J; Reep, R (2008). "Striatal projections from the rat lateral posterior thalamic nucleus". *Brain Research*. **1204**: 24–39. doi:10.1016/j.brainres.2008.01.094. PMID 18342841.

[10] Hikosaka, O; Takikawa, Y; Kawagoe, R (2000). "Role of the basal ganglia in the control of purposive saccadic eye movements". *Physiological reviews*. **80** (3): 953–78. PMID 10893428.

[11] Niv, Y.; Rivlin-Etzion, M. (2007). "Parkinson's Disease: Fighting the Will?". *Journal of Neuroscience*. **27** (44): 11777–9. doi:10.1523/JNEUROSCI.4010-07.2007. PMID 17978012.

[12] Kempton MJ, Salvador Z, Munafò MR, Geddes JR, Simmons A, Frangou S, Williams SC (2011). "Structural Neuroimaging Studies in Major Depressive Disorder: Meta-analysis and Comparison With Bipolar Disorder". *Arch Gen Psychiatry*. **68** (7): 675–90. doi:10.1001/archgenpsychiatry.2011.60. PMID 21727252. see also MRI database at www.depressiondatabase.org

[13] Radua, Joaquim; Mataix-Cols, David (November 2009). "Voxel-wise meta-analysis of grey matter changes in obsessive–compulsive disorder". *British Journal of Psychiatry*. **195** (5): 393–402. doi:10.1192/bjp.bp.108.055046. PMID 19880927.

[14] Radua, Joaquim; van den Heuvel, Odile A.; Surguladze, Simon; Mataix-Cols, David (5 July 2010). "Meta-analytical comparison of voxel-based morphometry studies in obsessive-compulsive disorder vs other anxiety disorders". *Archives of General Psychiatry*. **67** (7): 701–711. doi:10.1001/archgenpsychiatry.2010.70. PMID 20603451.

[15] Alm, Per A. (2004). "Stuttering and the basal ganglia circuits: a critical review of possible relations". *Journal of communication disorders*. **37** (4): 325–69. doi:10.1016/j.jcomdis.2004.03.001. PMID 15159193.

[16] Andrew Gilies, *A brief history of the basal ganglia*, retrieved on 27 June 2005

[17] Vieussens (1685)

[18] Percheron, G; Fénelon, G; Leroux-Hugon, V; Fève, A (1994). "History of the basal ganglia system. Slow development of a major cerebral system". *Revue neurologique*. **150** (8–9): 543–54. PMID 7754290.

[19] Soltanzadeh, Akbar (2004). *Neurologic Disorders*. Tehran: Jafari. ISBN 964-6088-03-1.

[20] Percheron, G; Filion, M (1991). "Parallel processing in the basal ganglia: up to a point". *Trends in Neurosciences*. **14** (2): 55–9. doi:10.1016/0166-2236(91)90020-U. PMID 1708537.

[21] Parent, Martin; Parent, Andre (2005). "Single-axon tracing and three-dimensional reconstruction of centre median-parafascicular thalamic neurons in primates". *The Journal of Comparative Neurology*. **481** (1): 127–44. doi:10.1002/cne.20348. PMID 15558721.

[22] Menasegovia, J; Bolam, J; Magill, P (2004). "Pedunculopontine nucleus and basal ganglia: distant relatives or part of the same family?". *Trends in Neurosciences*. **27** (10): 585–8. doi:10.1016/j.tins.2004.07.009. PMID 15374668.

[23] Parent A (1986). *Comparative Neurobiology of the Basal Ganglia*. Wiley. ISBN 978-0-471-80348-5.

[24] Grillner, S; Ekeberg, O; Elmanira, A; Lansner, A; Parker, D; Tegner, J; Wallen, P (1998). "Intrinsic function of a neuronal network — a vertebrate central pattern generator1". *Brain Research Reviews*. **26** (2–3): 184–97. doi:10.1016/S0165-0173(98)00002-2. PMID 9651523.

24.8 External links

- Imaging of Basal Ganglia at USUHS
- Models of Basal ganglia Jim Houk Scholarpedia 2(10):1633. doi:10.4249/scholarpedia.1633
- The International Basal Ganglia Society

24.8. EXTERNAL LINKS

- Basal ganglia – Official journal of LIMPE (Lega Italiana per la Lotta Contro la Malattia di Parkinson, le Sindromi Extrapiramidali e le Demenze, Italy), the German Parkinson Society (DPG, Deutsche Parkinson Gesellschaft), and the Japanese Basal Ganglia Society (JBAGS Japan Basal Ganglia Society)

Chapter 25

Olfactory bulb

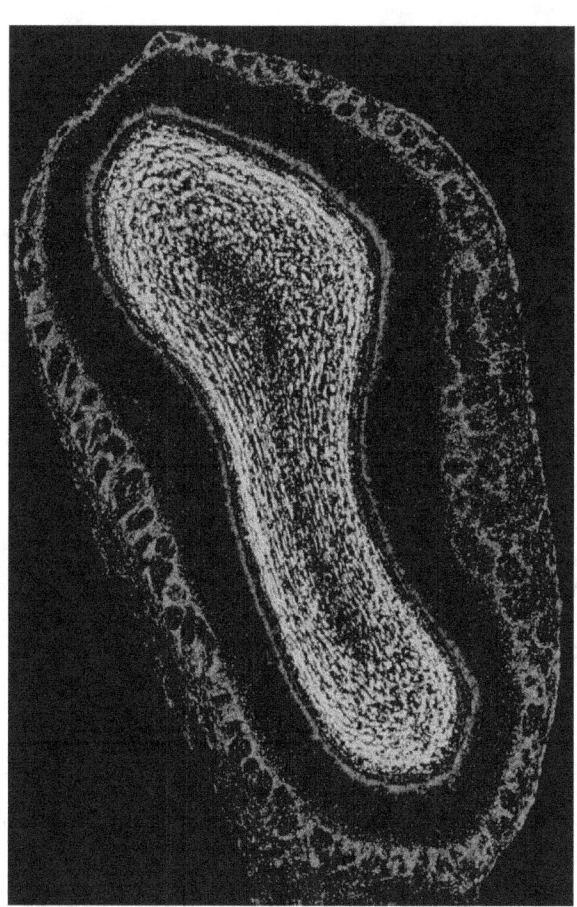

Coronal image of mouse main olfactory bulb cell nuclei.
Blue - Glomerular layer;
Red - External Plexiform and Mitral cell layer;
Green - Internal Plexiform and Granule cell layer.
Top of image is dorsal aspect, right of image is lateral aspect. Scale, ventral to dorsal, is approximately 2mm.

The **olfactory bulb** (bulbus olfactorius) is a neural structure of the vertebrate forebrain involved in olfaction, or the sense of smell.

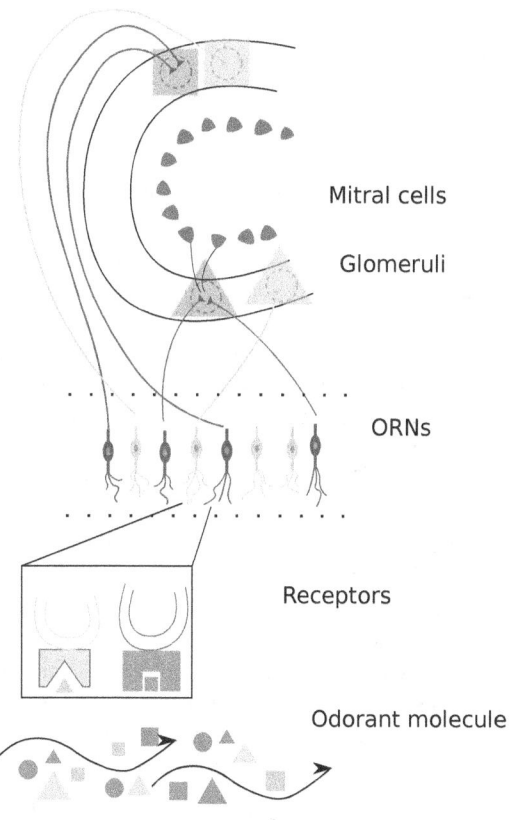

Flow of olfactory information from receptors to glomeruli layer

25.1 Structure

In most vertebrates, the olfactory bulb is the most rostral (forward) part of the brain, as seen in rats. In humans, however, the olfactory bulb is on the inferior (bottom) side of the brain. The olfactory bulb is supported and protected by the cribriform plate of the ethmoid bone, which in mammals separates it from the olfactory epithelium, and which is perforated by olfactory nerve axons. The bulb is divided into two distinct structures: the main olfactory bulb and the accessory olfactory bulb.

25.1.1 Layers

The main olfactory bulb has a multi-layered cellular architecture. In order from surface to the center the layers are:

- Glomerular layer
- External plexiform layer
- Mitral cell layer
- Internal plexiform layer
- Granule cell layer

The olfactory bulb transmits smell information from the nose to the brain, and is thus necessary for a proper sense of smell. As a neural circuit, the glomerular layer receives direct input from olfactory nerves, made up of the axons from approximately ten million olfactory receptor neurons in the olfactory mucosa, a region of the nasal cavity. The ends of the axons cluster in spherical structures known as glomeruli such that each glomerulus receives input primarily from olfactory receptor neurons that express the same olfactory receptor.The glomeruli layer of the olfactory bulb is the first level of synaptic processing.[1] The glomeruli layer represents a spatial odor map organized by chemical structure of odorants like functional group and carbon chain length. This spatial map is divided into zones and clusters, which represent similar glomeruli and therefore similar odors. One cluster in particular is associated with rank, spoiled smells which are represented by certain chemical characteristics. This classification may be evolutionary to help identify food that is no longer good to eat.

The spatial map of the glomeruli layer may be used for perception of odor in the olfactory cortex.[2] The next level of synaptic processing in the olfactory bulb occurs in the external plexiform layer, between the glomerular layer and the mitral cell layer. The external plexiform layer contains astrocytes, interneurons and some mitral cells. It does not contain many cell bodies, rather mostly dendrites of mitral cells and GABAergic granule cells [3] are also permeated by dendrites from neurons called mitral cells, which in turn output to the olfactory cortex. Numerous interneuron types exist in the olfactory bulb including periglomerular cells which synapse within and between glomeruli, and granule cells which synapse with mitral cells. The granule cell layer is the deepest layer in the olfactory bulb. It is made up of dendrodendritic granule cells that synapse to the mitral cell layer.[4]

25.2 Function

As a neural circuit, the olfactory bulb has one source of sensory input (axons from olfactory receptor neurons of the olfactory epithelium), and one output (mitral cell axons). As a result, it is generally assumed that it functions as a filter, as opposed to an associative circuit that has many inputs and many outputs. However, the olfactory bulb also receives "top-down" information from such brain areas as the amygdala, neocortex, hippocampus, locus coeruleus, and substantia nigra.[5] Its potential functions can be placed into four non-exclusive categories:

- discriminating among odors
- enhancing sensitivity of odor detection
- filtering out many background odors to enhance the transmission of a few select odors
- permitting higher brain areas involved in arousal and attention to modify the detection or the discrimination of odors.

While all of these functions could theoretically arise from the olfactory bulb's circuit layout, it is unclear which, if any, of these functions are performed exclusively by the olfactory bulb. By analogy to similar parts of the brain such as the retina, many researchers have focused on how the olfactory bulb filters incoming information from receptor neurons in space, or how it filters incoming information in time. At the core of these proposed filters are the two classes of interneurons; the periglomerular cells, and the granule cells. Processing occurs at each level of the main olfactory bulb, beginning with the spatial maps that categorize odors in the glomeruli layer.[2]

Interneurons in the external plexiform layer are responsive to pre-synaptic action potentials and exhibit both excitatory postsynaptic potentials and inhibitory postsynaptic potentials. Neural firing varies temporally, there are periods of fast, spontaneous firing and slow modulation of firing. These patterns may be related to sniffing or change in intensity and concentration of odorant.[3] Temporal patterns may have effect in later processing of spatial awareness of odorant. For example, synchronized mitral cell spike trains appear to help to discriminate similar odors better than when those spike trains are not synchronized.[6] Destruction to the olfactory bulb results in ipsilateral anosmia while irritative lesion of the uncus can result in olfactory and gustatory hallucinations.

25.2.1 Lateral inhibition

External plexiform layer

The interneurons in the external plexiform layer perform feedback inhibition on the mitral cells to control back propagation. They also participate in lateral inhibition of the mitral cells. This inhibition is an important part of olfaction as it aids in odor discrimination by decreasing firing in response to background odors and differentiating the responses of olfactory nerve inputs in the mitral cell layer.[1] Inhibition of the mitral cell layer by the other layers contributes to odor discrimination and higher level processing by modulating the output from the olfactory bulb. These hyperpolarizations during odor stimulation shape the responses of the mitral cells to make them more specific to an odor.[4]

There is a lack of information regarding the function of the internal plexiform layer which lies between the mitral cell layer and the granule cell layer.

Granule cell layer

The basal dendrites of mitral cells are connected to interneurons known as granule cells, which by some theories produce lateral inhibition between mitral cells. The synapse between mitral and granule cells is of a rare class of synapses that are "dendro-dendritic" which means that both sides of the synapse are dendrites that release neurotransmitter. In this specific case, mitral cells release the excitatory neurotransmitter glutamate, and granule cells release the inhibitory neurotransmitter Gamma-aminobutyric acid (GABA). As a result of its bi-directionality, the dendro-dendritic synapse can cause mitral cells to inhibit themselves (auto-inhibition), as well as neighboring mitral cells (lateral inhibition). More specifically, the granule cell layer receives excitatory glutamate signals from the basal dendrites of the mitral and tufted cells. The granule cell in turn releases GABA to cause an inhibitory effect on the mitral cell. More neurotransmitter is released from the activated mitral cell to the connected dendrite of the granule cell, making the inhibitory effect from the granule cell to the activated mitral cell stronger than the surrounding mitral cells.[4] It is not clear what the functional role of lateral inhibition would be, though it may be involved in boosting the signal-to-noise ratio of odor signals by silencing the basal firing rate of surrounding non-activated neurons. This in turn aids in odor discrimination.[1] Other research suggest that the lateral inhibition contributes to differentiated odor responses, which aids in the processing and perception of distinct odors.[4] There is also evidence of cholinergic effects on granule cells that enhance depolarization of granule cells making them more excitable which in turn increases inhibition of mitral cells. This may contribute to a more specific output from the olfactory bulb that would closer resemble the glomerular odor map.[7][8] Olfaction is distinct from the other sensory systems where peripheral sensory receptors have a relay in the diencephalon. Therefore, the olfactory bulb plays this role for the olfactory system.

25.2.2 Accessory Olfactory Bulb

The accessory olfactory bulb (AOB), which resides on the dorsal-posterior region of the main olfactory bulb, forms a parallel pathway independent from the main olfactory bulb. The vomeronasal organ sends projections to the accessory olfactory bulb[9] making it the second processing stage of the accessory olfactory system. As in the main olfactory bulb, axonal input to the accessory olfactory bulb forms synapses with mitral cells within glomeruli. The accessory olfactory bulb receives axonal input from the vomeronasal organ, a distinct sensory epithelium from the main olfactory epithelium that detects chemical stimuli relevant for social and reproductive behaviors, but probably also generic odorants.[10] It has been hypothesized that, in order for the vomernasal pump to turn on, the main olfactory epithelium must first detect the appropriate odor.[11] However, the possibility that the vomeronasal system works in parallel or independently from generic olfactory inputs has not been ruled out yet.

Vomeronasal sensory neurons provide direct excitatory inputs to AOB principle neurons called mitral cells[12] which are transmitted to the amygdala and hypothalamus and therefore are directly involved in sex hormone activity and may influence aggressiveness and mating behavior.[13] Axons of the vomeronasal sensory neurons express a given receptor type which, differently from what occurs in the main olfactory bulb, diverge between 6 and 30 AOB glomeruli. Mitral cell dendritic endings go through a dramatic period of targeting and clustering just after presynaptic unification of the sensory neuron axons. The connectivity of the vomernasal sensorglomery neurons to mitral cells is precise, with mitral cell dendrites targeting the glomeruli.[12] There is evidence against the presence of a functional accessory olfactory bulb in humans and other higher primates.[14]

The AOB is divided into two main subregions, anterior and posterior, which receive segregated synaptic inputs from two main categories of vomeronasal sensory neurons, V1R and V2R, respectively. This appears as a clear functional specialization, given the differential role of the two populations of sensory neurons in detecting chemical stimuli of different type and molecular weight. Although it doesn't seem to be maintained centrally, where mitral cell projections from both sides of the AOB converge. A clear difference of the AOB circuitry, compared to the rest of the bulb, is its heterogeneous connectivity between mitral cells and vomeronasal sensory afferents within neuropil glomeruli. AOB mitral cells indeed contact through apical dendritic processes glomeruli formed by afferents of different receptor neurons, thus breaking the one-receptor-one-neuron rule which generally holds for the main olfactory system. This implies that stimuli sensed through the VNO and elaborated in the AOB are subjected to a different and probably more complex level of elaboration. Accordingly, AOB mitral cells show clearly different firing patterns compared to other bulbar projection neurons.[15] Additionally, top down input to the olfactory bulb differentially affects olfactory outputs.[16]

25.2.3 Further processing

The olfactory bulb sends olfactory information to be further processed in the amygdala, the orbitofrontal cortex (OFC) and the hippocampus where it plays a role in emotion, memory and learning. The main olfactory bulb connects to the amygdala via the piriform cortex of the primary olfactory cortex and directly projects from the main olfactory bulb to specific amygdala areas.[17] The amygdala passes olfactory information on to the hippocampus. The orbitofrontal cortex, amygdala, hippocampus, thalamus, and olfactory bulb have many interconnections directly and indirectly through the cortices of the primary olfactory cortex. These connections are indicative of the association between the olfactory bulb and higher areas of processing, specifically those related to emotion and memory.[17]

Amygdala

Associative learning between odors and behavioral responses takes place in the amygdala. The odors serve as the reinforcers or the punishers during the associative learning process; odors that occur with positive states reinforce the behavior that resulted in the positive state while odors that occur with negative states do the opposite. Odor cues are coded by neurons in the amygdala with the behavioral effect or emotion that they produce. In this way odors reflect certain emotions or physiological states.[18] Odors become associated with pleasant and unpleasant responses, and eventually the odor becomes a cue and can cause an emotional response. These odor associations contribute to emotional states such as fear. Brain imaging shows amygdala activation correlated with pleasant and unpleasant odors, reflecting the association between odors and emotions.[18]

Hippocampus

The hippocampus aids in olfactory memory and learning as well. Several olfaction-memory processes occur in the hippocampus. Similar to the process in the amygdala, an odor is associated with a particular reward, i.e. the smell of food with receiving sustenance.[19] Odor in the hippocampus also contributes to the formation of episodic memory; the memories of events at a specific place or time. The time at which certain neurons fire in the hippocampus is associated by neurons with a stimulus such as an odor. Presentation of the odor at a different time may cause recall of the memory, therefore odor aids in recall of episodic memories.[19]

Olfactory coding in Habenula

In lower vertebrates (lampreys and teleost fishes), mitral cell (principal olfactory neurons) axons project exclusively to the right hemisphere of Habenula in an asymmetric manner. It is reported that dorsal Habenula (Hb) are functional asymmetric with predominant odor responses in right hemisphere. Interestingly, it was also shown that Hb neurons are spontaneous active even in absence of olfactory stimulation. These spontaneous active Hb neurons are organized into functional clusters which were proposed to govern olfactory responses. (Jetti, SK. et al. 2014, Current Biology)

Depression models

Further evidence of the link between the olfactory bulb and emotion and memory is shown through animal depression models. Olfactory bulb removal in rats effectively causes structural changes in the amygdala and hippocampus and behavioral changes similar to that of a person with depression. Researchers use rats with olfactory bulbectomies to research antidepressants.[20] Research has shown that removal of the olfactory bulb in rats leads to dendrite reorganization, disrupted cell growth in the hippocampus, and decreased neuroplasticity in the hippocampus. These hippocampal changes due to olfactory bulb removal are associated with behavioral changes characteristic of depression, demonstrating the correlation between the olfactory bulb and emotion.[21] The hippocampus and amygdala affect odor perception. During certain physiological states such as hunger a food odor may seem more pleasant and rewarding due to the associations in the amygdala and hippocampus of the food odor stimulus with the reward of eating.[18]

Orbitofrontal cortex

Olfactory information is sent to the primary olfactory cortex, where projections are sent to the orbitofrontal cortex. The OFC contributes to this odor-reward association as well as it assesses the value of a reward, i.e. the nutritional value of a food. The OFC receives projections from the piriform cortex, amygdala, and parahippocampal cortices.[18] Neurons in the OFC that encode food reward information activate the reward system when stimulated, associating the act of eating with reward. The OFC further projects to the anterior cingulate cortex where it plays a role in appetite.[22] The OFC also associates odors with other stimuli, such as taste.[18] Odor perception and discrimination also involve the OFC. The spatial odor map in the glomeruli layer of the olfactory bulb may contribute to these functions. The odor map begins processing of olfactory information by spatially organizing the glomeruli. This organizing aids the olfactory cortices in its functions of perceiving and discriminating odors.[2]

25.2.4 Adult neurogenesis

The olfactory bulb is, along with both the subventricular zone and the subgranular zone of the dentate gyrus of the hippocampus, one of only three structures in the brain observed to undergo continuing neurogenesis in adult mammals. In most mammals, new neurons are born from neural stem cells in the sub-ventricular zone and migrate rostrally towards the main [23] and accessory[24] olfactory bulbs. Within the olfactory bulb these immature neuroblasts develop into fully functional granule cell interneurons and periglomerular cell interneurons that reside in the granule cell layer and glomerular layers, respectively. The olfactory sensory neuron axons that form synapses in olfactory bulb glomeruli are also capable of regeneration following regrowth of an olfactory sensory neuron residing in the olfactory epithelium. Despite dynamic turnover of sensory axons and interneurons, the projection neurons (mitral and tufted neurons) that form synapses with these axons are not structurally plastic.

The function of adult neurogenesis in this region remains a matter of study. The survival of immature neurons as they enter the circuit is highly sensitive to olfactory activity and in particular associative learning tasks. This has led to the hypothesis that new neurons participate in learning processes.[25] No definitive behavioral effect has been observed in loss-of-function experiments suggesting that the function of this process, if at all related to olfactory processing, may be subtle.

25.3 Clinical significance

Destruction to the olfactory bulb results in ipsilateral anosmia The olfactory lobe is a neutral structure of the vertebrate forebrain involved in olfaction, or sense of smell.

25.4 Other animals

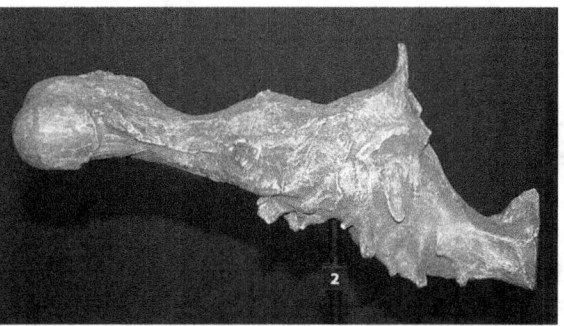

Fossil endocast of a Tyrannosaurus *cranial vault, showing extensive olfactory bulb (structure to the left)*

25.4.1 Evolution

Comparing the structure of the olfactory bulb among vertebrate species, such as the leopard frog and the lab mouse, reveals that they all share the same fundamental layout (five layers containing the nuclei of three major cell types; see "Anatomy" for details), despite being dissimilar in shape and size. A similar structure is shared by the analogous olfactory center in the fruit fly *Drosophila melanogaster*, the antennal lobe. One possibility is that vertebrate olfactory bulb and insect antennal lobe structure may be similar because they contain an optimal solution to a computational problem experienced by all olfactory systems and thus may have evolved independently in different phyla - a phenomenon generally known as convergent evolution.[26][27]

> "The increase of brain size relative to body size—encephalization—is intimately linked with human evolution. However, two genetically different evolutionary lineages, Neanderthals and modern humans, have produced similarly large-brained human species. Thus, understanding human brain evolution should include research into specific cerebral reorganization, possibly reflected by brain shape changes. Here we exploit developmental integration between the brain and its underlying skeletal base to test hypotheses about brain evolution in *Homo*. Three-dimensional geometric morphometric analyses of endobasicranial shape reveal previously undocumented details of evolutionary changes in *Homo sapiens*. Larger olfactory bulbs, relatively wider orbitofrontal cortex, relatively increased and forward projecting temporal lobe poles appear unique to modern humans. Such brain reorganization, beside physical consequences for overall skull shape, might have contributed to the evolution of *H. sapiens'* learning and social capacities, in which higher olfactory functions and its cognitive, neurological behavioral impli-

cations could have been hitherto underestimated factors."[28]

25.5 See also

- Olfactory ensheathing glia
- Phantosmia
- Nobiletin

25.6 References

[1] Hamilton, K.A.; Heinbockel, T.; Ennis, M.; Szabó, G.; Erdélyi, F.; Hayar, A. (2005). "Properties of external plexiform layer interneurons in mouse olfactory bulb slices". *Neuroscience.* **133** (3): 819–829. doi:10.1016/j.neuroscience.2005.03.008. ISSN 0306-4522. PMC 2383877. PMID 15896912.

[2] Mori K, Takahashi YK, Igarashi KM, Yamaguchi M (April 2006). "Maps of odorant molecular features in the Mammalian olfactory bulb". *Physiol. Rev.* **86** (2): 409–33. doi:10.1152/physrev.00021.2005. PMID 16601265.

[3] Spors, H.; Albeanu, D. F.; Murthy, V. N.; Rinberg, D.; Uchida, N.; Wachowiak, M.; Friedrich, R. W. (2012). "Illuminating Vertebrate Olfactory Processing". *Journal of Neuroscience.* **32** (41): 14102–14108a. doi:10.1523/JNEUROSCI.3328-12.2012. PMC 3752119. PMID 23055479.

[4] Scott JW, Wellis DP, Riggott MJ, Buonviso N (February 1993). "Functional organization of the main olfactory bulb". *Microsc. Res. Tech.* **24** (2): 142–56. doi:10.1002/jemt.1070240206. PMID 8457726.

[5] Prof. Leon Zurawicki (2 Sep 2010). *Neuromarketing: Exploring the Brain of the Consumer.* Springer Science & Business Media. p. 22. ISBN 978-3-540-77828-8. Retrieved 4 July 2015.

[6] Linster, Christiane; Cleland, Thomas (17 June 2013). "Spatiotemporal Coding in the Olfactory System". *20 Years of Computational Neuroscience.* **9**: 238. doi:10.1007/978-1-4614-1424-7_11. Retrieved 29 March 2016.

[7] Pressler, R. T.; Inoue, T.; Strowbridge, B. W. (2007). "Muscarinic Receptor Activation Modulates Granule Cell Excitability and Potentiates Inhibition onto Mitral Cells in the Rat Olfactory Bulb". *Journal of Neuroscience.* **27** (41): 10969–10981. doi:10.1523/JNEUROSCI.2961-07.2007. PMID 17928438.

[8] Smith, RS; Hu, R; DeSouza, A; Eberly, CL; Krahe, K; Chan, W; Araneda, RC (29 July 2015). "Differential Muscarinic Modulation in the Olfactory Bulb.". *The Journal of neuroscience : the official journal of the Society for Neuroscience.* **35** (30): 10773–85. doi:10.1523/JNEUROSCI.0099-15.2015. PMID 26224860.

[9] Taniguchi, K.; Saito, S.; Taniguchi, K. (Feb 2011). "Phylogenic outline of the olfactory system in vertebrates.". *J Vet Med Sci.* **73** (2): 139–47. doi:10.1292/jvms.10-0316. PMID 20877153.

[10] Trinh, K.; Storm DR. (2003). "Vomeronasal organ detects odorants in absence of signaling through main olfactory epithelium.". *Nat Neurosci.* **6** (5): 519–25. doi:10.1038/nn1039. PMID 12665798.

[11] Slotnick, B.; Restrepo, D.; Schellinck, H.; Archbold, G.; Price, S.; Lin, W. (Mar 2010). "Accessory olfactory bulb function is modulated by input from the main olfactory epithelium.". *Eur J Neurosci.* **31** (6): 1108–16. doi:10.1111/j.1460-9568.2010.07141.x. PMC 3745274. PMID 20377623.

[12] Hovis, KR.; Ramnath, R.; Dahlen, JE.; Romanova, AL.; LaRocca, G.; Bier, ME.; Urban, NN. (Jun 2012). "Activity regulates functional connectivity from the vomeronasal organ to the accessory olfactory bulb.". *J Neurosci.* **32** (23): 7907–16. doi:10.1523/JNEUROSCI.2399-11.2012. PMC 3483887. PMID 22674266.

[13] Trotier, D. (Sep 2011). "Vomeronasal organ and human pheromones.". *European Annals of Otorhinolaryngology Head Neck Diseases.* **128** (4): 184–90. doi:10.1016/j.anorl.2010.11.008. PMID 21377439.

[14] Brennan PA, Zufall F (November 2006). "Pheromonal communication in vertebrates". *Nature.* **444** (7117): 308–15. doi:10.1038/nature05404. PMID 17108955.

[15] Shpak, G.; Zylbertal, A.; Yarom, Y.; Wagner, S. (2012). "Calcium-Activated Sustained Firing Responses Distinguish Accessory from Main Olfactory Bulb Mitral Cells". *Journal of Neuroscience.* **32** (18): 6251–62. doi:10.1523/JNEUROSCI.4397-11.2012. PMID 22553031.

[16] Smith, RS; Hu, R; DeSouza, A; Eberly, CL; Krahe, K; Chan, W; Araneda, RC (29 July 2015). "Differential Muscarinic Modulation in the Olfactory Bulb.". *The Journal of neuroscience : the official journal of the Society for Neuroscience.* **35** (30): 10773–85. doi:10.1523/JNEUROSCI.0099-15.2015. PMID 26224860.

[17] Royet JP, Plailly J (October 2004). "Lateralization of olfactory processes" (PDF). *Chem. Senses.* **29** (8): 731–45. doi:10.1093/chemse/bjh067. PMID 15466819.

[18] Kadohisa M (2013). "Effects of odor on emotion, with implications". *Front Syst Neurosci.* **7**: 66. doi:10.3389/fnsys.2013.00066. PMC 3794443. PMID 24124415.

[19] Rolls ET (December 2010). "A computational theory of episodic memory formation in the hippocampus". *Behav. Brain Res.* **215** (2): 180–96. doi:10.1016/j.bbr.2010.03.027. PMID 20307583.

[20] Song, C.; Leonard, BE. (2005). "The olfactory bulbectomised rat as a model of depression". *Neuroscience Biobehavioral Reviews.* **29** (4–5): 627–47. doi:10.1016/j.neubiorev.2005.03.010. PMID 15925697.

[21] Morales-Medina, JC.; Juarez, I.; Venancio-García, E.; Cabrera, SN.; Menard, C.; Yu, W.; Flores, G.; Mechawar, N.; Quirion, R. (Apr 2013). "Impaired structural hippocampal plasticity is associated with emotional and memory deficits in the olfactory bulbectomized rat". *Neuroscience.* **236**: 233–43. doi:10.1016/j.neuroscience.2013.01.037. PMID 23357118.

[22] Rolls, ET (November 2012). "Taste, olfactory and food texture reward processing in the brain and the control of appetite.". *The Proceedings of the Nutrition Society.* **71** (4): 488–501. doi:10.1017/S0029665112000821. PMID 22989943.

[23] Lazarini, F.; Lledo, PM. (Jan 2011). "Is adult neurogenesis essential for olfaction?". *Trends Neuroscience.* **34** (1): 20–30. doi:10.1016/j.tins.2010.09.006. PMID 20980064.

[24] Oboti, L; Savalli G; Giachino C; De Marchis S; Panzica GC; Fasolo A; Peretto P (2009). "Integration and sensory experience-dependent survival of newly-generated neurons in the accessory olfactory bulb of female mice.". *Eur J Neurosci.* **29** (4): 679–92. doi:10.1111/j.1460-9568.2009.06614.x. PMID 19200078.

[25] Lepousez, G.; Valley, MT.; Lledo, PM. (2013). "The impact of adult neurogenesis on olfactory bulb circuits and computations.". *Annual Review of Physiology.* **75**: 339–63. doi:10.1146/annurev-physiol-030212-183731. PMID 23190074.

[26] Ache, BW. (Sep 2010). "Odorant-specific modes of signaling in mammalian olfaction". *Chem Senses.* **35** (7): 533–9. doi:10.1093/chemse/bjq045. PMID 20519266.

[27] Wang, JW. (Jan 2012). "Presynaptic modulation of early olfactory processing in Drosophila". *Dev Neurobiol.* **72** (1): 87–99. doi:10.1002/dneu.20936. PMID 21688402.

[28] Bastir, M.; Rosas, A.; Gunz, P.; Peña-Melian, A.; Manzi, G.; Harvati, K.; Kruszynski, R.; Stringer, C.; Hublin, JJ. (2011). "Evolution of the base of the brain in highly encephalized human species". *Nat Commun.* **2**: 588. doi:10.1038/ncomms1593. PMID 22158443.

25.7 Further reading

- Shepherd, G. *The Synaptic Organization of the Brain*, Oxford University Press, 5th edition (November, 2003). ISBN 0-19-515956-X

- Halpern, M; Martínez-Marcos, A (2003). "Structure and function of the vomeronasal system: An update". *Progress in neurobiology.* **70** (3): 245–318. doi:10.1016/S0301-0082(03)00103-5. PMID 12951145.

- Ache, BW; Young, JM (2005). "Olfaction: Diverse species, conserved principles". *Neuron.* **48** (3): 417–30. doi:10.1016/j.neuron.2005.10.022. PMID 16269360.

25.8 External links

- Stained brain slice images which include the "Olfactory bulb" at the BrainMaps project

- Anatomy diagram: 13048.000-1 at Roche Lexicon - illustrated navigator, Elsevier

- Glomerular Response Archive Leon & Johnson UC Irvine

- Olfactory Systems Laboratory at University of Utah

Chapter 26

Hippocampus

This article is about the section in the brain. For the fish genus *Hippocampus*, see Seahorse. For other uses, see Hippocampus (disambiguation).

The **hippocampus** (named after its resemblance to the

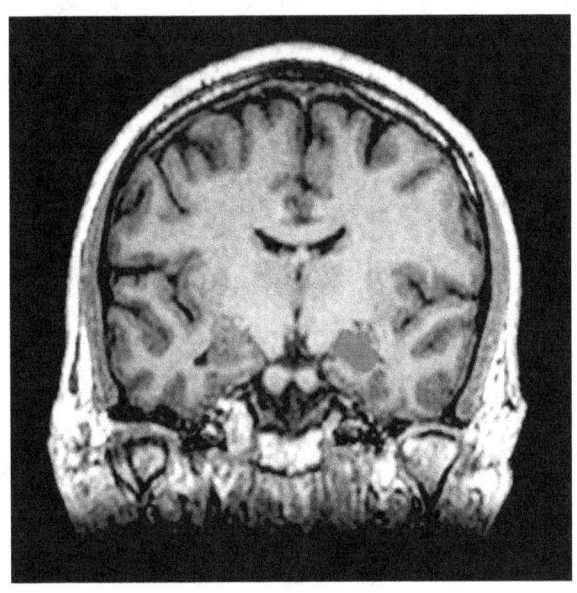

MRI coronal view of a hippocampus shown in red

seahorse, from the Greek ἱππόκαμπος, "seahorse" from ἵππος *hippos*, "horse" and κάμπος *kampos*, "sea monster") is a major component of the brains of humans and other vertebrates. Humans and other mammals have two hippocampi, one in each side of the brain. It belongs to the limbic system and plays important roles in the consolidation of information from short-term memory to long-term memory and spatial navigation. The hippocampus is located under the cerebral cortex;[1] and in primates it is located in the medial temporal lobe, underneath the cortical surface. It contains two main interlocking parts: the hippocampus proper (also called Ammon's horn)[2] and the dentate gyrus.

In Alzheimer's disease, the hippocampus is one of the first regions of the brain to suffer damage; memory loss and disorientation are included among the early symptoms.

Damage to the hippocampus can also result from oxygen starvation (hypoxia), encephalitis, or medial temporal lobe epilepsy. People with extensive, bilateral hippocampal damage may experience anterograde amnesia—the inability to form and retain new memories.

In rodents, the hippocampus has been studied extensively as part of a brain system responsible for spatial memory and navigation. Many neurons in the rat and mouse hippocampus respond as place cells: that is, they fire bursts of action potentials when the animal passes through a specific part of its environment. Hippocampal place cells interact extensively with head direction cells, whose activity acts as an inertial compass, and conjecturally with grid cells in the neighboring entorhinal cortex.

Since different neuronal cell types are neatly organized into layers in the hippocampus, it has frequently been used as a model system for studying neurophysiology. The form of neural plasticity known as long-term potentiation (LTP) was first discovered to occur in the hippocampus and has often been studied in this structure. LTP is widely believed to be one of the main neural mechanisms by which memory is stored in the brain.

26.1 Name

The earliest description of the ridge running along the floor of the temporal horn of the lateral ventricle comes from the Venetian anatomist Julius Caesar Aranzi (1587), who likened it first to a silkworm and then to a seahorse (Latin: *hippocampus* from Greek: ἵππος, "horse" and κάμπος, "sea monster"). The German anatomist Duvernoy (1729), the first to illustrate the structure, also wavered between "seahorse" and "silkworm." "Ram's horn" was proposed by the Danish anatomist Jacob Winsløw in 1732; and a decade later his fellow Parisian, the surgeon de Garengeot, used "cornu Ammonis" – horn of (the ancient Egyptian god) Amun.[3]

Another mythological reference appeared with the term

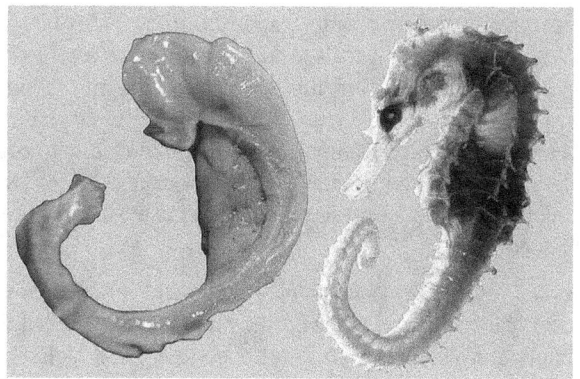

The human hippocampus and fornix compared with a seahorse (preparation by László Seress in 1980)

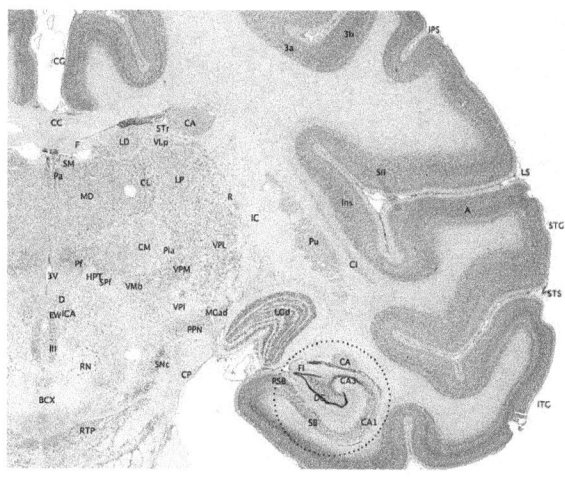

Nissl-stained coronal section of the brain of a macaque monkey, showing hippocampus (circled). Source: brainmaps.org.

pes hippocampi, which may date back to Diemerbroeck in 1672, introducing a comparison with the shape of the folded back forelimbs and webbed feet of the Classical hippocampus (Greek: ἱππόκαμπος), a sea monster with a horse's forequarters and a fish's tail. The hippocampus was then described as *pes hippocampi major*, with an adjacent bulge in the occipital horn, the calcar avis, being named *pes hippocampi minor*.[3] The renaming of the hippocampus as hippocampus major, and the calcar avis as hippocampus minor, has been attributed to Félix Vicq-d'Azyr systematising nomenclature of parts of the brain in 1786. Mayer mistakenly used the term hippopotamus in 1779, and was followed by some other authors until Karl Friedrich Burdach resolved this error in 1829. In 1861 the hippocampus minor became the centre of a dispute over human evolution between Thomas Henry Huxley and Richard Owen, satirised as the Great Hippocampus Question. The term hippocampus minor fell from use in anatomy textbooks, and was officially removed in the Nomina Anatomica of 1895.[4]

Today, the structure is called the hippocampus rather than hippocampus major, with *pes hippocampi* often being regarded as synonymous with De Garengeot's "cornu Ammonis",[3] a term that survives in the names of the four main histological divisions of the hippocampus: CA1, CA2, CA3, and CA4.[5]

26.2 Anatomy

Main article: Hippocampus anatomy

In terms of anatomy, the hippocampus is an elaboration of the edge of the cerebral cortex.[6] The structures that line the edge of the cortex make up the so-called limbic system (Latin *limbus* = border): These include the hippocampus, cingulate cortex, olfactory cortex, and amygdala. Paul MacLean once suggested, as part of his triune brain theory, that the limbic structures comprise the neural basis of emotion. Some neuroscientists no longer believe that the concept of a unified "limbic system" is valid, however.[7] Yet, the hippocampus is anatomically connected to parts of the brain that are involved with emotional behavior—the septum, the hypothalamic mammillary body, and the anterior nuclear complex in the thalamus—therefore its role as a limbic structure cannot be completely dismissed.

The hippocampus as a whole has the shape of a curved tube, which has been variously compared to a seahorse, a ram's horn (*Cornu Ammonis*, hence the subdivisions CA1 through CA4), or a banana.[6] It can be distinguished as a zone where the cortex narrows into a single layer of densely packed pyramidal neurons 3 to 6 cells deep in rats, which curl into a tight U shape; one edge of the "U," field CA4, is embedded into a backward-facing, strongly flexed, V-shaped cortex, the dentate gyrus. It consists of ventral and dorsal portions, both of which are of similar composition but are parts of different neural circuits.[8] This general layout holds across the full range of mammalian species, from hedgehog to human, although the details vary. In the rat, the two hippocampi resemble a pair of bananas, joined at the stems by the hippocampal commissure that crosses the midline under the anterior corpus callosum. In human or monkey brains, the portion of the hippocampus down at the bottom, near the base of the temporal lobe, is much broader than the part at the top. One of the consequences of this complex geometry is that cross-sections through the hippocampus can show a variety of shapes, depending on the angle and location of the cut.

The entorhinal cortex (EC), located in the parahippocampal gyrus, is considered to be part of the hippocampal region because of its anatomical connections. The EC is strongly and reciprocally connected with many other parts of the cerebral cortex. In addition, the medial septal nucleus, the

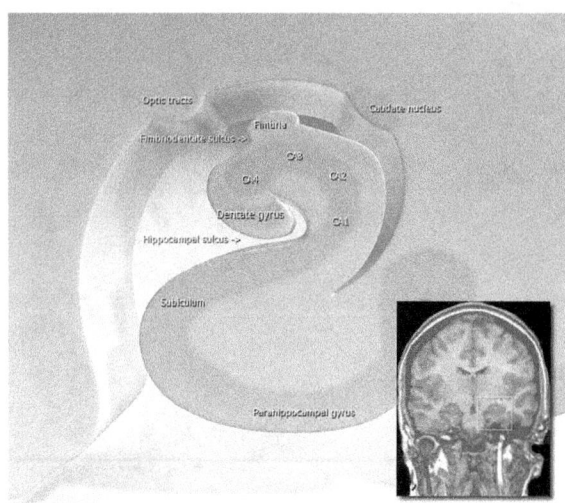

Hippocampal formation schematic

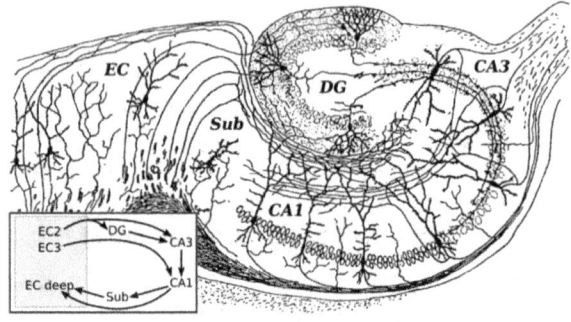

Basic circuit of the hippocampus, as drawn by Santiago Ramon y Cajal. DG: dentate gyrus. Sub: subiculum. EC: entorhinal cortex.

anterior nuclear complex and nucleus reuniens of the thalamus and the supramammillary nucleus of the hypothalamus, as well as the raphe nuclei and locus coeruleus in the brainstem send axons to the EC. The main output pathway (perforant path, first described by Ramon y Cajal) of EC axons comes from the large pyramidal cells in layer II that "perforate" the subiculum and project densely to the granule cells in the dentate gyrus, apical dendrites of CA3 get a less dense projection, and the apical dendrites of CA1 get a sparse projection. Thus, the perforant path establishes the EC as the main "interface" between the hippocampus and other parts of the cerebral cortex. The dentate granule cell axons (called mossy fibers) pass on the information from the EC on thorny spines that exit from the proximal apical dendrite of CA3 pyramidal cells. Then, CA3 axons exit from the deep part of the cell body and loop up into the region where the apical dendrites are located, then extend all the way back into the deep layers of the entorhinal cortex—the Schaffer collaterals completing the reciprocal circuit; field CA1 also sends axons back to the EC, but these are more sparse than the CA3 projection. Within the hippocampus, the flow of information from the EC is largely unidirectional, with signals propagating through a series of tightly packed cell layers, first to the dentate gyrus, then to the CA3 layer, then to the CA1 layer, then to the subiculum, then out of the hippocampus to the EC, mainly due to collateralization of the CA3 axons. Each of these layers also contains complex intrinsic circuitry and extensive longitudinal connections.[6]

Several other connections play important roles in hippocampal function.[6] Beyond the output to the EC, additional output pathways go to other cortical areas including the prefrontal cortex. A very important large output goes to the lateral septal area and to the mammillary body of the hypothalamus. The hippocampus receives modulatory input from the serotonin, norepinephrine, and dopamine systems, and from nucleus reuniens of the thalamus to field CA1. A very important projection comes from the medial septal area, which sends cholinergic and GABAergic fibers to all parts of the hippocampus. The inputs from the septal area play a key role in controlling the physiological state of the hippocampus; destruction of the septal area abolishes the hippocampal theta rhythm and severely impairs certain types of memory.[9]

The cortical region adjacent to the hippocampus is known collectively as the parahippocampal gyrus (or parahippocampus).[10] It includes the EC and also the perirhinal cortex, which derives its name from the fact that it lies next to the rhinal sulcus. The perirhinal cortex plays an important role in visual recognition of complex objects. There is also substantial evidence that it makes a contribution to memory, which can be distinguished from the contribution of the hippocampus. It is apparent that complete amnesia occurs only when both the hippocampus and the parahippocampus are damaged.[10]

26.3 Functions

Historically, the earliest widely held hypothesis was that the hippocampus is involved in olfaction. This idea was cast into doubt by a series of anatomical studies that did not find any direct projections to the hippocampus from the olfactory bulb.[11] However, later work did confirm that the olfactory bulb does project into the ventral part of the lateral entorhinal cortex, and field CA1 in the ventral hippocampus sends axons to the main olfactory bulb,[12] the anterior olfactory nucleus, and to the primary olfactory cortex. There continues to be some interest in hippocampal olfactory responses, in particular the role of the hippocampus in memory for odors, but few specialists today believe that olfaction is its primary function.[13][14]

Over the years, three main ideas of hippocampal function

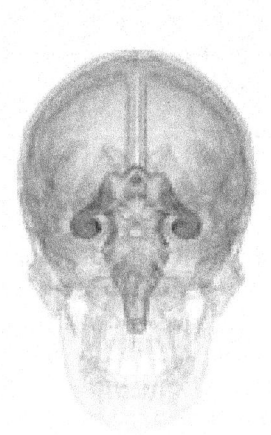

Hippocampus (animation)

have dominated the literature: inhibition, memory, and space. The behavioral inhibition theory (caricatured by O'Keefe and Nadel as "slam on the brakes!")[15] was very popular up to the 1960s. It derived much of its justification from two observations: first, that animals with hippocampal damage tend to be hyperactive; second, that animals with hippocampal damage often have difficulty learning to inhibit responses that they have previously been taught, especially if the response requires remaining quiet as in a passive avoidance test. Jeffrey Gray developed this line of thought into a full-fledged theory of the role of the hippocampus in anxiety.[16] The inhibition theory is currently the least popular of the three.[17]

The second major line of thought relates the hippocampus to memory. Although it had historical precursors, this idea derived its main impetus from a famous report by William Beecher Scoville and Brenda Milner[18] describing the results of surgical destruction of the hippocampi (in an attempt to relieve epileptic seizures), in Henry Molaison,[19] known until his death in 2008 as "Patient H.M." The unexpected outcome of the surgery was severe anterograde and partial retrograde amnesia; Molaison was unable to form new episodic memories after his surgery and could not remember any events that occurred just before his surgery, but he did retain memories of events that occurred many years earlier extending back into his childhood. This case attracted such widespread professional interest that Molaison became the most intensively studied subject in medical history.[20] In the ensuing years, other patients with similar levels of hippocampal damage and amnesia (caused by accident or disease) have also been studied, and thousands of experiments have studied the physiology of activity-driven changes in synaptic connections in the hippocampus. There is now universal agreement that the hippocampi play some sort of important role in memory; however, the precise nature of this role remains widely debated.[21][22]

The third important theory of hippocampal function relates the hippocampus to space. The spatial theory was originally championed by O'Keefe and Nadel, who were influenced by E.C. Tolman's theories about "cognitive maps" in humans and animals. O'Keefe and his student Dostrovsky in 1971 discovered neurons in the rat hippocampus that appeared to them to show activity related to the rat's location within its environment.[23] Despite skepticism from other investigators, O'Keefe and his co-workers, especially Lynn Nadel, continued to investigate this question, in a line of work that eventually led to their very influential 1978 book *The Hippocampus as a Cognitive Map*.[24] There is now almost universal agreement that hippocampal function plays an important role in spatial coding, but the details are widely debated.[25]

26.3.1 Role in memory

See also: Amnesia

Psychologists and neuroscientists generally agree that the hippocampus plays an important role in the formation of new memories about experienced events (episodic or autobiographical memory).[22][26] Part of this function is hippocampal involvement in the detection of novel events, places and stimuli.[27] Some researchers regard the hippocampus as part of a larger medial temporal lobe memory system responsible for general declarative memory (memories that can be explicitly verbalized—these would include, for example, memory for facts in addition to episodic memory).[21]

Due to bilateral symmetry the brain has a hippocampus in each cerebral hemisphere, so every normal brain has two of them. If damage to the hippocampus occurs in only one hemisphere, leaving the structure intact in the other hemisphere, the brain can retain near-normal memory functioning.[28] Severe damage to the hippocampi in both hemispheres results in profound difficulties in forming new memories (anterograde amnesia) and often also affects memories formed before the damage occurred (retrograde amnesia). Although the retrograde effect normally extends many years back before the brain damage, in some cases older memories remain. This retention of older memories leads to the idea that consolidation over time involves the transfer of memories out of the hippocampus to other parts of the brain.[29]

Damage to the hippocampus does not affect some types of

memory, such as the ability to learn new skills (playing a musical instrument or solving certain types of puzzles, for example). This fact suggests that such abilities depend on different types of memory (procedural memory) and different brain regions. Furthermore, amnesic patients frequently show "implicit" memory for experiences even in the absence of conscious knowledge. For example, patients asked to guess which of two faces they have seen most recently may give the correct answer most of the time in spite of stating that they have never seen either of the faces before. Some researchers distinguish between conscious *recollection*, which depends on the hippocampus, and *familiarity*, which depends on portions of the medial temporal cortex.[30]

26.3.2 Role in spatial memory and navigation

Main article: Place cell

Studies conducted on freely moving rats and mice have

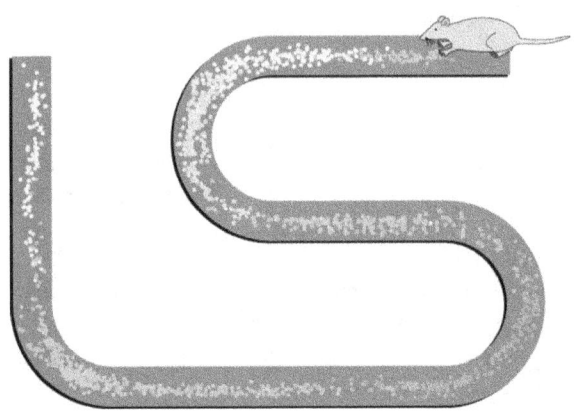

Spatial firing patterns of 8 place cells recorded from the CA1 layer of a rat. The rat ran back and forth along an elevated track, stopping at each end to eat a small food reward. Dots indicate positions where action potentials were recorded, with color indicating which neuron emitted that action potential.

shown that many hippocampal neurons have "place fields", that is, they fire bursts of action potentials when a rat passes through a particular part of the environment. Evidence for place cells in primates is limited, perhaps in part because it is difficult to record brain activity from freely moving monkeys. Place-related hippocampal neural activity has been reported in monkeys moving around inside a room while seated in a restraint chair;[31] on the other hand, Edmund Rolls and his colleagues instead described hippocampal cells that fire in relation to the place a monkey is looking at, rather than the place where its body is located.[32] In humans, cells with location-specific firing patterns have been reported in a study of patients with drug-resistant epilepsy who were undergoing an invasive procedure to localize the source of their seizures, with a view to surgical resection. The patients had diagnostic electrodes implanted in their hippocampus and then used a computer to move around in a virtual reality town.[33]

Place responses in rats and mice have been studied in hundreds of experiments over four decades, yielding a large quantity of information.[25] Place cell responses are shown by pyramidal cells in the hippocampus proper, and granule cells in the dentate gyrus. These constitute the great majority of neurons in the densely packed hippocampal layers. Inhibitory interneurons, which make up most of the remaining cell population, frequently show significant place-related variations in firing rate that are much weaker than those displayed by pyramidal or granule cells. There is little if any spatial topography in the representation; in general, cells lying next to each other in the hippocampus have uncorrelated spatial firing patterns. Place cells are typically almost silent when a rat is moving around outside the place field but reach sustained rates as high as 40 Hertz when the rat is near the center. Neural activity sampled from 30 to 40 randomly chosen place cells carries enough information to allow a rat's location to be reconstructed with high confidence. The size of place fields varies in a gradient along the length of the hippocampus, with cells at the dorsal end showing the smallest fields, cells near the center showing larger fields, and cells at the ventral tip fields that cover the entire environment.[25] In some cases, the firing rate of rat hippocampal cells depends not only on place but also on the direction a rat is moving, the destination toward which it is traveling, or other task-related variables.[34]

The discovery of place cells in the 1970s led to a theory that the hippocampus might act as a cognitive map—a neural representation of the layout of the environment.[35] Several lines of evidence support the hypothesis. It is a frequent observation that without a fully functional hippocampus, humans may not remember where they have been and how to get where they are going: Getting lost is one of the most common symptoms of amnesia.[36] Studies with animals have shown that an intact hippocampus is required for initial learning and long-term retention of some spatial memory tasks, in particular ones that require finding the way to a hidden goal.[37][38][39][40] The "cognitive map hypothesis" has been further advanced by recent discoveries of head direction cells, grid cells, and border cells in several parts of the rodent brain that are strongly connected to the hippocampus.[25][41]

Brain imaging shows that people have more active hippocampi when correctly navigating, as tested in a computer-simulated "virtual" navigation task.[42] Also, there is evidence that the hippocampus plays a role in finding shortcuts and new routes between familiar places. For example, London's taxi drivers must learn a large number of places and the most direct routes between them (they have to pass a strict test, The Knowledge, before being licensed to drive the famous black cabs). A study at University College London by Maguire, et al.. (2000)[43] showed that part of the hippocampus is larger in taxi drivers than in the general public, and that more experienced drivers have bigger hippocampi. Whether having a bigger hippocampus helps an individual to become a better cab driver, or if finding shortcuts for a living makes an individual's hippocampus grow is yet to be elucidated. However, in that study, Maguire et al. examined the correlation between size of the grey matter and length of time that had been spent as a taxi driver, and found a positive correlation between the length of time an individual had spent as a taxi driver and the volume of the right hippocampus. It was found that the total volume of the hippocampus remained constant, from the control group vs. taxi drivers. That is to say that the posterior portion of a taxi driver's hippocampus is indeed increased, but at the expense of the anterior portion. There have been no known detrimental effects reported from this disparity in hippocampal proportions.[43]

26.3.3 Hippocampal formation

Various sections of the hippocampal formation are shown to be functionally and anatomically distinct. The dorsal (DH), ventral (VH) and intermediate regions of the hippocampal formation serve different functions, project with differing pathways, and have varying degrees of place field neurons.[44] The dorsal region of the hippocampal formation serves for spatial memory, verbal memory, and learning of conceptual information. Using the radial arm maze, Pothuizen et al. (2004) found lesions in the DH to cause spatial memory impairment while VH lesions did not. Its projecting pathways include the medial septal complex and supramammillary nucleus.[45] The dorsal hippocampal formation also has more place field neurons than both the ventral and intermediate hippocampal formations.[46]

The intermediate hippocampus has overlapping characteristics with both the ventral and dorsal hippocampus.[44] Using PHAL anterograde tracing methods, Cenquizca and Swanson (2007) located the moderate projections to two primary olfactory cortical areas and prelimbic areas of the mPFC. This region has the smallest number of place field neurons. The ventral hippocampus functions in fear conditioning and affective processes.[47] Anagnostaras et al. (2002) showed that alterations to the ventral hippocampus reduced the amount of information sent to the amygdala by the dorsal and ventral hippocampus, consequently altering fear conditioning in rats.[48]

26.4 Physiology

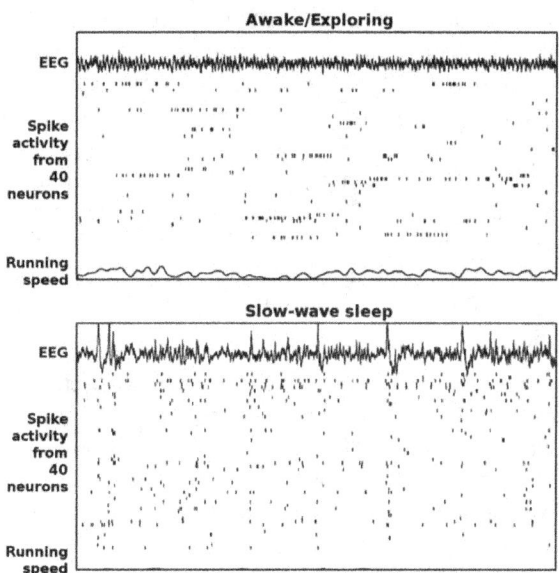

Examples of rat hippocampal EEG and CA1 neural activity in the theta (awake/behaving) and LIA (slow-wave sleep) modes. Each plot shows 20 seconds of data, with a hippocampal EEG trace at the top, spike rasters from 40 simultaneously recorded CA1 pyramidal cells in the middle (each raster line represents a different cell), and a plot of running speed at the bottom. The top plot represents a time period during which the rat was actively searching for scattered food pellets. For the bottom plot the rat was asleep.

The hippocampus shows two major "modes" of activity, each associated with a distinct pattern of neural population activity and waves of electrical activity as measured by an electroencephalogram (EEG). These modes are named after the EEG patterns associated with them: theta and large irregular activity (LIA). The main characteristics described below are for the rat, which is the animal most extensively studied.[49]

The theta mode appears during states of active, alert behavior (especially locomotion), and also during REM (dreaming) sleep.[50] In the theta mode, the EEG is dominated by large regular waves with a frequency range of 6 to 9 Hertz, and the main groups of hippocampal neurons (pyramidal cells and granule cells) show sparse population activity, which means that in any short time interval, the great majority of cells are silent, while the small remaining fraction fire at relatively high rates, up to 50 spikes in one second for the most active of them. An active cell typically stays active for half a second to a few seconds. As the rat behaves, the active cells fall silent and new cells become active, but the overall percentage of active cells remains more or less constant. In many situations, cell activity is determined largely by the spatial location of the animal, but other behavioral

26.4. PHYSIOLOGY

variables also clearly influence it.

The LIA mode appears during slow-wave (non-dreaming) sleep, and also during states of waking immobility such as resting or eating.[50] In the LIA mode, the EEG is dominated by sharp waves that are randomly timed large deflections of the EEG signal lasting for 25–50 milliseconds. Sharp waves are frequently generated in sets, with sets containing up to 5 or more individual sharp waves and lasting up to 500 ms. The spiking activity of neurons within the hippocampus is highly correlated with sharp wave activity. Most neurons decrease their firing rate between sharp waves; however, during a sharp wave, there is a dramatic increase of firing rate in up to 10% of the hippocampal population

These two hippocampal activity modes can be seen in primates as well as rats, with the exception that it has been difficult to see robust theta rhythmicity in the primate hippocampus. There are, however, qualitatively similar sharp waves and similar state-dependent changes in neural population activity.[51]

26.4.1 Theta rhythm

Main article: Theta rhythm

Because of its densely packed neural layers, the hippocampus generates some of the largest EEG signals of any brain structure. In some situations the EEG is dominated by regular waves at 3 to 10 Hertz, often continuing for many seconds. These reflect subthreshold membrane potentials and strongly modulate the spiking of hippocampal neurons and synchronise across the hippocampus in a travelling wave pattern.[52] This EEG pattern is known as a theta rhythm.[53] Theta rhythmicity is very obvious in rabbits and rodents and also clearly present in cats and dogs. Whether theta can be seen in primates is a vexing question.[54] In rats (the animals that have been the most extensively studied), theta is seen mainly in two conditions: first, when an animal is walking or in some other way actively interacting with its surroundings; second, during REM sleep.[55] The function of theta has not yet been convincingly explained although numerous theories have been proposed.[49] The most popular hypothesis has been to relate it to learning and memory. An example would be the phase with which theta rhythms, at the time of stimulation of a neuron, shape the effect of that stimulation upon its synapses. What is meant here is that theta rhythms may affect those aspects of learning and memory that are dependent upon synaptic plasticity.[56] It is well established that lesions of the medial septum—the central node of the theta system—cause severe disruptions of memory. However, the medial septum is more than just the controller of theta; it is also the main source of cholinergic projections to the hippocampus.[6] It has not been established that septal lesions exert their effects specifically by eliminating the theta rhythm.[57]

26.4.2 Sharp waves

Main article: Sharp wave–ripple complexes

During sleep or during waking states when an animal is resting or otherwise not engaged with its surroundings, the hippocampal EEG shows a pattern of irregular slow waves, somewhat larger in amplitude than theta waves. This pattern is occasionally interrupted by large surges called *sharp waves*.[58] These events are associated with bursts of spike activity lasting 50 to 100 milliseconds in pyramidal cells of CA3 and CA1. They are also associated with short-lived high-frequency EEG oscillations called "ripples", with frequencies in the range 150 to 200 Hertz in rats. Sharp waves are most frequent during sleep when they occur at an average rate of around 1 per second (in rats) but in a very irregular temporal pattern. Sharp waves are less frequent during inactive waking states and are usually smaller. Sharp waves have also been observed in humans and monkeys. In macaques, sharp waves are robust but do not occur as frequently as in rats.[51]

One of the most interesting aspects of sharp waves is that they appear to be associated with memory. Wilson and McNaughton 1994,[59] and numerous later studies, reported that when hippocampal place cells have overlapping spatial firing fields (and therefore often fire in near-simultaneity), they tend to show correlated activity during sleep following the behavioral session. This enhancement of correlation, commonly known as *reactivation*, has been found to occur mainly during sharp waves.[60] It has been proposed that sharp waves are, in fact, reactivations of neural activity patterns that were memorized during behavior, driven by strengthening of synaptic connections within the hippocampus.[61] This idea forms a key component of the "two-stage memory" theory, advocated by Buzsáki and others, which proposes that memories are stored within the hippocampus during behavior and then later transferred to the neocortex during sleep. Sharp waves are suggested to drive Hebbian synaptic changes in the neocortical targets of hippocampal output pathways.[62]

26.4.3 Long-term potentiation

Main article: Long-term potentiation

Since at least the time of Ramon y Cajal, psychologists have speculated that the brain stores memory by altering the strength of connections between neurons that are simultaneously active.[63] This idea was formalized by Donald Hebb in 1948,[64] but for many years thereafter, attempts to find a brain mechanism for such changes failed. In 1973, Tim Bliss and Terje Lømo described a phenomenon in the rabbit hippocampus that appeared to meet Hebb's specifications: a change in synaptic responsiveness induced by brief strong activation and lasting for hours or days or longer.[65] This phenomenon was soon referred to as *long-term potentiation*, abbreviated *LTP*. As a candidate mechanism for memory, LTP has since been studied intensively, and a great deal has been learned about it.

The hippocampus is a particularly favorable site for studying LTP because of its densely packed and sharply defined layers of neurons, but similar types of activity-dependent synaptic change have now been observed in many other brain areas.[66] The best-studied form of LTP occurs at synapses that terminate on dendritic spines and use the transmitter glutamate. Several of the major pathways within the hippocampus fit this description and exhibit LTP.[67] The synaptic changes depend on a special type of glutamate receptor, the NMDA receptor, which has the special property of allowing calcium to enter the postsynaptic spine only when presynaptic activation and postsynaptic depolarization occur at the same time.[68] Drugs that interfere with NMDA receptors block LTP and have major effects on some types of memory, especially spatial memory. Transgenic mice, genetically modified in ways that disable the LTP mechanism, also generally show severe memory deficits.[68]

26.5 Pathology

26.5.1 Aging

See also: Neurobiological effects of physical exercise § Structural growth, and Aging brain

Age-related conditions such as Alzheimer's disease (for which hippocampal disruption is one of the earliest signs[69]) have a severe impact on many types of cognition, but even normal aging is associated with a gradual decline in some types of memory, including episodic memory and working memory (or short-term memory). Because the hippocampus is thought to play a central role in memory, there has been considerable interest in the possibility that age-related declines could be caused by hippocampal deterioration.[70] Some early studies reported substantial loss of neurons in the hippocampus of elderly people, but later studies using more precise techniques found only minimal differences.[70] Similarly, some MRI studies have reported shrinkage of the hippocampus in elderly people, but other studies have failed to reproduce this finding. There is, however, a reliable relationship between the size of the hippocampus and memory performance — meaning that not all elderly people show hippocampal shrinkage, but those who do tend to perform less well on some memory tasks.[71] There are also reports that memory tasks tend to produce less hippocampal activation in elderly than in young subjects.[71] Furthermore, a randomized-control study published in 2011 found that aerobic exercise could increase the size of the hippocampus in adults aged 55 to 80 and also improve spatial memory.[72]

26.5.2 Stress

The hippocampus contains high levels of glucocorticoid receptors, which make it more vulnerable to long-term stress than most other brain areas.[73] Stress-related steroids affect the hippocampus in at least three ways: first, by reducing the excitability of some hippocampal neurons; second, by inhibiting the genesis of new neurons in the dentate gyrus; third, by causing atrophy of dendrites in pyramidal cells of the CA3 region. There is evidence that humans having experienced severe, long-lasting traumatic stress show atrophy of the hippocampus more than of other parts of the brain.[74] These effects show up in post-traumatic stress disorder,[75] and they may contribute to the hippocampal atrophy reported in schizophrenia[76] and severe depression.[77] A recent study has also revealed atrophy as a result of depression, but this can be stopped with anti-depressants even if they are not effective in relieving other symptoms.[78] Hippocampal atrophy is also frequently seen in Cushing's syndrome, a disorder caused by high levels of cortisol in the bloodstream. At least some of these effects appear to be reversible if the stress is discontinued. There is, however, evidence derived mainly from studies using rats that stress occurring shortly after birth can affect hippocampal function in ways that persist throughout life.[79]

Sex-specific responses to stress have also been demonstrated to have an effect on the hippocampus. During situations in which adult male and female rats were exposed to chronic stress the females were shown to be better able to cope.[80]

26.5.3 Epilepsy

The hippocampus is often the focus of epileptic seizures: hippocampal sclerosis is the most commonly visible type of tissue damage in temporal lobe epilepsy.[81] It is not yet clear, however, whether the epilepsy is usually caused by hippocampal abnormalities or whether the hippocampus is damaged by cumulative effects of seizures.[82] In experimental settings where repetitive seizures are artificially induced in animals, hippocampal damage is a frequent result. This may be a consequence of the hippocampus's being one of the most electrically excitable parts of the brain. It may also have something to do with the fact that the hippocampus is one of very few brain regions where new neurons continue to be created throughout life.[83]

26.5.4 Schizophrenia

The causes of schizophrenia are not at all well understood, but numerous abnormalities of brain structure have been reported. The most thoroughly investigated alterations involve the cerebral cortex, but effects on the hippocampus have also been described. Many reports have found reductions in the size of the hippocampus in schizophrenic subjects.[84] The changes probably result from altered development rather than tissue damage and show up even in subjects never having been medicated. Several lines of evidence implicate changes in synaptic organization and connectivity.[84] It is unclear whether hippocampal alterations play any role in causing the psychotic symptoms that are the most important feature of schizophrenia. Anthony Grace and his co-workers have suggested, on the basis of experimental work using animals, that hippocampal dysfunction might produce an alteration of dopamine release in the basal ganglia, thereby indirectly affecting the integration of information in the prefrontal cortex.[85] Others have suggested that hippocampal dysfunction might account for disturbances in long-term memory frequently observed in people with schizophrenia.[86]

26.5.5 Transient global amnesia

A current hypothesis as to one cause of transient global amnesia—a dramatic, sudden, temporary, near-total loss of short-term memory—is that it may be due to venous congestion of the brain,[87] leading to ischemia of structures such as the hippocampus that are involved in memory.[88]

26.6 Evolution

The hippocampus has a generally similar appearance across the range of mammal species, from monotremes such as the echidna to primates such as humans.[89] The hippocampal-size-to-body-size ratio broadly increases, being about twice as large for primates as for the echidna. It does not, however, increase at anywhere close to the rate of the neocortex-to-body-size ratio. Therefore, the hippocampus takes up a much larger fraction of the cortical mantle in rodents than in primates. In adult humans the volume of the hippocampus

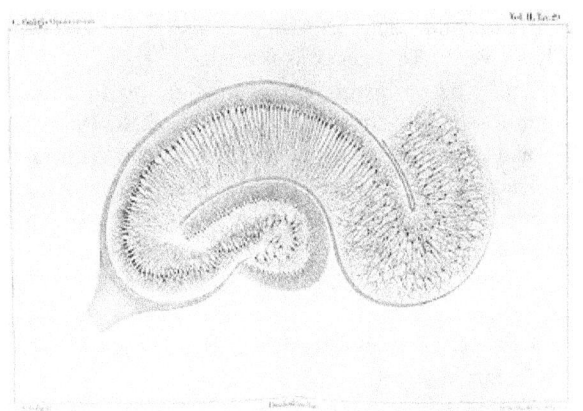

Drawing by Camillo Golgi of a hippocampus stained using the silver nitrate method

on each side of the brain is about 3.0 to 3.5 cm^3 as compared to 320 to 420 cm^3 for the volume of the neocortex.[90]

There is also a general relationship between the size of the hippocampus and spatial memory. When comparisons are made between similar species, those that have a greater capacity for spatial memory tend to have larger hippocampal volumes.[91] This relationship also extends to sex differences; in species where males and females show strong differences in spatial memory ability they also tend to show corresponding differences in hippocampal volume.[92]

Non-mammalian species do not have a brain structure that looks like the mammalian hippocampus, but they have one that is considered homologous to it. The hippocampus, as pointed out above, is in essence the medial edge of the cortex. Only mammals have a fully developed cortex, but the structure it evolved from, called the pallium, is present in all vertebrates, even the most primitive ones such as the lamprey or hagfish.[93] The pallium is usually divided into three zones: medial, lateral and dorsal. The medial pallium forms the precursor of the hippocampus. It does not resemble the hippocampus visually because the layers are not warped into an S shape or enfolded by the dentate gyrus, but the homology is indicated by strong chemical and functional affinities. There is now evidence that these hippocampal-like structures are involved in spatial cognition in birds, reptiles, and fish.[94]

In birds, the correspondence is sufficiently well established that most anatomists refer to the medial pallial zone as the "avian hippocampus".[95] Numerous species of birds have strong spatial skills, in particular those that cache food. There is evidence that food-caching birds have a larger hippocampus than other types of birds and that damage to the hippocampus causes impairments in spatial memory.[96]

The story for fish is more complex. In teleost fish (which make up the great majority of existing species), the forebrain is distorted in comparison to other types of vertebrates: Most neuroanatomists believe that the teleost forebrain is in essence everted, like a sock turned inside-out, so that structures that lie in the interior, next to the ventricles, for most vertebrates, are found on the outside in teleost fish, and vice versa.[97] One of the consequences of this is that the medial pallium ("hippocampal" zone) of a typical vertebrate is thought to correspond to the lateral pallium of a typical fish. Several types of fish (particularly goldfish) have been shown experimentally to have strong spatial memory abilities, even forming "cognitive maps" of the areas they inhabit.[91] There is evidence that damage to the lateral pallium impairs spatial memory.[98][99]

Thus, the role of the hippocampal region in navigation appears to begin far back in vertebrate evolution, predating splits that occurred hundreds of millions of years ago.[100] It is not yet known whether the medial pallium plays a similar role in even more primitive vertebrates, such as sharks and rays, or even lampreys and hagfish. Some types of insects, and molluscs such as the octopus, also have strong spatial learning and navigation abilities, but these appear to work differently from the mammalian spatial system, so there is as yet no good reason to think that they have a common evolutionary origin; nor is there sufficient similarity in brain structure to enable anything resembling a "hippocampus" to be identified in these species. Some have proposed, however, that the insect's mushroom bodies may have a function similar to that of the hippocampus.[101]

26.7 See also

- Trisynaptic circuit

26.8 Notes

[1] Wright, Anthony. Chapter 5: Limbic System: Hippocampus. Department of Neurobiology and Anatomy, The UT Medical School at Houston

[2] Pearce, 2001

[3] Duvernoy, 2005

[4] Gross, 1993

[5] Wechsler, 2004

[6] Amaral and Lavenex, 2006

[7] Kötter & Stephan, 1997

[8] Moser and Moser, 1998

[9] Winson, 1978

[10] Eichenbaum et al, 2007

[11] Finger, p. 183

[12] "Extrinsic projections from area CA1 of the rat hippocampus: olfactory, cortical, subcortical, and bilateral hippocampal formation projections". *Journal of Comparative Neurology*. 1990. doi:10.1002/cne.903020308.

[13] Eichenbaum et al, 1991

[14] Vanderwolf, 2001

[15] Nadel et al., 1975

[16] Gray and McNaughton, 2000

[17] Best & White, 1999

[18] Scoville and Milner, 1957

[19] New York Times, 12-06-2008

[20] Squire, 2009

[21] Squire, 1992

[22] Eichenbaum and Cohen, 1993

[23] O'Keefe and Dostrovsky, 1971

[24] O'Keefe and Nadel, 1978

[25] Moser et al., 2008

[26] Squire and Schacter, 2002

[27] VanElzakker et al., 2008

[28] Di Gennaro G, Grammaldo LG, Quarato PP, Esposito V, Mascia A, Sparano A, Meldolesi GN, Picardi A (Jun 2006). "Severe amnesia following bilateral medial temporal lobe damage occurring on two distinct occasions". *Neurological Sciences*. **27** (2): 129–33. doi:10.1007/s10072-006-0614-y. PMID 16816912.

[29] Squire and Schacter, 2002, Ch. 1

[30] Diana et al., 2007

[31] Matsumara et al., 1999

[32] Rolls and Xiang, 2006

[33] Ekstrom et al., 2003

[34] Smith and Mizumori, 2006

[35] O'Keefe and Nadel

[36] Chiu et al., 2004

[37] Morris et al., 1982

[38] Sutherland et al., 1982

[39] Sutherland et al., 2001

26.8. NOTES

[40] Clark et al., 2005

[41] Solstad et al., 2008

[42] Maguire et al., 1998

[43] Maguire et al., 2000

[44] Fanselow, 2010

[45] Pothuizen *et al.*, 2004

[46] Jung *et al.*, 1994

[47] Cenquizca *et al.*, 2007

[48] Anagnostaras *et al.*, 2002

[49] Buzsáki, 2006

[50] Buzsáki et al., 1990

[51] Skaggs et al., 2007

[52] Lubenov & Siapas, 2009

[53] Buzsáki, 2002

[54] Cantero et al., 2003

[55] Vanderwolf, 1969

[56] Huerta & Lisman, 1993

[57] Kahana et al., 2001

[58] Buzsáki, 1986

[59] Wilson & McNaughton, 1994

[60] Jackson et al., 2006

[61] Sutherland & McNaughton, 2000

[62] Buzsáki, 1989

[63] Ramon y Cajal, 1894

[64] Hebb, 1948

[65] Bliss & Lømo, 1973

[66] Cooke & Bliss, 2006

[67] Malenka & Bear, 2004

[68] Nakazawa et al., 2004

[69] Hampel et al., 2008

[70] Prull et al., 2000, p. 105

[71] Prull et al., 2000, p. 107

[72] Erickson et al., 2011

[73] Joels, 2008

[74] Fu et al, 2010

[75] Karl A, Schaefer M, Malta LS, Dörfel D, Rohleder N, Werner A (2006). "A meta-analysis of structural brain abnormalities in PTSD". *Neuroscience and Biobehavioral Reviews.* **30** (7): 1004–31. doi:10.1016/j.neubiorev.2006.03.004. PMID 16730374.

[76] Wright IC, Rabe-Hesketh S, Woodruff PW, David AS, Murray RM, Bullmore ET (January 2000). "Meta-analysis of regional brain volumes in schizophrenia". *The American Journal of Psychiatry.* **157** (1): 16–25. doi:10.1176/ajp.157.1.16. PMID 10618008.

[77] Kempton MJ, Salvador Z, Munafò MR, Geddes JR, Simmons A, Frangou S, Williams SC (Jul 2011). "Structural neuroimaging studies in major depressive disorder. Meta-analysis and comparison with bipolar disorder". *Archives of General Psychiatry.* **68** (7): 675–90. doi:10.1001/archgenpsychiatry.2011.60. PMID 21727252. see also MRI database at www.depressiondatabase.org

[78] Campbell & MacQueen, 2004

[79] Garcia-Segura, pp. 170–71

[80] Conrad CD (2008). "Chronic stress-induced hippocampal vulnerability: the glucocorticoid vulnerability hypothesis". *Reviews in the Neurosciences.* **19** (6): 395–411. doi:10.1515/revneuro.2008.19.6.395. PMC 2746750. PMID 19317179.

[81] Chang and Lowenstein, 2003

[82] Sloviter, 2005

[83] Kuruba et al., 2009

[84] Harrison, 2004

[85] Goto & Grace, 2008

[86] Boyer et al., 2007

[87] Lewis SL (Aug 1998). "Aetiology of transient global amnesia". *Lancet.* **352** (9125): 397–9. doi:10.1016/S0140-6736(98)01442-1. PMID 9717945.

[88] Chung CP, Hsu HY, Chao AC, Chang FC, Sheng WY, Hu HH (Jun 2006). "Detection of intracranial venous reflux in patients of transient global amnesia". *Neurology.* **66** (12): 1873–77. doi:10.1212/01.wnl.0000219620.69618.9d. PMID 16801653.

[89] West, 1990

[90] Suzuki et al, 2005

[91] Jacobs, 2003

[92] Jacobs et al., 1990

[93] Aboitiz et al., 2003

[94] Rodríguez et al., 2002

[95] Colombo and Broadbent, 2000

[96] Shettleworth, 2003

[97] Nieuwenhuys, 1982

[98] Portavella et al., 2002

[99] Vargas et al., 2006

[100] Broglio et al., 2005

[101] Mizunami et al., 1998

26.9 References

- Aboitiz F, Morales D, Montiel J (Oct 2003). "The evolutionary origin of the mammalian isocortex: towards an integrated developmental and functional approach". *The Behavioral and Brain Sciences*. **26** (5): 535–52. doi:10.1017/S0140525X03000128. PMID 15179935.

- Amaral D, Lavenex P (2006). "Ch 3. Hippocampal Neuroanatomy". In Andersen P, Morris R, Amaral D, Bliss T, O'Keefe J. *The Hippocampus Book*. Oxford University Press. ISBN 978-0-19-510027-3.

- Anagnostaras SG, Gale GD, Fanselow MS (2002). "The hippocampus and Pavlovian fear conditioning: reply to Bast et al" (PDF). *Hippocampus*. **12** (4): 561–565. doi:10.1002/hipo.10071. PMID 12201641.

- Best PJ, White AM (1999). "Placing hippocampal single-unit studies in a historical context". *Hippocampus*. **9** (4): 346–51. doi:10.1002/(SICI)1098-1063(1999)9:4<346::AID-HIPO2>3.0.CO;2-3. PMID 10495017.

- Bliss TV, Lomo T (Jul 1973). "Long-lasting potentiation of synaptic transmission in the dentate area of the anaesthetized rabbit following stimulation of the perforant path". *The Journal of Physiology*. **232** (2): 331–56. doi:10.1113/jphysiol.1973.sp010273. PMC 1350458. PMID 4727084.

- Boyer P, Phillips JL, Rousseau FL, Ilivitsky S (Apr 2007). "Hippocampal abnormalities and memory deficits: new evidence of a strong pathophysiological link in schizophrenia". *Brain Research Reviews*. **54** (1): 92–112. doi:10.1016/j.brainresrev.2006.12.008. PMID 17306884.

- Broglio C, Gómez A, Durán E, Ocaña FM, Jiménez-Moya F, Rodríguez F, Salas C (Sep 2005). "Hallmarks of a common forebrain vertebrate plan: specialized pallial areas for spatial, temporal and emotional memory in actinopterygian fish". *Brain Research Bulletin*. **66** (4-6): 397–99. doi:10.1016/j.brainresbull.2005.03.021. PMID 16144602.

- Burke SN, Barnes CA (Jan 2006). "Neural plasticity in the ageing brain". *Nature Reviews. Neuroscience*. **7** (1): 30–40. doi:10.1038/nrn1809. PMID 16371948.

- Buzsáki G (Nov 1986). "Hippocampal sharp waves: their origin and significance". *Brain Research*. **398** (2): 242–52. doi:10.1016/0006-8993(86)91483-6. PMID 3026567.

- Buzsáki G (1989). "Two-stage model of memory trace formation: a role for "noisy" brain states". *Neuroscience*. **31** (3): 551–70. doi:10.1016/0306-4522(89)90423-5. PMID 2687720.

- Buzsáki G, Chen LS, Gage FH (1990). "Spatial organization of physiological activity in the hippocampal region: relevance to memory formation". *Progress in Brain Research*. Progress in Brain Research. **83**: 257–68. doi:10.1016/S0079-6123(08)61255-8. ISBN 9780444811493. PMID 2203100.

- Buzsáki G (Jan 2002). "Theta oscillations in the hippocampus" (PDF). *Neuron*. **33** (3): 325–40. doi:10.1016/S0896-6273(02)00586-X. PMID 11832222.

- Buzsáki G (2006). *Rhythms of the Brain*. Oxford University Press. ISBN 0-19-530106-4.

- Ramón y Cajal S (1894). "The Croonian Lecture: La Fine Structure des Centres Nerveux". *Proceedings of the Royal Society*. **55** (331–335): 444–68. doi:10.1098/rspl.1894.0063.

- Campbell S, Macqueen G (Nov 2004). "The role of the hippocampus in the pathophysiology of major depression". *Journal of Psychiatry & Neuroscience*. **29** (6): 417–26. PMC 524959. PMID 15644983.

- Cantero JL, Atienza M, Stickgold R, Kahana MJ, Madsen JR, Kocsis B (Nov 2003). "Sleep-dependent theta oscillations in the human hippocampus and neocortex". *The Journal of Neuroscience*. **23** (34): 10897–903. PMID 14645485.

- Carey B (2008-12-04). "H. M., an Unforgettable Amnesiac, Dies at 82". *The New York Times*. Retrieved 2009-04-27.

- Chiu YC, Algase D, Whall A, Liang J, Liu HC, Lin KN, Wang PN (2004). "Getting lost: directed attention and executive functions in early Alzheimer's disease patients". *Dementia and Geriatric Cognitive Disorders*. **17** (3): 174–80. doi:10.1159/000076353. PMID 14739541.

26.9. REFERENCES

- Chang BS, Lowenstein DH (Sep 2003). "Epilepsy". *The New England Journal of Medicine*. **349** (13): 1257–66. doi:10.1056/NEJMra022308. PMID 14507951.

- Cho RY, Gilbert A, Lewis DA (2005). "Ch 22. The neurobiology of schizophrenia". In Charney DS, Nestler EJ. *Neurobiology of Mental Illness*. Oxford University Press US. ISBN 978-0-19-518980-3.

- Cenquizca LA, Swanson LW (Nov 2007). "Spatial organization of direct hippocampal field CA1 axonal projections to the rest of the cerebral cortex". *Brain Research Reviews*. **56** (1): 1–26. doi:10.1016/j.brainresrev.2007.05.002. PMC 2171036. PMID 17559940.

- Clark RE, Broadbent NJ, Squire LR (2005). "Hippocampus and remote spatial memory in rats". *Hippocampus*. **15** (2): 260–72. doi:10.1002/hipo.20056. PMC 2754168. PMID 15523608.

- Colombo M, Broadbent N (Jun 2000). "Is the avian hippocampus a functional homologue of the mammalian hippocampus?". *Neuroscience and Biobehavioral Reviews*. **24** (4): 465–84. doi:10.1016/S0149-7634(00)00016-6. PMID 10817844.

- Cooke SF, Bliss TV (Jul 2006). "Plasticity in the human central nervous system". *Brain*. **129** (Pt 7): 1659–73. doi:10.1093/brain/awl082. PMID 16672292.

- de Olmos J, Hardy H, Heimer L (Sep 1978). "The afferent connections of the main and the accessory olfactory bulb formations in the rat: an experimental HRP-study". *The Journal of Comparative Neurology*. **181** (2): 213–244. doi:10.1002/cne.901810202. PMID 690266.

- Diana RA, Yonelinas AP, Ranganath C (Sep 2007). "Imaging recollection and familiarity in the medial temporal lobe: a three-component model". *Trends in Cognitive Sciences*. **11** (9): 379–86. doi:10.1016/j.tics.2007.08.001. PMID 17707683.

- Duvernoy HM (2005). "Introduction". *The Human Hippocampus* (3rd ed.). Berlin: Springer-Verlag. p. 1. ISBN 3-540-23191-9.

- Eichenbaum H, Otto TA, Wible CG, Piper JM (1991). "Ch 7. Building a model of the hippocampus in olfaction and memory". In Davis JL, Eichenbaum H,. *Olfaction*. MIT Press. ISBN 978-0-262-04124-9.

- Eichenbaum H, Cohen NJ (1993). *Memory, Amnesia, and the Hippocampal System*. MIT Press.

- Eichenbaum H, Yonelinas AP, Ranganath C (2007). "The medial temporal lobe and recognition memory". *Annual Review of Neuroscience*. **30**: 123–52. doi:10.1146/annurev.neuro.30.051606.094328. PMC 2064941. PMID 17417939.

- Ekstrom AD, Kahana MJ, Caplan JB, Fields TA, Isham EA, Newman EL, Fried I (Sep 2003). "Cellular networks underlying human spatial navigation" (PDF). *Nature*. **425** (6954): 184–88. Bibcode:2003Natur.425..184E. doi:10.1038/nature01964. PMID 12968182.

- Erickson KI, et al. (Feb 2011). "Exercise training increases size of hippocampus and improves memory". *Proceedings of the National Academy of Sciences of the United States of America*. **108** (7): 3017–3022. Bibcode:2011PNAS..108.3017E. doi:10.1073/pnas.1015950108. PMC 3041121. PMID 21282661.

- Fanselow MS, Dong HW (Jan 2010). "Are the dorsal and ventral hippocampus functionally distinct structures?" (PDF). *Neuron*. **65** (1): 7–19. doi:10.1016/j.neuron.2009.11.031. PMC 2822727. PMID 20152109.

- Finger, S (2001). *Origins of Neuroscience: A History of Explorations Into Brain Function*. Oxford University Press US. ISBN 978-0-19-514694-3.

- Garcia-Segura LM (2009). *Hormones and Brain Plasticity*. Oxford University Press US. ISBN 978-0-19-532661-1.

- Woon FL, Sood S, Hedges DW (Oct 2010). "Hippocampal volume deficits associated with exposure to psychological trauma and posttraumatic stress disorder in adults: a meta-analysis". *Progress in Neuro-Psychopharmacology & Biological Psychiatry*. **34** (7): 1181–1188. doi:10.1016/j.pnpbp.2010.06.016. PMID 20600466.

- Gorwood P, Corruble E, Falissard B, Goodwin GM (Jun 2008). "Toxic effects of depression on brain function: impairment of delayed recall and the cumulative length of depressive disorder in a large sample of depressed outpatients". *The American Journal of Psychiatry*. **165** (6): 731–9. doi:10.1176/appi.ajp.2008.07040574. PMID 18381906.

- Goto Y, Grace AA (Nov 2008). "Limbic and cortical information processing in the nucleus accumbens". *Trends in Neurosciences*. **31** (11): 552–8. doi:10.1016/j.tins.2008.08.002. PMC 2884964. PMID 18786735.

- Gray, JA; McNaughton N (2000). *The Neuropsychology of Anxiety: An Enquiry into the Functions of the Septo-Hippocampal System*. Oxford University Press.

- Gross CG (Oct 1993). "Hippocampus minor and man's place in nature: a case study in the social construction of neuroanatomy". *Hippocampus*. **3** (4): 403–416. doi:10.1002/hipo.450030403. PMID 8269033.

- Hampel H, Bürger K, Teipel SJ, Bokde AL, Zetterberg H, Blennow K (Jan 2008). "Core candidate neurochemical and imaging biomarkers of Alzheimer's disease". *Alzheimer's & Dementia*. **4** (1): 38–48. doi:10.1016/j.jalz.2007.08.006. PMID 18631949.

- Harrison PJ (Jun 2004). "The hippocampus in schizophrenia: a review of the neuropathological evidence and its pathophysiological implications". *Psychopharmacology*. **174** (1): 151–62. doi:10.1007/s00213-003-1761-y. PMID 15205886.

- Hebb DO (1949). *Organization of Behavior: a Neuropsychological Theory*. New York: John Wiley. ISBN 0-471-36727-3.

- Huerta PT, Lisman JE (Aug 1993). "Heightened synaptic plasticity of hippocampal CA1 neurons during a cholinergically induced rhythmic state". *Nature*. **364** (6439): 723–5. Bibcode:1993Natur.364..723H. doi:10.1038/364723a0. PMID 8355787.

- Jackson JC, Johnson A, Redish AD (Nov 2006). "Hippocampal sharp waves and reactivation during awake states depend on repeated sequential experience". *The Journal of Neuroscience*. **26** (48): 12415–26. doi:10.1523/JNEUROSCI.4118-06.2006. PMID 17135403.

- Jacobs LF, Gaulin SJ, Sherry DF, Hoffman GE (Aug 1990). "Evolution of spatial cognition: sex-specific patterns of spatial behavior predict hippocampal size". *Proceedings of the National Academy of Sciences of the United States of America*. **87** (16): 6349–52. Bibcode:1990PNAS...87.6349J. doi:10.1073/pnas.87.16.6349. PMC 54531. PMID 2201026.

- Jacobs LF (2003). "The evolution of the cognitive map". *Brain, Behavior and Evolution*. **62** (2): 128–39. doi:10.1159/000072443. PMID 12937351.

- Jung MW, Wiener SI, McNaughton BL (Dec 1994). "Comparison of spatial firing characteristics of units in dorsal and ventral hippocampus of the rat" (PDF). *The Journal of Neuroscience*. **14** (12): 7347–7356. PMID 7996180.

- Kahana MJ, Seelig D, Madsen JR (Dec 2001). "Theta returns". *Current Opinion in Neurobiology*. **11** (6): 739–44. doi:10.1016/S0959-4388(01)00278-1. PMID 11741027.

- Kötter R, Stephan KE (1997). "Useless or helpful? The "limbic system" concept". *Reviews in the Neurosciences*. **8** (2): 139–45. doi:10.1515/REVNEURO.1997.8.2.139. PMID 9344183.

- Joëls M (Apr 2008). "Functional actions of corticosteroids in the hippocampus". *European Journal of Pharmacology*. **583** (2-3): 312–321. doi:10.1016/j.ejphar.2007.11.064. PMID 18275953.

- Kuruba R, Hattiangady B, Shetty AK (Jan 2009). "Hippocampal neurogenesis and neural stem cells in temporal lobe epilepsy". *Epilepsy & Behavior*. 14 Suppl 1: 65–73. doi:10.1016/j.yebeh.2008.08.020. PMC 2654382. PMID 18796338.

- Lubenov EV, Siapas AG (May 2009). "Hippocampal theta oscillations are travelling waves". *Nature*. **459** (7246): 534–9. doi:10.1038/nature08010. PMID 19489117.

- Maguire EA, Burgess N, Donnett JG, Frackowiak RS, Frith CD, O'Keefe J (May 1998). "Knowing where and getting there: a human navigation network". *Science*. **280** (5365): 921–24. Bibcode:1998Sci...280..921M. doi:10.1126/science.280.5365.921. PMID 9572740.

- Maguire EA, Gadian DG, Johnsrude IS, Good CD, Ashburner J, Frackowiak RS, Frith CD (Apr 2000). "Navigation-related structural change in the hippocampi of taxi drivers". *Proceedings of the National Academy of Sciences of the United States of America*. **97** (8): 4398–403. Bibcode:2000PNAS...97.4398M. doi:10.1073/pnas.070039597. PMC 18253. PMID 10716738.

- Malenka RC, Bear MF (Sep 2004). "LTP and LTD: an embarrassment of riches". *Neuron*. **44** (1): 5–21. doi:10.1016/j.neuron.2004.09.012. PMID 15450156.

- Matsumura N, Nishijo H, Tamura R, Eifuku S, Endo S, Ono T (Mar 1999). "Spatial- and task-dependent neuronal responses during real and virtual translocation in the monkey hippocampal formation". *The Journal of Neuroscience*. **19** (6): 2381–93. PMID 10066288.

- McNaughton BL, Battaglia FP, Jensen O, Moser EI, Moser MB (Aug 2006). "Path integration and the neural basis of the 'cognitive map'". *Nature Reviews. Neuroscience.* **7** (8): 663–78. doi:10.1038/nrn1932. PMID 16858394.

- Mizunami M, Weibrecht JM, Strausfeld NJ (Dec 1998). "Mushroom bodies of the cockroach: their participation in place memory". *The Journal of Comparative Neurology.* **402** (4): 520–37. doi:10.1002/(SICI)1096-9861(19981228)402:4<520::AID-CNE6>3.0.CO;2-K. PMID 9862324.

- Morris RG, Garrud P, Rawlins JN, O'Keefe J (Jun 1982). "Place navigation impaired in rats with hippocampal lesions". *Nature.* **297** (5868): 681–83. Bibcode:1982Natur.297..681M. doi:10.1038/297681a0. PMID 7088155.

- Moser MB, Moser EI (1998). "Functional differentiation in the hippocampus". *Hippocampus.* **8** (6): 608–19. doi:10.1002/(SICI)1098-1063(1998)8:6<608::AID-HIPO3>3.0.CO;2-7. PMID 9882018.

- Moser EI, Kropff E, Moser MB (2008). "Place cells, grid cells, and the brain's spatial representation system". *Annual Review of Neuroscience.* **31**: 69. doi:10.1146/annurev.neuro.31.061307.090723. PMID 18284371.

- Nadel L, O'Keefe J, Black A (Jun 1975). "Slam on the brakes: a critique of Altman, Brunner, and Bayer's response-inhibition model of hippocampal function". *Behavioral Biology.* **14** (2): 151–62. doi:10.1016/S0091-6773(75)90148-0. PMID 1137539.

- Nakazawa K, McHugh TJ, Wilson MA, Tonegawa S (May 2004). "NMDA receptors, place cells and hippocampal spatial memory". *Nature Reviews. Neuroscience.* **5** (5): 361–72. doi:10.1038/nrn1385. PMID 15100719.

- Nieuwenhuys, R (1982). "An Overview of the Organization of the Brain of Actinopterygian Fishes". *Am. Zool.* **22** (2): 287–310. doi:10.1093/icb/22.2.287.

- O'Kane G, Kensinger EA, Corkin S (2004). "Evidence for semantic learning in profound amnesia: an investigation with patient H.M". *Hippocampus.* **14** (4): 417–25. doi:10.1002/hipo.20005. PMID 15224979.

- O'Keefe J, Dostrovsky J (Nov 1971). "The hippocampus as a spatial map. Preliminary evidence from unit activity in the freely-moving rat". *Brain Research.* **34** (1): 171–75. doi:10.1016/0006-8993(71)90358-1. PMID 5124915.

- O'Keefe, J; Nadel L (1978). *The Hippocampus as a Cognitive Map.* Oxford University Press.

- Portavella M, Vargas JP, Torres B, Salas C (2002). "The effects of telencephalic pallial lesions on spatial, temporal, and emotional learning in goldfish". *Brain Research Bulletin.* **57** (3-4): 397–99. doi:10.1016/S0361-9230(01)00699-2. PMID 11922997.

- Pearce JM (Sep 2001). "Ammon's horn and the hippocampus". *Journal of Neurology, Neurosurgery, and Psychiatry.* **71** (3): 351. doi:10.1136/jnnp.71.3.351. PMC 1737533. PMID 11511709.

- Pothuizen HH, Zhang WN, Jongen-Rêlo AL, Feldon J, Yee BK (Feb 2004). "Dissociation of function between the dorsal and the ventral hippocampus in spatial learning abilities of the rat: a within-subject, within-task comparison of reference and working spatial memory". *The European Journal of Neuroscience.* **19** (3): 705–712. doi:10.1111/j.0953-816X.2004.03170.x. PMID 14984421.

- Prull MW, Gabrieli JD, Bunge SA (2000). "Ch 2. Age-related changes in memory: A cognitive neuroscience perspective". In Craik FI, Salthouse TA. *The handbook of aging and cognition.* Erlbaum. ISBN 978-0-8058-2966-2.

- Rodríguez F, López JC, Vargas JP, Broglio C, Gómez Y, Salas C (2002). "Spatial memory and hippocampal pallium through vertebrate evolution: insights from reptiles and teleost fish". *Brain Research Bulletin.* **57** (3-4): 499–503. doi:10.1016/S0361-9230(01)00682-7. PMID 11923018.

- Rolls ET, Xiang JZ (2006). "Spatial view cells in the primate hippocampus and memory recall". *Reviews in the Neurosciences.* **17** (1-2): 175–200. doi:10.1515/REVNEURO.2006.17.1-2.175. PMID 16703951.

- Rosenzweig ES, Barnes CA (Feb 2003). "Impact of aging on hippocampal function: plasticity, network dynamics, and cognition". *Progress in Neurobiology.* **69** (3): 143–79. doi:10.1016/S0301-0082(02)00126-0. PMID 12758108.

- Scoville WB, Milner B (Feb 1957). "Loss of recent memory after bilateral hippocampal lesions". *Journal of Neurology, Neurosurgery, and Psychiatry.* **20** (1): 11–21. doi:10.1136/jnnp.20.1.11. PMC 497229. PMID 13406589.

- Shettleworth SJ (2003). "Memory and hippocampal specialization in food-storing birds: challenges for research on comparative cognition".

- *Brain, Behavior and Evolution.* **62** (2): 108–16. doi:10.1159/000072441. PMID 12937349.

- Skaggs WE, McNaughton BL, Wilson MA, Barnes CA (1996). "Theta phase precession in hippocampal neuronal populations and the compression of temporal sequences". *Hippocampus.* **6** (2): 149–76. doi:10.1002/(SICI)1098-1063(1996)6:2<149::AID-HIPO6>3.0.CO;2-K. PMID 8797016.

- Skaggs WE, McNaughton BL, Permenter M, Archibeque M, Vogt J, Amaral DG, Barnes CA (Aug 2007). "EEG sharp waves and sparse ensemble unit activity in the macaque hippocampus". *Journal of Neurophysiology.* **98** (2): 898–910. doi:10.1152/jn.00401.2007. PMID 17522177.

- Sloviter RS (Feb 2005). "The neurobiology of temporal lobe epilepsy: too much information, not enough knowledge". *Comptes Rendus Biologies.* **328** (2): 143–53. doi:10.1016/j.crvi.2004.10.010. PMID 15771000.

- Smith DM, Mizumori SJ (2006). "Hippocampal place cells, context, and episodic memory". *Hippocampus.* **16** (9): 716–29. doi:10.1002/hipo.20208. PMID 16897724.

- Solstad T, Boccara CN, Kropff E, Moser MB, Moser EI (Dec 2008). "Representation of geometric borders in the entorhinal cortex". *Science.* **322** (5909): 1865–68. Bibcode:2008Sci...322.1865S. doi:10.1126/science.1166466. PMID 19095945.

- Squire LR (Apr 1992). "Memory and the hippocampus: a synthesis from findings with rats, monkeys, and humans". *Psychological Review.* **99** (2): 195–231. doi:10.1037/0033-295X.99.2.195. PMID 1594723.

- Squire, LR; Schacter DL (2002). *The Neuropsychology of Memory.* Guilford Press.

- Squire LR (Jan 2009). "The legacy of patient H.M. for neuroscience". *Neuron.* **61** (1): 6–9. doi:10.1016/j.neuron.2008.12.023. PMC 2649674. PMID 19146808.

- Sutherland GR, McNaughton B (Apr 2000). "Memory trace reactivation in hippocampal and neocortical neuronal ensembles". *Current Opinion in Neurobiology.* **10** (2): 180–86. doi:10.1016/S0959-4388(00)00079-9. PMID 10753801.

- Sutherland RJ, Kolb B, Whishaw IQ (Aug 1982). "Spatial mapping: definitive disruption by hippocampal or medial frontal cortical damage in the rat". *Neuroscience Letters.* **31** (3): 271–6. doi:10.1016/0304-3940(82)90032-5. PMID 7133562.

- Sutherland RJ, Weisend MP, Mumby D, Astur RS, Hanlon FM, Koerner A, Thomas MJ, Wu Y, Moses SN, Cole C, Hamilton DA, Hoesing JM (2001). "Retrograde amnesia after hippocampal damage: recent vs. remote memories in two tasks". *Hippocampus.* **11** (1): 27–42. doi:10.1002/1098-1063(2001)11:1<27::AID-HIPO1017>3.0.CO;2-4. PMID 11261770.

- Suzuki M, Hagino H, Nohara S, Zhou SY, Kawasaki Y, Takahashi T, Matsui M, Seto H, Ono T, Kurachi M (Feb 2005). "Male-specific volume expansion of the human hippocampus during adolescence". *Cerebral Cortex.* **15** (2): 187–93. doi:10.1093/cercor/bhh121. PMID 15238436.

- Vanderwolf CH (Dec 2001). "The hippocampus as an olfacto-motor mechanism: were the classical anatomists right after all?". *Behavioural Brain Research.* **127** (1-2): 25–47. doi:10.1016/S0166-4328(01)00354-0. PMID 11718883.

- Vargas JP, Bingman VP, Portavella M, López JC (Nov 2006). "Telencephalon and geometric space in goldfish". *The European Journal of Neuroscience.* **24** (10): 2870–78. doi:10.1111/j.1460-9568.2006.05174.x. PMID 17156211.

- VanElzakker M, Fevurly RD, Breindel T, Spencer RL (Dec 2008). "Environmental novelty is associated with a selective increase in Fos expression in the output elements of the hippocampal formation and the perirhinal cortex". *Learning & Memory.* **15** (12): 899–908. doi:10.1101/lm.1196508. PMC 2632843. PMID 19050162.

- Wechsler RT, Morss, AM, Wustoff, CJ, & Caughey, AB (2004). *Blueprints notes & cases: Neuroscience.* Oxford: Blackwell Publishing. p. 37. ISBN 1-4051-0349-3.

- West MJ (1990). "Stereological studies of the hippocampus: a comparison of the hippocampal subdivisions of diverse species including hedgehogs, laboratory rodents, wild mice and men". *Progress in Brain Research.* Progress in Brain Research. **83**: 13–36. doi:10.1016/S0079-6123(08)61238-8. ISBN 9780444811493. PMID 2203095.

- Wilson MA, McNaughton BL (Jul 1994). "Reactivation of hippocampal ensemble memories during sleep". *Science.* **265** (5172): 676–79. Bibcode:1994Sci...265..676W. doi:10.1126/science.8036517. PMID 8036517.

- Winson J (Jul 1978). "Loss of hippocampal theta rhythm results in spatial memory deficit in the rat". *Science*. **201** (4351): 160–63. Bibcode:1978Sci...201..160W. doi:10.1126/science.663646. PMID 663646.

26.10 Further reading

26.10.1 Journals

- *Hippocampus* (Wiley)

26.10.2 Books

- Per Andersen; Richard Morris; David Amaral; Tim Bliss; John O'Keefe, eds. (2007). *The Hippocampus Book*. Oxford University Press. ISBN 978-0-19-510027-3.

- Dori Derdikman; James J. Knierim, eds. (2014). *Space, Time and Memory in the Hippocampal Formation*. Springer. ISBN 978-3-7091-1292-2.

- Henri M. Duvernoy; F. Cattin (2005). *The Human Hippocampus: Functional Anatomy, Vascularization, and Serial Sections with MRI*. Springer. ISBN 978-3-540-23191-2.

- Howard Eichenbaum (2002). *The Cognitive Neuroscience of Memory*. Oxford University Press US. ISBN 978-0-19-514175-7.

- edited by Patricia E. Sharp. (2002). Patricia E. Sharp, ed. *The Neural Basis of Navigation: Evidence from Single Cell Recording*. Springer. ISBN 978-0-7923-7579-1.

- Philippe Taupin (2007). *The Hippocampus: Neurotransmission and Plasticity in the Nervous System*. Nova Publishers. ISBN 978-1-60021-914-6.

- John H Byrne, ed. (2008). *Learning and Memory: A comprehensive reference*. Elsevier. ISBN 978-0-12-370509-9.

26.11 External links

- Stained brain slice images which include the "hippocampus" at the BrainMaps project
- Diagram of a Hippocampal Brain Slice
- Hippocampus – Cell Centered Database
- Temporal-lobe.com An interactive diagram of the rat parahippocampal-hippocampal region
- Search Hippocampus on BrainNavigator via BrainNavigator
- Gyorgy Buzsaki (2010) Hippocampus. Scholarpedia. 6(1):1468.

Chapter 27

Amygdala

For other uses, see Amygdala (disambiguation).

The **amygdalae** (singular: **amygdala**; /əˈmɪɡdələ/; also

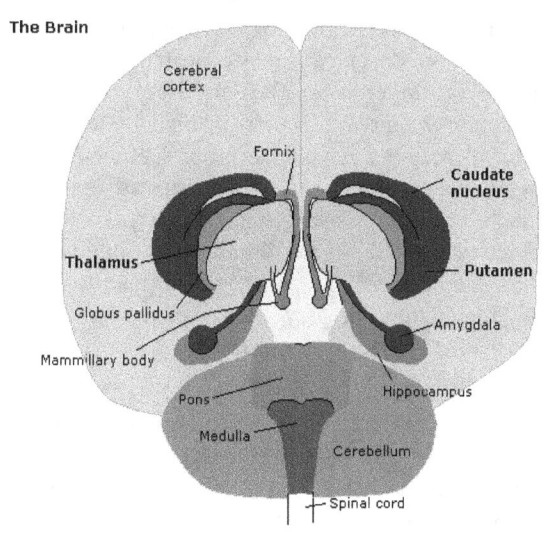
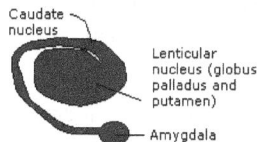

Human brain in the coronal orientation. Amygdalae are shown in dark red.

corpus amygdaloideum; Latin, from Greek ἀμυγδαλή, *amygdalē*, 'almond', 'tonsil'[1]) are two almond-shaped groups of nuclei located deep and medially within the temporal lobes of the brain in complex vertebrates, including humans.[2] Shown in research to perform a primary role in the processing of memory, decision-making, and emotional reactions, the amygdalae are considered part of the limbic system.[3]

27.1 Structure

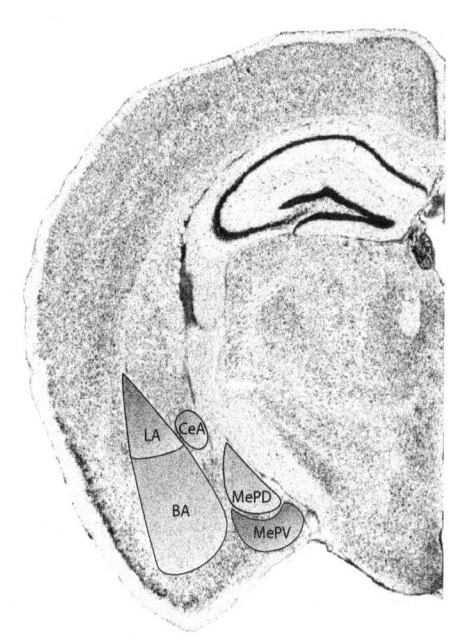

Subdivisions of the mouse amygdala

The regions described as amygdala nuclei encompass several structures with distinct connectional and functional characteristics in humans and other animals.[4] Among these nuclei are the basolateral complex, the cortical nucleus, the medial nucleus, the central nucleus, and the intercalated cell clusters (ITCs). The basolateral complex can be further subdivided into the lateral, the basal, and the accessory basal nuclei.[3][5][6]

Anatomically, the amygdala[7] and more particularly, its central and medial nuclei,[8] have sometimes been classified as a part of the basal ganglia.

27.1. STRUCTURE

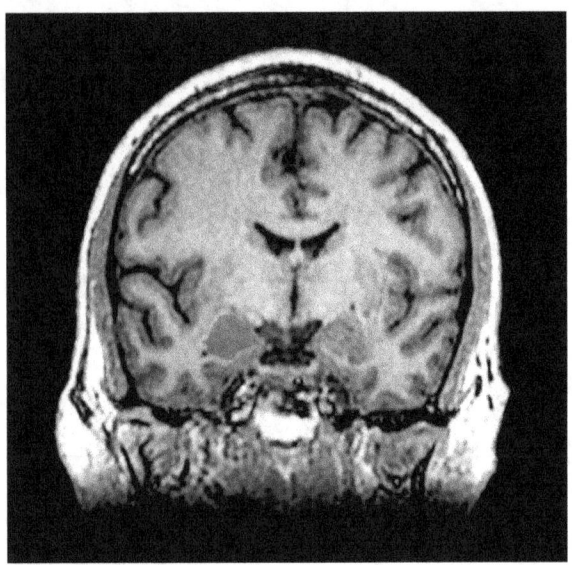

MRI coronal view of the right amygdala

27.1.1 Hemispheric specializations

There are functional differences between the right and left amygdala. In one study, electrical stimulations of the right amygdala induced negative emotions, especially fear and sadness. In contrast, stimulation of the left amygdala was able to induce either pleasant (happiness) or unpleasant (fear, anxiety, sadness) emotions.[9] Other evidence suggests that the left amygdala plays a role in the brain's reward system.[10]

Each side holds a specific function in how we perceive and process emotion. The right and left portions of the amygdala have independent memory systems, but work together to store, encode, and interpret emotion.

The right hemisphere is associated with negative emotion. It plays a role in the expression of fear and in the processing of fear-inducing stimuli. Fear conditioning, which is when a neutral stimulus acquires aversive properties, occurs within the right hemisphere. When an individual is presented with a conditioned, aversive stimulus, it is processed within the right amygdala, producing an unpleasant or fearful response. This emotional response conditions the individual to avoid fear-inducing stimuli.

The right hemisphere is also linked to declarative memory, which consists of facts and information from previously experienced events and must be consciously recalled. It also plays a significant role in the retention of episodic memory. Episodic memory consists of the autobiographical aspects of memory, permitting you to recall your personal emotional and sensory experience of an event. This type of memory does not require conscious recall. The right amygdala plays a role in the association of time and places with emotional properties.[11]

27.1.2 Amygdalar development

There is considerable growth within the first few years of structural development in both male and female amygdalae. Within this early period, female limbic structures grow at a more rapid pace than do males. Amongst female subjects, the amygdala reaches its full growth potential approximately 1.5 years before the peak of male development. The structural development of the male amygdala occurs over a longer period than in women. Despite the early development of female amygdalae, they reach their growth potential sooner than males, whose amygdalae continue to develop. The larger relative size of the male amygdala may be attributed to this extended developmental period.

In addition to longer periods of development, other neurological and hormonal factors may contribute to sex-specific developmental differences. The amygdala is rich in androgen receptors – nuclear receptors that bind to testosterone. Androgen receptors play a role in the DNA binding that regulates gene expression. Though testosterone is present within the female hormonal systems, women have lower levels of testosterone than men. The abundance of testosterone in the male hormonal system may contribute to development. In addition, the grey matter volume on the amygdala is predicted by testosterone levels, which may also contribute to the increased mass of the male amygdala.

In addition to sex differences, there are observable developmental differences between the right and left amygdala in both males and females. The left amygdala reaches its developmental peak approximately 1.5–2 years prior to the right amygdala. Despite the early growth of the left amygdala, the right increases in volume for a longer period of time. The right amygdala is associated with response to fearful stimuli as well as face recognition. It is inferred that the early development of the left amygdala functions to provide infants the ability to detect danger.[12]

In childhood, the amygdala is found to react differently to same-sex versus opposite-sex individuals. This reactivity decreases until a person enters adolescence, where it increases dramatically at puberty.[13]

27.1.3 Gender distinction

The amygdala is one of the best-understood brain regions with regard to differences between the sexes. The amygdala is larger in males than females in children ages 7–11,[14] in adult humans,[15] and in adult rats.[16]

In addition to size, other differences between men and women exist with regards to the amygdala. Subjects' amygdala activation was observed when watching a horror film and subliminal stimuli. The results of the study showed a different lateralization of the amygdala in men and women. Enhanced memory for the film was related to enhanced activity of the left, but not the right, amygdala in women, whereas it was related to enhanced activity of the right, but not the left, amygdala in men.[17] One study found evidence that on average, women tend to retain stronger memories for emotional events than men.[18]

The right amygdala is also linked with taking action as well as being linked to negative emotions,[19] which may help explain why males tend to respond to emotionally stressful stimuli physically. The left amygdala allows for the recall of details, but it also results in more thought rather than action in response to emotionally stressful stimuli, which may explain the absence of physical response in women.

27.2 Function

27.2.1 Connections

The amygdala sends projections to the hypothalamus, the dorsomedial thalamus, the thalamic reticular nucleus, the nuclei of the trigeminal nerve and the facial nerve, the ventral tegmental area, the locus coeruleus, and the laterodorsal tegmental nucleus.[5]

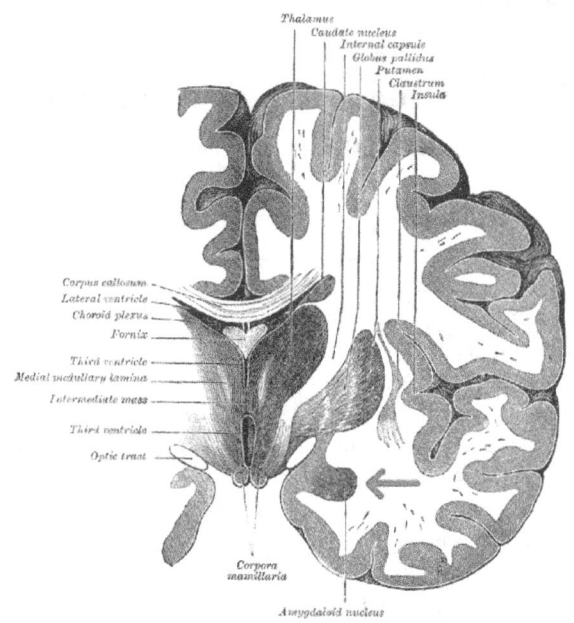

Coronal section of brain through intermediate mass of third ventricle. Amygdala is shown in purple.

The medial nucleus is involved in the sense of smell and pheromone-processing. It receives input from the olfactory bulb and olfactory cortex.[20] The lateral amygdalae, which send impulses to the rest of the basolateral complexes and to the centromedial nuclei, receive input from the sensory systems. The centromedial nuclei are the main outputs for the basolateral complexes, and are involved in emotional arousal in rats and cats.[5][6][21]

27.2.2 Emotional learning

In complex vertebrates, including humans, the amygdalae perform primary roles in the formation and storage of memories associated with emotional events. Research indicates that, during fear conditioning, sensory stimuli reach the basolateral complexes of the amygdalae, particularly the lateral nuclei, where they form associations with memories of the stimuli. The association between stimuli and the aversive events they predict may be mediated by long-term potentiation,[22][23] a sustained enhancement of signaling between affected neurons.[24] There have been studies that show that damage to the amygdala can interfere with memory that is strengthened by emotion. One study examined a patient with bilateral degeneration of the amygdala. He was told a violent story accompanied by matching pictures and was observed based on how much he could recall from the story. The patient had less recollection of the story than patients with functional amygdala, showing that the amygdala has a strong connection with emotional learning.[25]

Memories of emotional experiences imprinted in reactions of synapses in the lateral nuclei elicit fear behavior through neuronal connections with the central nucleus of the amygdalae and the bed nuclei of the stria terminalis (BNST). The axon terminals from sensory neurons form synapses with dendritic spines on neurons from the central nucleus.[26] The central nuclei are involved in the genesis of many fear responses such as defensive behavior (freezing or escape responses), autonomic nervous system responses (changes in blood pressure and heart rate/tachycardia), neuroendocrine responses (stress-hormone release), etc. Damage to the amygdalae impairs both the acquisition and expression of Pavlovian fear conditioning, a form of classical conditioning of emotional responses.[24]

The amygdalae are also involved in appetitive (positive) conditioning. It seems that distinct neurons respond to positive and negative stimuli, but there is no clustering of these distinct neurons into clear anatomical nuclei.[27][28] However, lesions of the central nucleus in the amygdala have been shown to reduce appetitive learning in rats. Lesions of the basolateral regions do not exhibit the same effect.[29] Research like this indicates that different nuclei within the amygdala have different functions in appetitive conditioning.[30][31]

27.2.3 Memory modulation

The amygdala is also involved in the modulation of memory consolidation. Following any learning event, the long-term memory for the event is not formed instantaneously. Rather, information regarding the event is slowly assimilated into long-term (potentially lifelong) storage over time, possibly via long-term potentiation. Recent studies suggest that the amygdala regulates memory consolidation in other brain regions. Also, fear conditioning, a type of memory that is impaired following amygdala damage, is mediated in part by long-term potentiation.[22][23]

During the consolidation period, the memory can be modulated. In particular, it appears that emotional arousal following the learning event influences the strength of the subsequent memory for that event. Greater emotional arousal following a learning event enhances a person's retention of that event. Experiments have shown that administration of stress hormones to mice immediately after they learn something enhances their retention when they are tested two days later.[32]

The amygdala, especially the basolateral nuclei, are involved in mediating the effects of emotional arousal on the strength of the memory for the event, as shown by many laboratories including that of James McGaugh. These laboratories have trained animals on a variety of learning tasks and found that drugs injected into the amygdala after training affect the animals' subsequent retention of the task. These tasks include basic classical conditioning tasks such as inhibitory avoidance, where a rat learns to associate a mild footshock with a particular compartment of an apparatus, and more complex tasks such as spatial or cued water maze, where a rat learns to swim to a platform to escape the water. If a drug that activates the amygdalae is injected into the amygdalae, the animals had better memory for the training in the task.[33] If a drug that inactivates the amygdalae is injected, the animals had impaired memory for the task.

Buddhist monks who do compassion meditation have been shown to modulate their amygdala, along with their temporoparietal junction and insula, during their practice.[34] In an fMRI study, more intensive insula activity was found in expert meditators than in novices.[35] Increased activity in the amygdala following compassion-oriented meditation may contribute to social connectedness.[36]

Amygdala activity at the time of encoding information correlates with retention for that information. However, this correlation depends on the relative "emotionalness" of the information. More emotionally arousing information increases amygdalar activity, and that activity correlates with retention. Amygdala neurons show various types of oscillation during emotional arousal, such as theta activity. These synchronized neuronal events could promote synaptic plasticity (which is involved in memory retention) by increasing interactions between neocortical storage sites and temporal lobe structures involved in declarative memory.[37]

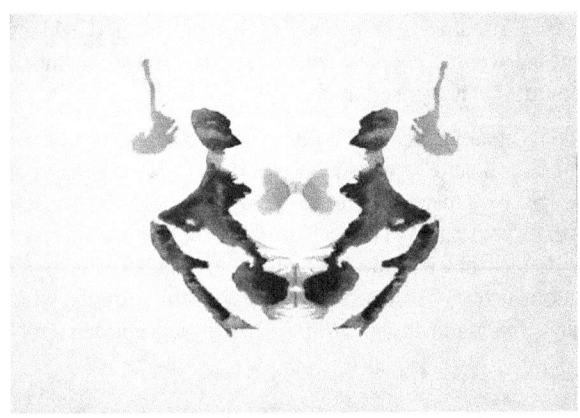

Rorschach test blot 03

Research using Rorschach test blot 03 finds that the number of unique responses to this random figure links to larger sized amygdalae. The researchers note, "Since previous reports have indicated that unique responses were observed at higher frequency in the artistic population than in the nonartistic normal population, this positive correlation suggests that amygdalar enlargement in the normal population might be related to creative mental activity."[38]

27.3 Neuropsychological correlates of amygdala activity

Early research on primates provided explanations as to the functions of the amygdala, as well as a basis for further research. As early as 1888, rhesus monkeys with a lesioned temporal cortex (including the amygdala) were observed to have significant social and emotional deficits.[39] Heinrich Klüver and Paul Bucy later expanded upon this same observation by showing that large lesions to the anterior temporal lobe produced noticeable changes, including overreaction to all objects, hypoemotionality, loss of fear, hypersexuality, and hyperorality, a condition in which inappropriate objects are placed in the mouth. Some monkeys also displayed an inability to recognize familiar objects and would approach animate and inanimate objects indiscriminately, exhibiting a loss of fear towards the experimenters. This behavioral disorder was later named Klüver-Bucy syndrome accordingly,[40] and later research proved it was specifically due to amygdala lesions. Monkey mothers who had amygdala damage showed a reduction in maternal behaviors towards their infants, often physically abusing or neglecting them.[41] In 1981, researchers found that selective radio frequency lesions of the whole amygdala caused Klüver-Bucy syndrome.[42]

With advances in neuroimaging technology such as MRI, neuroscientists have made significant findings concerning the amygdala in the human brain. A variety of data shows the amygdala has a substantial role in mental states, and is related to many psychological disorders. Some studies have shown children with anxiety disorders tend to have a smaller left amygdala. In the majority of the cases, there was an association between an increase in the size of the left amygdala with the use of SSRIs (antidepressant medication) or psychotherapy. The left amygdala has been linked to social anxiety, obsessive and compulsive disorders, and post traumatic stress, as well as more broadly to separation and general anxiety.[43] In a 2003 study, subjects with borderline personality disorder showed significantly greater left amygdala activity than normal control subjects. Some borderline patients even had difficulties classifying neutral faces or saw them as threatening.[44] Individuals with psychopathy show reduced autonomic responses, relative to comparison individuals, to instructed fear cues.[45] In 2006, researchers observed hyperactivity in the amygdala when patients were shown threatening faces or confronted with frightening situations. Patients with severe social phobia showed a correlation with increased response in the amygdala.[46] Similarly, depressed patients showed exaggerated left amygdala activity when interpreting emotions for all faces, and especially for fearful faces. Interestingly, this hyperactivity was normalized when patients were administered antidepressant medication.[47] By contrast, the amygdala has been observed to respond differently in people with bipolar disorder. A 2003 study found that adult and adolescent bipolar patients tended to have considerably smaller amygdala volumes and somewhat smaller hippocampal volumes.[48] Many studies have focused on the connections between the amygdala and autism.[49]

Studies in 2004 and 2006 showed that normal subjects exposed to images of frightened faces or faces of people from another race will show increased activity of the amygdala, even if that exposure is subliminal.[50][51] However, the amygdala is not necessary for the processing of fear-related stimuli, since persons in whom it is bilaterally damaged show rapid reactions to fearful faces, even in the absence of a functional amygdala.[52]

Recent research suggests that parasites, in particular toxoplasma, form cysts in the brain of rats, often taking up residence in the amygdala. This may provide clues as to how specific parasites may contribute to the development of disorders, including paranoia.[53]

Future studies have been proposed to address the role of the amygdala in positive emotions, and the ways in which the amygdala networks with other brain regions.[54]

27.3.1 Sexual orientation

Recent studies have suggested possible correlations between brain structure, including differences in hemispheric ratios and connection patterns in the amygdala, and sexual orientation. Homosexual men tend to exhibit more female-like patterns in the amygdala than heterosexual males do, just as homosexual females tend to show more male-like patterns in the amygdala than heterosexual women do. It was observed that amygdala connections were more widespread from the left amygdala in homosexual males, as is also found in heterosexual females. Amygdala connections were more widespread from the right amygdala in homosexual females, as in heterosexual males.[55][55][56]

27.3.2 Social interaction

Amygdala volume correlates positively with both the size (the number of contacts a person has) and the complexity (the number of different groups to which a person belongs) of social networks.[57][58] Individuals with larger amygdalae had larger and more complex social networks. They were also better able to make accurate social judgments about other persons' faces.[59] The amygdala's role in the analysis of social situations stems specifically from its ability to identify and process changes in facial features. It does not, however, process the direction of the gaze of the person being perceived.[60][61]

The amygdala is also thought to be a determinant of the level of a person's emotional intelligence. It is particularly hypothesized that larger amygdalae allow for greater emotional intelligence, enabling greater societal integration and cooperation with others.[62]

The amygdala processes reactions to violations concerning personal space. These reactions are absent in persons in whom the amygdala is damaged bilaterally.[63] Furthermore, the amygdala is found to be activated in fMRI when people observe that others are physically close to them, such as when a person being scanned knows that an experimenter is standing immediately next to the scanner, versus standing at a distance.[63]

27.3.3 Aggression

Animal studies have shown that stimulating the amygdala appears to increase both sexual and aggressive behavior. Likewise, studies using brain lesions have shown that harm to the amygdala may produce the opposite effect. Thus, it appears that this part of the brain may play a role in the display and modulation of aggression.[64]

27.3.4 Fear

There are cases of human patients with focal bilateral amygdala lesions, due to the rare genetic condition Urbach-Wiethe disease.[65][66] Such patients fail to exhibit fear-related behaviors, leading one, Patient S.M., to be dubbed the "woman with no fear". This finding reinforces the conclusion that the amygdala "plays a pivotal role in triggering a state of fear".[67]

27.3.5 Alcoholism and binge drinking

The amygdala appears to play a role in binge drinking, being damaged by repeated episodes of intoxication and withdrawal.[68] Alcoholism is associated with dampened activation in brain networks responsible for emotional processing, including the amygdala.[69] Protein kinase C-epsilon in the amygdala is important for regulating behavioral responses to morphine, ethanol, and controlling anxiety-like behavior. The protein is involved in controlling the function of other proteins and plays a role in development of the ability to consume a large amount of ethanol.[70][71]

27.3.6 Anxiety

There may also be a link between the amygdala and anxiety.[72] In particular, there is a higher prevalence of females that are affected by anxiety disorders. In an experiment, degu pups were removed from their mother but allowed to hear her call. In response, the males produced increased serotonin receptors in the amygdala but females lost them. This led to the males being less affected by the stressful situation.

The clusters of the amygdala are activated when an individual expresses feelings of fear or aggression. This occurs because the amygdala is the primary structure of the brain responsible for flight or fight response. Anxiety and panic attacks can occur when the amygdala senses environmental stressors that stimulate fight or flight response.

The amygdala is directly associated with conditioned fear. Conditioned fear is the framework used to explain the behavior produced when an originally neutral stimulus is consistently paired with a stimulus that evokes fear. The amygdala represents a core fear system in the human body, which is involved in the expression of conditioned fear. Fear is measured by changes in autonomic activity including increased heart rate, increased blood pressure, as well as in simple reflexes such as flinching or blinking.

The central nucleus of the amygdala has direct correlations to the hypothalamus and brainstem – areas directly related to fear and anxiety. This connection is evident from studies of animals that have undergone amygdalae removal. Such studies suggest that animals lacking an amygdala have less fear expression and indulge in non-species-like behavior. Many projection areas of the amygdala are critically involved in specific signs that are used to measure fear and anxiety.

Mammals have very similar ways of processing and responding to danger. Scientists have observed similar areas in the brain – specifically in the amygdala – lighting up or becoming more active when a mammal is threatened or beginning to experience anxiety. Similar parts of the brain are activated when rodents and when humans observe a dangerous situation, the amygdala playing a crucial role in this assessment. By observing the amygdala's functions, people can determine why one rodent may be much more anxious than another. There is a direct relationship between the activation of the amygdala and the level of anxiety the subject feels.

Feelings of anxiety start with a catalyst – an environmental stimulus that provokes stress. This can include various smells, sights, and internal feelings that result in anxiety. The amygdala reacts to this stimuli by preparing to either stand and fight or to turn and run. This response is triggered by the release of adrenaline into the bloodstream. Consequently, blood sugar rises, becoming immediately available to the muscles for quick energy. Shaking may occur in an attempt to return blood to the rest of the body. A better understanding of the amygdala and its various functions may lead to a new way of treating clinical anxiety.[73]

27.3.7 Posttraumatic stress disorder

There seems to be a connection with the amygdalae and how the brain processes posttraumatic stress disorder. Multiple studies have found that the amygdalae may be responsible for the emotional reactions of PTSD patients. One study in particular found that when PTSD patients are shown pictures of faces with fearful expressions, their amygdalae tended to have a higher activation than someone without PTSD.[74]

27.3.8 Bipolar disorder

Amygdala dysfunction during face emotion processing is well-documented in bipolar disorder. Individuals with bipolar disorder showed greater amygdala activity (especially the amygdala/medial-prefrontal-cortex circuit).[75]

[76]

27.3.9 Political orientation

Amygdala size has been correlated with cognitive styles with regard to political thinking. A study found that "greater liberalism was associated with increased gray matter volume in the anterior cingulate cortex, whereas greater conservatism was associated with increased volume of the right amygdala."[77]

27.4 See also

- Amygdala hijack
- BELBIC
- List of regions in the human brain
- Triune brain
- Intercalated cells of the amygdala

27.5 Further reading

- Amygdala Joseph E. LeDoux, Scholarpedia, 3(4):2698. doi:10.4249/scholarpedia.2698

27.6 References

[1] amygdala – Definitions from Dictionary.com

[2] University of Idaho College of Science (2004). "amygdala". Archived from the original on 31 March 2007. Retrieved 15 March 2007.

[3] Amunts K, Kedo O, Kindler M, Pieperhoff P, Mohlberg H, Shah N, Habel U, Schneider F, Zilles K (2005). "Cytoarchitectonic mapping of the human amygdala, hippocampal region and entorhinal cortex: intersubject variability and probability maps". *Anat Embryol (Berl)*. **210** (5–6): 343–52. doi:10.1007/s00429-005-0025-5. PMID 16208455.

[4] Bzdok D, Laird A, Zilles K, Fox PT, Eickhoff S.: An investigation of the structural, connectional and functional subspecialization in the human amygdala. Human Brain Mapping, 2012.

[5] Ben Best (2004). "The Amygdala and the Emotions". Archived from the original on 9 March 2007. Retrieved 15 March 2007.

[6] Solano-Castiella E, Anwander A, Lohmann G, Weiss M, Docherty C, Geyer S, Reimer E, Friederici AD, Turner R (2010). "Diffusion tensor imaging segments the human amygdala in vivo". *NeuroImage*. **49** (4): 2958–65. doi:10.1016/j.neuroimage.2009.11.027. PMID 19931398.

[7] See *Amygdala* in the BrainInfo database

[8] Larry W. Swanson; Gorica D. Petrovich (August 1998). "What is the amygdala?". *Trends in Neurosciences*. **21** (8): 323–331. doi:10.1016/S0166-2236(98)01265-X.

[9] Lanteaume, L.; et al. (Jun 2007). "Emotion induction after direct intracerebral stimulations of human amygdala". *Cerebral Cortex*. **17** (6): 1307–13. doi:10.1093/cercor/bhl041. PMID 16880223.

[10] Murray, Elizabeth A.; et al. (2009). "Amygdala function in positive reinforcement". *The Human Amygdala*. Guilford Press.

[11] Markowitsch, H. (1998). Differential contribution of right and left amygdala to affective information processing. IOS Press. 11(4), 233–244.

[12] Uematsu, A., Matsui, M., Tanaka C., Takahashi, T., Noguchi K., Suzuki M., Nishijo H. (2012). Developmental trajectories of amygdala and hippocampus from infancy to early adulthood in healthy individuals. PLOS One Journal. doi:10.1371/journal.pone.0046970

[13] Telzer, E. H., Flannery, J., Humphreys, K. L., Goff, B., Gabard-Durman, L., Gee, D. G., & Tottenham, N. (2015). 'The cooties effect': Amygdala reactivity to opposite- versus same-sex faces declines from childhood to adolescence. Journal Of Cognitive Neuroscience, 27(9), 1685-1696. doi:10.1162/jocn_a_00813

[14] Caviness, V. S.; Kennedy, D. N.; Richelme, C.; Rademacher, J.; Filipek, P. A. (1996). "The Human Brain Age 7–11 Years: A Volumetric Analysis Based on Magnetic Resonance Images". *Cerebral Cortex*. **6** (5): 726–36. doi:10.1093/cercor/6.5.726. PMID 8921207.

[15] Goldstein, J. M.; Seidman, LJ; Horton, NJ; Makris, N; Kennedy, DN; Caviness Jr, VS; Faraone, SV; Tsuang, MT (2001). "Normal Sexual Dimorphism of the Adult Human Brain Assessed by in Vivo Magnetic Resonance Imaging". *Cerebral Cortex*. **11** (6): 490–7. doi:10.1093/cercor/11.6.490. PMID 11375910.

[16] Hines, Melissa; Allen, Laura S.; Gorski, Roger A. (1992). "Sex differences in subregions of the medial nucleus of the amygdala and the bed nucleus of the stria terminalis of the rat". *Brain Research*. **579** (2): 321–6. doi:10.1016/0006-8993(92)90068-K. PMID 1352729.

[17] Cahill, L; Haier, RJ; White, NS; Fallon, J; Kilpatrick, L; Lawrence, C; Potkin, SG; Alkire, MT (2001). "Sex-Related Difference in Amygdala Activity during Emotionally Influenced Memory Storage". *Neurobiology of Learning and Memory*. **75** (1): 1–9. doi:10.1006/nlme.2000.3999. PMID 11124043.

27.6. REFERENCES

[18] Hamann, Stephan (2005). "Sex Differences in the Responses of the Human Amygdala". *Neuroscience*. **11** (4): 288–93. doi:10.1177/1073858404271981. PMID 16061516.

[19] Lanteaume, L.; Khalfa, S.; Régis, J.; Marquis, P.; Chauvel, P.; Bartolomei, F. (2006). "Emotion Induction After Direct Intracerebral Stimulations of Human Amygdala". *Cerebral Cortex*. **17** (6): 1307–13. doi:10.1093/cercor/bhl041. PMID 16880223.

[20] Carlson, Neil (12 January 2012). *Physiology of behavior*. Pearson. p. 336. ISBN 978-0205239399.

[21] Groshek, Frank; Kerfoot, Erin; McKenna, Vanessa; Polackwich, Alan S.; Gallagher, Michela; Holland, Peter C. (2005). "Amygdala Central Nucleus Function is Necessary for Learning, but Not Expression, of Conditioned Auditory Orienting". *Behavioral Neuroscience*. **119** (1): 202–12. doi:10.1037/0735-7044.119.1.202. PMC 1255918. PMID 15727525.

[22] Maren (Dec 1999). "Long-term potentiation in the amygdala: a mechanism for emotional learning and memory". *Trends Neurosci*. **22** (12): 561–7. doi:10.1016/S0166-2236(99)01465-4. PMID 10542437.

[23] Blair, H. T. (2001). "Synaptic Plasticity in the Lateral Amygdala: A Cellular Hypothesis of Fear Conditioning". *Learning & Memory*. **8** (5): 229–242. doi:10.1101/lm.30901.

[24] Ressler, Kerry; Davis, Michael (2003). "Genetics of Childhood Disorders: L. Learning and Memory, Part 3: Fear Conditioning". *Journal of the American Academy of Child & Adolescent Psychiatry*. **42** (5): 612–5. doi:10.1097/01.CHI.0000046835.90931.32. PMID 12707566.

[25] Carlson, Neil R. (12 January 2012). *Physiology of Behavior*. Pearson. p. 364. ISBN 978-0205239399.

[26] Carlson, Neil R. (12 January 2012). *Physiology of Behavior*. Pearson. p. 453. ISBN 978-0205239399.

[27] Paton, Joseph J.; Belova, Marina A.; Morrison, Sara E.; Salzman, C. Daniel (2006). "The primate amygdala represents the positive and negative value of visual stimuli during learning". *Nature*. **439** (7078): 865–70. doi:10.1038/nature04490. PMC 2396495. PMID 16482160.

[28] Redondo, RL; Kim, J; Arons, AL; Ramirez, S; Liu, X; Tonegawa, S (2014). "Bidirectional switch of the valence associated with a hippocampal contextual memory engram". *Nature*. **513**: 426–30. doi:10.1038/nature13725.

[29] Parkinson, John A.; Robbins, Trevor W.; Everitt, Barry J. (2000). "Dissociable roles of the central and basolateral amygdala in appetitive emotional learning". *European Journal of Neuroscience*. **12** (1): 405–13. doi:10.1046/j.1460-9568.2000.00960.x. PMID 10651899.

[30] See recent TINS article by Balleine and Killcross (2006)

[31] Killcross S, Robbins T, Everitt B (1997). "Different types of fear-conditioned behaviour mediated by separate nuclei within amygdala". *Nature*. **388** (6640): 377–80. doi:10.1038/41097. PMID 9237754.

[32] "Researchers Prove A Single Memory Is Processed In Three Separate Parts Of The Brain" http://www.sciencedaily.com/releases/2006/02/060202182107.htm

[33] Ferry B, Roozendaal B, McGaugh J (1999). "Role of norepinephrine in mediating stress hormone regulation of long-term memory storage: a critical involvement of the amygdala". *Biol Psychiatry*. **46** (9): 1140–52. doi:10.1016/S0006-3223(99)00157-2. PMID 10560021.

[34] "Cultivating compassion: Neuroscientific and behavioral approaches" a talk given by Richard J. Davidson found online at http://ccare.stanford.edu/node/25

[35] Lutz, Antoine; Brefczynski-Lewis, Julie; Johnstone, Tom; Davidson, Richard J. (2008). Baune, Bernhard, ed. "Regulation of the Neural Circuitry of Emotion by Compassion Meditation: Effects of Meditative Expertise". *PLoS ONE*. **3** (3): e1897. doi:10.1371/journal.pone.0001897. PMC 2267490. PMID 18365029.

[36] Hutcherson, Cendri A.; Seppala, Emma M.; Gross, James J. (2008). "Loving-kindness meditation increases social connectedness". *Emotion*. **8** (5): 720–4. doi:10.1037/a0013237. PMID 18837623.

[37] Paré D.; Collins D.R.; Pelletier J.G. (2002). "Amygdala oscillations and the consolidation of emotional memories". *Trends in Cognitive Sciences*. **6** (7): 306–314. doi:10.1016/S1364-6613(02)01924-1. PMID 12110364.

[38] Asari T, Konishi S, Jimura K, Chikazoe J, Nakamura N, Miyashita Y (2010). "Amygdalar enlargement associated with unique perception". *Cortex*. **46** (1): 94–99. doi:10.1016/j.cortex.2008.08.001. PMID 18922517.

[39] Brown, S.; Shafer, E. (1888). "An investigation into the functions of the occipital and temporal lobes of the monkey's brain". *Philosophical Transactions of the Royal Society B*. **179**: 303–327. doi:10.1098/rstb.1888.0011.

[40] Kluver, H.; Bucy, P. (1939). "Preliminary analysis of function of the temporal lobe in monkeys". *Archives of Neurology*. **42** (6): 979–1000. doi:10.1001/archneurpsyc.1939.02270240017001.

[41] Bucher, K.; Myersn, R.; Southwick, C. (1970). "Anterior temporal cortex and maternal behaviour in monkey". *Neurology*. **20** (4): 415. doi:10.1212/wnl.20.4.402. PMID 4998075.

[42] Aggleton, JP.; Passingham, RE. (1981). "Syndrome produced by lesions of the amygdala in monkeys (Macaca mulatta)". *Journal of Comparative and Physiological Psychology*. **95** (6): 961–977. doi:10.1037/h0077848. PMID 7320283.

[43] http://pn.psychiatryonline.org/content/40/9/37.full[][]

[44] Donegan, Nelson H; Sanislow, CA; Blumberg, HP; Fulbright, RK; Lacadie, C; Skudlarski, P; Gore, JC; Olson, IR; McGlashan, TH; et al. (2003). "Amygdala hyperreactivity in borderline personality disorder: implications for emotional dysregulation". *Biological Psychiatry.* **54** (11): 1284–1293. doi:10.1016/S0006-3223(03)00636-X. PMID 14643096.

[45] R. J. R. Blair (23 April 2008). "The amygdala and ventromedial prefrontal cortex: functional contributions and dysfunction in psychopathy". *Philosophical Transactions of the Royal Society B: Biological Sciences.* **363** (1503): 2557–2565. doi:10.1098/rstb.2008.0027. PMC 2606709. PMID 18434283.

[46] Studying Brain Activity Could Aid Diagnosis Of Social Phobia. Monash University. 19 January 2006.

[47] Sheline; Barch, DM; Donnelly, JM; Ollinger, JM; Snyder, AZ; Mintun, MA; et al. (2001). "Increased amygdala response to masked emotional faces in depressed subjects resolves with antidepressant treatment: an fMRI study". *Biological Psychiatry.* **50** (9): 651–658. doi:10.1016/S0006-3223(01)01263-X. PMID 11704071.

[48] Blumberg; Kaufman, J; Martin, A; Whiteman, R; Zhang, JH; Gore, JC; Charney, DS; Krystal, JH; Peterson, BS; et al. (2003). "Amygdala and hippocampal volumes in adolescents and adults with bipolar disorder". *Arch Gen Psychiatry.* **60** (12): 1201–8. doi:10.1001/archpsyc.60.12.1201. PMID 14662552.

[49] Schultz RT (2005). "Developmental deficits in social perception in autism: the role of the amygdala and fusiform face area". *Int J Dev Neurosci.* **23** (2–3): 125–41. doi:10.1016/j.ijdevneu.2004.12.012. PMID 15749240.

[50] Williams, Leanne M.; Belinda J. Liddell; Andrew H. Kemp; Richard A. Bryant; Russell A. Meares; Anthony S. Peduto; Evian Gordon (2006). "Amygdala-prefrontal dissociation of subliminal and supraliminal fear". *Human Brain Mapping.* **27** (8): 652–661. doi:10.1002/hbm.20208. PMID 16281289.

[51] Brain Activity Reflects Complexity Of Responses To Other-race Faces, *Science Daily*, 14 December 2004

[52] Tsuchiya N, Moradi F, Felsen C, Yamazaki M, Adolphs R (2009). "Intact rapid detection of fearful faces in the absence of the amygdala". *Nature Neuroscience.* **12** (10): 1224–12225. doi:10.1038/nn.2380. PMC 2756300. PMID 19718036.

[53] Vyas; Kim, SK; Giacomini, N; Boothroyd, JC; Sapolsky, RM; et al. (2007). "Behavioral changes induced by Toxoplasma infection of rodents are highly specific to aversion of cat odors". *Proceedings of the National Academy of Sciences of the United States of America.* **104** (15): 6442–7. doi:10.1073/pnas.0608310104. PMC 1851063. PMID 17404235.

[54] Gazzaniga, M.S., Ivry, R.B., & Mangun, G.R. (2009). Cognitive neuroscience: the biology of the mind. NY: W.W.Norton&Company.

[55] Swaab, D. F. (2008). "Sexual orientation and its basis in brain structure and function". *Proceedings of the National Academy of Sciences of the United States of America.* **105** (30): 10273–4. doi:10.1073/pnas.0805542105. PMC 2492513. PMID 18653758.

[56] Swaab, Dick F. (2007). "Sexual differentiation of the brain and behavior". *Best Practice & Research Clinical Endocrinology & Metabolism.* **21** (3): 431–44. doi:10.1016/j.beem.2007.04.003. PMID 17875490.

[57] Bickart, Kevin C; Wright, Christopher I; Dautoff, Rebecca J; Dickerson, Bradford C; Barrett, Lisa Feldman (2010). "Amygdala volume and social network size in humans". *Nature Neuroscience.* **14** (2): 163–4. doi:10.1038/nn.2724. PMC 3079404. PMID 21186358.

[58] Szalavitz, Maia (28 December 2010). "How to Win Friends: Have a Big Amygdala?". *Time*. Retrieved 30 December 2010.

[59] Bzdok, D.; Langner, R.; Caspers, S.; Kurth, F.; Habel, U.; Zilles, K.; Laird, A.; Eickhoff, Simon B. (2010). "ALE meta-analysis on facial judgments of trustworthiness and attractiveness". *Brain Structure and Function.* **215** (3–4): 209–23. doi:10.1007/s00429-010-0287-4. PMID 20978908.

[60] Mormann, F.; Niediek, J.; Tudusciuc, O.; Quesada, C. M.; Coenen, V. A.; Elger, C. E.; Adolphs, R. (2015). "Neurons in the human amygdala encode face identity, but not gaze direction". *Nature Neuroscience.* **18** (11): 1568–1570. doi:10.1038/nn.4139.

[61] Huijgen, J.; Dinkelacker, V.; Lachat, F.; Yahia-Cherif, L.; El Karoui, I.; Lemaréchal, J.; George, N. (2015). "Amygdala processing of social cues from faces: An intracrebral EEG study". *Social Cognitive And Affective Neuroscience.* **10** (11): 1568–1576.

[62] Buchanan, T.W., Tranel, D. & Adolphs, R. in The Human Amygdala (eds. Whalen, P.J. & Phelps, E.A.) 289–318 (Guilford, New York, 2009).

[63] Kennedy DP, Gläscher J, Tyszka JM, Adolphs R (2009). "Personal space regulation by the human amygdala". *Nature Neuroscience.* **12** (10): 1226–1227. doi:10.1038/nn.2381. PMC 2753689. PMID 19718035.

[64] T.L. Brink. (2008) Psychology: A Student Friendly Approach. "Unit 4: The Nervous System." pp 61

[65] Feinstein, Justin S.; Adolphs, Ralph; Damasio, Antonio; Tranel, Daniel (2011). "The Human Amygdala and the Induction and Experience of Fear". *Current Biology.* **21** (1): 34–8. doi:10.1016/j.cub.2010.11.042. PMC 3030206. PMID 21167712.

[66] Staut, C. C. V.; Naidich, T. P. (1998). "Urbach-Wiethe Disease(Lipoid Proteinosis)". *Pediatric Neurosurgery*. **28** (4): 212–214. doi:10.1159/000028653. PMID 9732251.

[67] http://bps-research-digest.blogspot.com/2013/02/extreme-fear-experienced-without.html[]

[68] Stephens, D. N; Duka, T. (2008). "Cognitive and emotional consequences of binge drinking: Role of amygdala and prefrontal cortex". *Philosophical Transactions of the Royal Society B*. **363** (1507): 3169–79. doi:10.1098/rstb.2008.0097. PMC 2607328. PMID 18640918.

[69] Marinkovic, Ksenija; Oscar-Berman, Marlene; Urban, Trinity; o'Reilly, Cara E.; Howard, Julie A.; Sawyer, Kayle; Harris, Gordon J. (2009). "Alcoholism and Dampened Temporal Limbic Activation to Emotional Faces". *Alcoholism: Clinical and Experimental Research*. **33** (11): 1880–92. doi:10.1111/j.1530-0277.2009.01026.x. PMC 3543694. PMID 19673745.

[70] Newton, P; Ron, D (2007). "Protein kinase C and alcohol addiction". *Pharmacological Research*. **55** (6): 570–7. doi:10.1016/j.phrs.2007.04.008. PMID 17566760.

[71] Lesscher, H. M. B.; Wallace, M. J.; Zeng, L.; Wang, V.; Deitchman, J. K.; McMahon, T.; Messing, R. O.; Newton, P. M. (2009). "Amygdala protein kinase C epsilon controls alcohol consumption". *Genes, Brain and Behavior*. **8** (5): 493–9. doi:10.1111/j.1601-183X.2009.00485.x. PMC 2714877. PMID 19243450.

[72] Ziabreva, Irina; Poeggel, Gerd; Schnabel, Reinhild; Braun, Katharina (2003). "Separation-induced receptor changes in the hippocampus and amygdala of Octodon degus: Influence of maternal vocalizations". *The Journal of Neuroscience*. **23** (12): 5329–36. PMID 12832558.

[73] Davis, M (1992). "The role of the amygdala in fear and anxiety". *Annual Review of Neuroscience*. **15**: 353–375. doi:10.1146/annurev.ne.15.030192.002033. PMID 1575447.

[74] Carlson, Neil R. (12 January 2012). *Physiology of Behavior*. Pearson. p. 608. ISBN 978-0205239399.

[75] Laura A; Thomas; et al. (2013). "Elevated amygdala responses to emotional faces in youths with chronic irritability or bipolar disorder.". *Neuroimage Clinical*. **2** (2): 637–645. doi:10.1016/j.nicl.2013.04.007. PMC 3746996. PMID 23977455.

[76] M. T. Keener; et al. (2012). "Dissociable patterns of medial prefrontal and amygdala activity to face identity versus emotion in bipolar disorder.". *Psychological Medicine*. **42** (9): 1913–1924. doi:10.1017/S0033291711002935. PMC 3685204. PMID 22273442.

[77] http://www.cell.com/current-biology/abstract/S0960-9822%2811%2900289-2

27.7 External links

- Media related to amygdala at Wikimedia Commons
- Stained brain slice images which include the "amygdala" at the BrainMaps project
- international committee for amygdala and health studies

Chapter 28

Pallium (neuroanatomy)

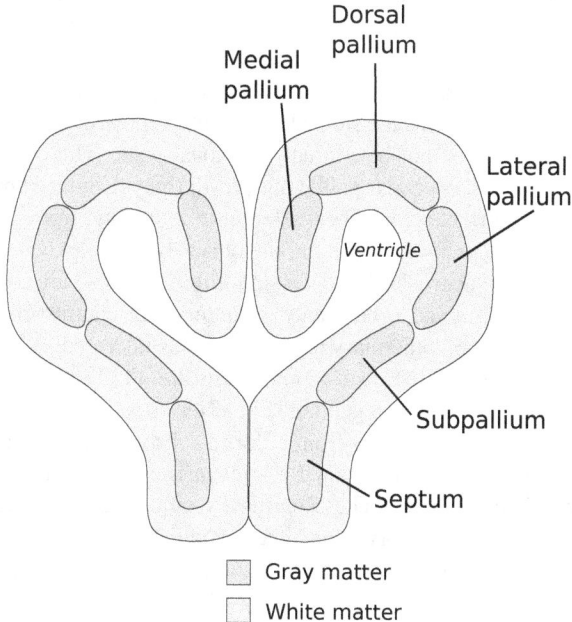

Schematic cross-section of the forebrain of a basal vertebrate such as a shark.

In neuroanatomy, **pallium** refers to the layers of gray and white matter that cover the upper surface of the cerebrum in vertebrates. The non-pallial part of the telencephalon builds the subpallium. In basal vertebrates the pallium is a relatively simple three-layered structure, encompassing 3-4 histogenetically distinct domains, plus the olfactory bulb. It used to be thought that pallium equals cortex and subpallium equals telencephalic nuclei, but it has turned out, according to comparative evidence provided by molecular markers, that the pallium develops both cortical structures (allocortex and isocortex) and pallial nuclei (claustroamygdaloid complex), whereas the subpallium develops striatal, pallidal, diagonal-innominate and preoptic nuclei, plus the corticoid structure of the olfactory tuberculum. In mammals, the cortical part of the pallium registers a definite evolutionary step-up in complexity, forming the cerebral cortex, most of which consists of a progressively expanded six-layered portion isocortex, with simpler three-layered cortical regions allocortex at the margins. The allocortex subdivides into hippocampal allocortex, medially, and olfactory allocortex, laterally (including rostrally the olfactory bulb and anterior olfactory areas).

28.1 Structure

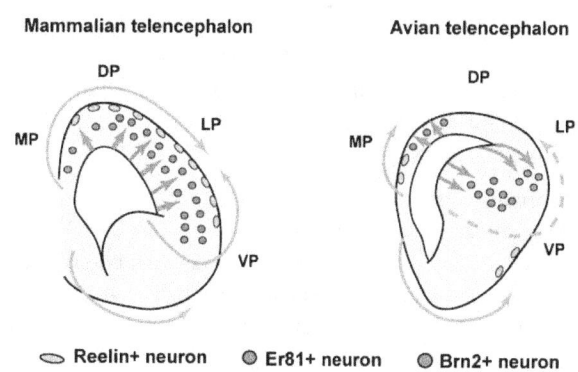

Schematic illustration of differences in neuronal specification and migration patterns between the mammalian and avian pallium

The general layout or body plan of the pallium is already clearly defined in animals with relatively simple brains, including lampreys, sharks and amphibians. In teleost fish, reptiles, birds, and mammals, the pallial architecture is greatly modified (sharply divergently in fish), with differential growth and specialization of diverse sectors of the conserved pallial Bauplan. In all vertebrate brains, the telencephalic forebrain consists of two hemispheres, joined at the midline by a region called the septum. The septum is continuous with the preoptic area across the plane defined by the anterior commissure; it is largely subpallial, but also contains a small pallial portion, where the hippocampal commissure forms, which is contiguous to the medial pallium. The telencephalic part of the rostral choroidal tela (roof plate continuous caudally with a diencephalic part) is inserted at the back of this commissure at a locus where mammals show the subfornical circumventricular organ, and extends laterally over the interventricular foramen into a wing-shaped medial telencephalic territory, the so-called choroidal fissure. Here the choroidal tissue is attached to the fimbria of the hippocampus (also known as the cortical hem area), bordering lengthwise the medial pallium. At its rostral and caudal ends, the medial pallium contacts the ventral pallium, which builds the pallial portion that contacts the subpallium across the pallio subpallial boundary, observed at the lateral telencephalic wall. Inside the ring formed thus by the medial and ventral pallium there is a sort of island that contains the dorsal and lateral pallial portions. In older literature the pallium used to be subdivided only into three zones, called the medial pallium, dorsal (or dorsolateral) pallium, and lateral pallium. The old lateral pallium encompassed the modern lateral and ventral parts of the pallium. The medial pallium is the progenitor of the mammalian hippocampus, and is thought to be involved in spatial cognitive mapping and memory formation across a broad range of species. The lateral and ventral pallium is the progenitor of the mammalian piriform cortex, and has an olfactory function in every species in which it has been studied. The evolutionary diversifications and specialization in functions of the dorsal pallium have been more difficult to decipher. It is widely believed to be the progenitor of the bulk of the mammalian cerebral cortex, although the evidence for this is considered by some anatomists not yet to be conclusive.

Importantly, the lateral and ventral parts of the pallium produce also deep to their respective sectors of subpial olfactory cortex sets of pallial nuclei, the neurons entering the claustrum, rostrally, and the pallial amygdala, caudally. The concept of hypopallium refers to this histogenetically unitary complex of olfactory (piriform) cortex and deep pallial nuclei. In reptiles and birds the hypopallium becomes differentially enlarged (largest in crocodiles and birds, whose olfactory cortex gets nevertheless reduced), whereas in mammals it becomes reduced to the claustroamygdaloid complex and relatively enlarged olfactory (prepiriform and piriform) cortex.

The pallial amygdala contains mainly the so-called basolateral amygdala, encompassing the lateral, basolateral (basal) and basomedial (accessory basal) nuclei, plus the anterior, amygdalopiriform and posterolateral corticoid areas at its surface. The medial pallium also may contribute to the pallial amygdala, forming the amygdalohippocampal nucleus and the posteromedial corticoid area. It has been postulated that the neurons forming the nucleus of the lateral olfactory tract derive from the dorsal pallium and migrate tangentially into its final position caudal to the olfactory tuberculum. Situated ventral to the pallium in the basic vertebrate forebrain plan (though representing a topologically rostral field in neural plate fate maps) is another region of telencephalic gray matter known as the subpallium, which is the progenitor area for the basal ganglia, a set of structures that play a crucial role in the executive control of behavior. The subpallium region has distinct striatal, pallidal, diagonal and preoptic subregions, which are stretched obliquely between the septal midline and the amygdala at the posterior pole of the telencephalon. At least the striatum, pallidum and diagonal domains extend into the amygdala, representing there the subpallial amygdala, forming its central and medial nucleis, as well as the amygdaloid end of the bed nucleus stria terminalis complex.

The amygdala thus encompasses an heterogeneous group of subpallial nuclei and hypopallial olfactory and amygdalohippocampal corticonuclear cell masses which are on the whole heavily involved in emotion and motivation. The pallial portions build the analytic or perceptual end of this complex, whereas the subpallial portions represent the corresponding output or efferent functional pole. The olfactory bulb is a peculiar pallial outgrowth (maybe induced by the primary olfactory fibers afferent to it, coming from the sensory neurons developed in the olfactory placode) whose projection neurons (the mitral and tufted neurons) are pallial in origin and accordingly excitatory. In contrast, the superfial periglomerulary neurons, various intermediate interneurons and the deep granule cells are all of subpallial origin and migrate tangentially out of the striatal part of the subpallium (apparently from a dorsal subsector of this domain) through the so-called rostral migratory stream into the olfactory bulb. These extremely numerous subpallial cells are all inhibitory. The olfactory bulb is thus singularly formed by a minority of autochthonous pallial neurons and a majority of immigrated inhibitory subpallial cells (it is nevertheless classified as a part of the ventral pallium). There is also a modified accessory olfactory bulb at the base of the principal one, which is associated specifically to incoming afferents from Jacobson's organ found at the nasal septum. The accessory olfactory pathway is maximally developed in some reptiles (e.g., snakes) and is lost in birds.

28.2 Evolution

The evolution of the dorsal pallium is not fully understood yet. Some authors hold that it largely contributes to the mammalian hippocampal allocortical and parahippocampal mesocortical (transitional) areas. Others postulate it directly transforms into the six-layered isocortex (neocortex) characteristic of mammals, and still others suppose that medial and lateral parts of the dorsal pallium contribute (perhaps with some contributions from the lateral pallium) to the alternative allocortical and isocortical fates.

28.3 In humans

The human pallium (*cloak* in Latin) envelops most of the telencephalon, due to extensive surface expansion of the isocortex. The telencephalic pallium has been described classically as having three parts: the archipallium, the paleopallium and the neopallium, but these concepts are now considered obsolete, having been substituted by the concept of medial pallium, dorsal pallium, lateral pallium and ventral pallium mentioned above under pallial Bauplan. It used to be said in anatomy textbooks that pallium equals cortex and subpallium equals telencephalic nuclei, but it has turned out, according to molecular markers, that the pallium develops both cortical structures (allocortex and isocortex) and pallial nuclei (claustroamygdaloid complex), whereas the subpallium develops striatal, pallidal, diagonal-innominate and preoptic nuclei, plus the corticoid structure of the olfactory tuberculum.

28.4 In amphibians and other anamniotes

In amphibians, the telencephalon distinctly shows medial, dorsal, lateral and ventral parts of the pallium, plus striatal, pallidal, diagonal and preoptic parts of the basal nuclei. However, the pallial portions do not show a visible lamination. They already have a mixture of glutamatergic (excitatory) and GABAergic (inhibitory) neurons, whereas the subpallium is largely populated by inhibitory neurons. This structure is very similar to that found generally in anamniotes, though cartilaginous fishes do show a layered arrangement of their pallial neurons.

28.5 In reptiles and birds

Reptiles developed a distinct three-layered structure of medial and dorsal portions of their pallium, a morphological schema referred to with the concept of allocortex. In contrast, the lateral and ventral pallium sectors of reptiles adopted hypopallial structure (superficial olfactory cortex, covering deep pallial nuclei). The hypopallial region is also known as the dorsal ventricular ridge, described as having anterior and posterior (amygdaloid) regions. Birds essentially show much increased cellularity, keeping within the reptilian morphological schema, which leads to the apparent disappearance of layering within its medial and dorsal pallial sectors. The olfactory cortex is much reduced, whereas the hypopallial or dorsal ventricular ridge nuclei increase significantly in size and relative differentiation.

28.6 See also

- Avian pallium
- Arcopallium
- Nidopallium

Chapter 29

Gyrus

For the video game, see Gyruss.
"Gyral" redirects here. For the album, see Gyral (album).
In neuroanatomy, a **gyrus** (pl. *gyri*) is a ridge on the

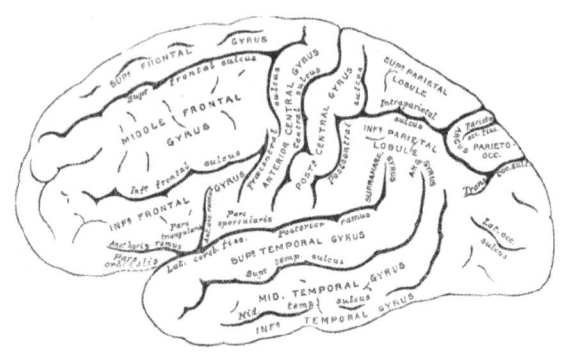

Gray's FIG. 726 – Lateral surface of left cerebral hemisphere, viewed from the side.

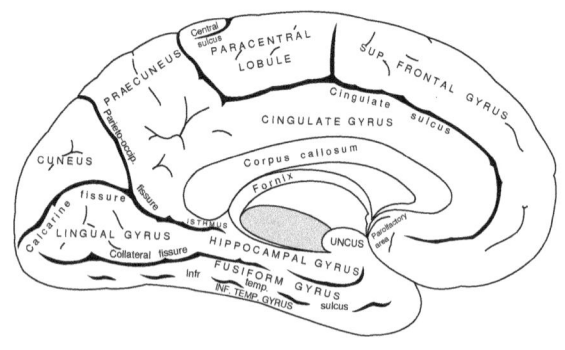

Gray's Fig. 727 – Medial surface of left cerebral hemisphere.

cerebral cortex. It is generally surrounded by one or more sulci (depressions or furrows; sg. *sulcus*).[1] Gyri and sulci create the folded appearance of the brain in humans and other mammals.

29.1 Structure

The gyri are part of a system of folds and ridges that create a larger surface area for the human brain and other mammalian brains.[2] Because the brain is confined to the skull, brain size is limited. Ridges and depressions create folds allowing a larger cortical surface area, and greater cognitive function, to exist in the confines of a smaller cranium.[3]

29.1.1 Development

The human brain undergoes gyrification during fetal and neonatal development. In embryonic development, all mammalian brains begin as smooth structures derived from the neural tube. A cerebral cortex without surface convolutions is lissencephalic, meaning 'smooth-brained'.[4] As development continues, gyri and sulci begin to take shape on the fetal brain, with deepening indentations and ridges developing on the surface of the cortex.[5]

29.2 Clinical significance

Changes in the structure of gyri in the cerebral cortex are associated with various diseases and disorders. Pachygyria, lissencephaly, and polymicrogyria are all the results of abnormal cell migration associated with a disorganized cellular architecture, failure to form six layers of cortical neurons (a four-layer cortex is common), and functional problems.[6] The abnormal formation is commonly associated with epilepsy and mental dysfunctions.[7]

Pachygyria (meaning "thick" or "fat" gyri) is a congenital malformation of the cerebral hemisphere, resulting in unusually thick gyri in the cerebral cortex.[8] Pachygyria is used to describe brain characteristics in association with several neuronal migration disorders; most commonly relating to lissencephaly.

Lissencephaly (*smooth brain*) is a rare congenital brain malformation caused by defective neuronal migration during the 12th to 24th weeks of fetal gestation resulting in a lack of development of gyri and sulci.[9]

Polymicrogyria (meaning "many small gyri") is a developmental malformation of the human brain characterized by excessive folding of the gyri and a thickening of the cerebral cortex,[10] It may be generalized, affecting the whole surface of the cerebral cortex or may be focal, affecting only parts of the surface.

29.3 Notable gyri

- Superior frontal gyrus, *lat.* gyrus frontalis superior
- Middle frontal gyrus, *lat.* gyrus frontalis medius
- Inferior frontal gyrus, *lat.* gyrus frontalis inferior with 3 parts: pars opercularis, pars triangularis, and pars orbitalis
- Superior temporal gyrus, *lat.* gyrus temporalis superior
- Middle temporal gyrus, *lat.* gyrus temporalis medius
- Inferior temporal gyrus, *lat.* gyrus temporalis inferior
- Fusiform gyrus, *lat.* gyrus occipitotemporalis medialis
- Parahippocampal gyrus, *lat.* gyrus parahippocampalis
- Transverse temporal gyrus
- Lingual gyrus *lat.* gyrus lingualis
- Precentral gyrus, *lat.* gyrus praecentralis
- Postcentral gyrus, *lat.* gyrus postcentralis
- Supramarginal gyrus, *lat.* gyrus supramarginalis
- Angular gyrus, *lat.* gyrus angularis
- Cingulate gyrus *lat.* gyrus cinguli
- Fornicate gyrus
- Cuneus
- Precuneus

29.4 References

[1] Deng, Fan; Jiang, Xi; Zhu, Dajiang; Zhang, Tuo; Li, Kaiming; Guo, Lei; Liu, Tianming (2013). "A functional model of cortical gyri and sulci". *Brain Structure and Function.* **219** (4): 1473–1491. doi:10.1007/s00429-013-0581-z. ISSN 1863-2653.

[2] Marieb, Elaine N.; Hoehn, Katja (2012). *Human Anatomy & Physiology* (9th ed.). Pearson. ISBN 0321852125.

[3] Cusack, Rhodri (April 2005). "The Intraparietal Sulcus and Perceptual Organization". *Journal of Cognitive Neuroscience.* **17** (4): 641–651. doi:10.1162/0898929053467541.

[4] Armstrong, E; Schleicher, A; Omran, H; Curtis, M; Zilles, K (1991). "The ontogeny of human gyrification.". *Cerebral cortex (New York, N.Y. : 1991).* **5** (1): 56–63. PMID 7719130.

[5] Rajagopalan, V; Scott, J; Habas, PA; Kim, K; Corbett-Detig, J; Rousseau, F; Barkovich, AJ; Glenn, OA; Studholme, C (23 February 2011). "Local tissue growth patterns underlying normal fetal human brain gyrification quantified in utero.". *The Journal of neuroscience : the official journal of the Society for Neuroscience.* **31** (8): 2878–87. doi:10.1523/jneurosci.5458-10.2011. PMID 21414909.

[6] Barkovich, A. J.; Guerrini, R.; Kuzniecky, R. I.; Jackson, G. D.; Dobyns, W. B. (2012). "A developmental and genetic classification for malformations of cortical development: update 2012". *Brain.* **135** (5): 1348–1369. doi:10.1093/brain/aws019. ISSN 0006-8950.

[7] Pang, Trudy; Atefy, Ramin; Sheen, Volney (2008). "Malformations of Cortical Development". *The Neurologist.* **14** (3): 181–191. doi:10.1097/NRL.0b013e31816606b9. ISSN 1074-7931.

[8] Guerrini R (2005). "Genetic malformations of the cerebral cortex and epilepsy". *Epilepsia.* 46 Suppl 1: 32–37. doi:10.1111/j.0013-9580.2005.461010.x. PMID 15816977.

[9] Dobyns WB (1987). "Developmental aspects of lissencephaly and the lissencephaly syndromes". *Birth Defects Orig. Artic. Ser.* **23** (1): 225–41. PMID 3472611.

[10] Chang, B; Walsh, CA; Apse, K; Bodell, A; Pagon, RA; Adam, TD; Bird, CR; Dolan, K; Fong, MP; Stephens, K (1993). "Polymicrogyria Overview". *GeneReviews.* PMID 20301504.

29.5 See also

- Gyrification
- Lissencephaly

29.5. SEE ALSO

- Sulcus
- Ulegyria

Chapter 30

Sulcus (neuroanatomy)

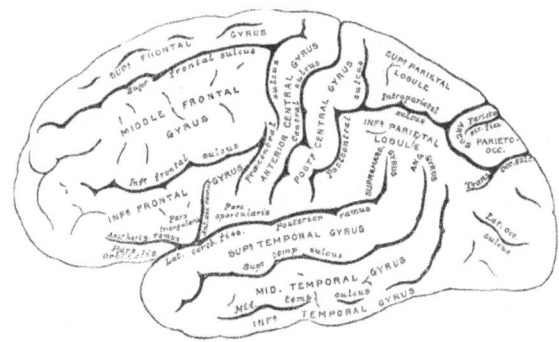

Gray's Fig. 726– Lateral surface of left cerebral hemisphere, viewed from the side.

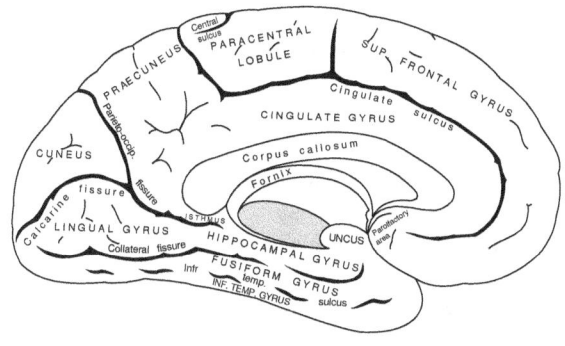

Gray's Fig. 727 - Medial surface of left cerebral hemisphere.

In neuroanatomy, a **sulcus** (Latin: "furrow", pl. *sulci*) is a depression or groove in the cerebral cortex. It surrounds a gyrus (pl. gyri), creating the characteristic folded appearance of the brain in humans and other mammals.

30.1 Structure

Sulci are one of three parts of the cerebral cortex, the others being the gyri and the fissures. The three different parts create a larger surface area for the human brain and other mammalian brains. When looking at the human brain, two-

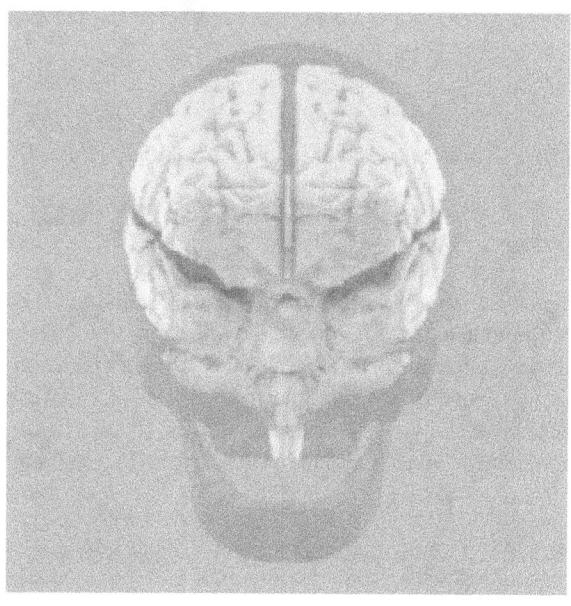

Rotating image of human brain, illustrating the Lateral sulcus

thirds of the surface are hidden in the grooves. The sulci and fissures are both grooves in the cortex but they are differentiated by size. A sulcus is a shallower groove that surrounds a gyrus. A fissure is a large furrow that divides the brain into lobes, and also into the two hemispheres as the medial longitudinal fissure does.[1]

30.1.1 Importance of expanded surface area

As the surface area of the brain increases more functions are made possible. A smooth-surfaced brain is only able to grow to a certain extent. A depression, sulcus, in the surface area allows for continued growth. This in turn allows for the functions of the brain to continue growing.[2]

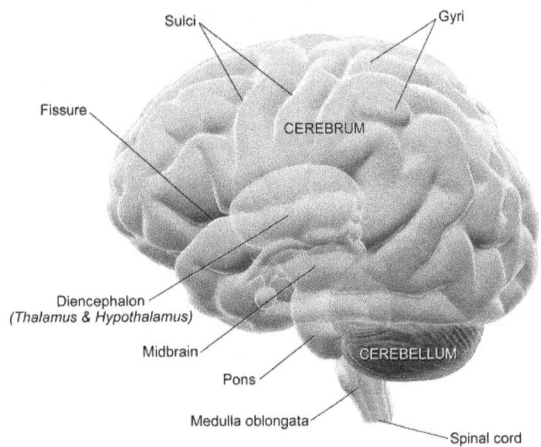

Illustration depicting general brain structures including sulci

30.1.2 Variation

The sulcal pattern varies between human individuals, and the most elaborate overview on this variation is probably an atlas by Ono, Kubick and Abernathey: *Atlas of the Cerebral Sulci*.[3] Some of the more prominent sulci are, however, seen across individuals - and even species - making a common nomenclature across individuals and species possible.

30.1.3 Development

In humans, cerebral convolutions appear at about 5 months and take at least into the first year after birth to fully develop.[4][5][6] Development varies greatly between individuals. The potential influences of genetic, epigenetic and environmental factors are not fully understood.[7] It has been found that the width of cortical sulci not only increases with age,[8] but also with cognitive decline in the elderly.[9]

30.2 Notable sulci

- Calcarine sulcus
- Central sulcus
- Central sulcus of insula
- Cingulate sulcus
- Circular sulcus of insula
- Collateral sulcus
- Fimbrodentate sulcus
- Hippocampal sulcus
- Inferior frontal sulcus
- Inferior temporal sulcus
- Intraparietal sulcus
- Lateral sulcus
- Lunate sulcus
- Occipitotemporal sulcus
- Olfactory sulcus
- Paracentral sulcus
- Parieto-occipital sulcus
- Postcentral sulcus
- Precentral sulcus
- Rhinal sulcus
- Subparietal sulcus
- Sulcus of corpus callosum
- Superior frontal sulcus
- Superior temporal sulcus
- Transverse occipital sulcus
- Transverse temporal sulcus

30.3 Other animals

The variation in the amount of fissures in the brain (gyrification) between species is related to the size of the animal and the size of the brain. Mammals that have smooth-surfaced or nonconvoluted brains are called lissencephalics and those that have folded or convoluted brains gyrencephalics.[4][5] The division between the two groups occurs when cortical surface area is about 10 cm^2 and the brain has a volume of 3–4 cm^3. Large rodents such as beavers (40 pounds (18 kg)) and capybaras (150 pounds (68 kg)) are gyrencephalic and smaller rodents such as rats and mice lissencephalic.[10]

30.3.1 Macaque

A macaque has a more simple sulcal pattern. In a monograph Bonin and Bailey list the following as the primary sulci:[11]

- Calcarine fissure (ca)
- Central sulcus (ce)
- Sulcus cinguli (ci)
- Hippocampal fissure (h)
- Sulcus intraparitalis (ip)
- Lateral fissure (or Sylvian fissure) (la)
- Sulcus olfactorius (olf)
- Medial parieto-occipital fissure (pom)
- fissura rhinalis (rh)
- Sulcus temporalis superior (ts) - this sulcus runs parallel to the lateral fissure and extends to the temporal pole and often superficially merges with it.

30.4 See also

This article uses anatomical terminology; for an overview, see Anatomical terminology.

- Sulcus (anatomy)

30.5 References

[1] Carlson, N. R. (2013). Physiology of Behavior. Upper Saddle River, NJ: Pearson Education Inc.

[2] Cusack, R. (2005). The intraparietal sulcus and perceptual organization. Journal of Cognitive Neuroscience, 17(4), 641-651. doi: 10.1162/0898929053467541

[3] Ono, Kubick, Abernathey, *Atlas of the Cerebral Sulci*, Thieme Medical Publishers, 1990. ISBN 0-86577-362-9. ISBN 3-13-732101-8.

[4] Hofman MA. (1985). Size and shape of the cerebral cortex in mammals. I. The cortical surface. Brain Behav Evol. 27(1):28-40. PMID 3836731

[5] Hofman MA. (1989).On the evolution and geometry of the brain in mammals. Prog Neurobiol.32(2):137-58. PMID 2645619

[6] Caviness VS Jr. (1975). Mechanical model of brain convolutional development. Science. 189(4196):18-21. PMID 1135626

[7] Dubois, J., & Benders, M. (2007). Mapping the early cortical folding process in preterm newborn brain. Oxford Journals, 18, 1444-1454. dpi: 10.1093/cercor/bhm180

[8] Tao Liu, Wei Wen, Wanlin Zhu, Julian Trollor, Simone Reppermund, John Crawford, Jesse S Jin, Suhuai Luo, Henry Brodaty, Perminder Sachdev (2010) The effects of age and sex on cortical sulci in the elderly. Neuroimage 51:1. 19–27 May. PMID 20156569

[9] Tao Liu, Wei Wen, Wanlin Zhu, Nicole A Kochan, Julian N Trollor, Simone Reppermund, Jesse S Jin, Suhuai Luo, Henry Brodaty, Perminder S Sachdev (2011) The relationship between cortical sulcal variability and cognitive performance in the elderly. Neuroimage 56:3. 865-873 Jun. PMID 21397704

[10] Martin I. Sereno, Roger B. H. Tootell, "From Monkeys to humans: what do we now know about brain homologies," *Current Opinion in Neurobiology* **15**:135-144, (2005).

[11] Gerhardt von Bonin, Percival Bailey, *The Neocortex of Macaca Mulatta*, The University of Illinois Press, Urbana, Illinois, 1947

30.6 External links

- Visual explanation of gyri, sulci, and fissures

Chapter 31

Development of the nervous system in humans

This article is about low-level development of anatomy. For development of brain functionality, see Cognitive development.

The study of **neural development in humans** draws on both neuroscience and developmental biology to describe the cellular and molecular mechanisms by which complex nervous systems emerge during embryonic development and throughout life.

Some landmarks of embryonic neural development include the birth and differentiation of neurons from stem cell precursors, the migration of immature neurons from their birthplaces in the embryo to their final positions, outgrowth of axons from neurons and guidance of the motile growth cone through the embryo towards postsynaptic partners, the generation of synapses between these axons and their postsynaptic partners, the neuron pruning that occurs in adolescence, and finally the lifelong changes in synapses which are thought to underlie learning and memory.

Typically, these neurodevelopmental processes can be broadly divided into two classes: activity-independent mechanisms and activity-dependent mechanisms. Activity-independent mechanisms are generally believed to occur as hardwired processes determined by genetic programs played out within individual neurons. These include differentiation, migration and axon guidance to their initial target areas. These processes are thought of as being independent of neural activity and sensory experience. Once axons reach their target areas, activity-dependent mechanisms come into play. Neural activity and sensory experience will mediate formation of new synapses, as well as synaptic plasticity, which will be responsible for refinement of the nascent neural circuits.

31.1 Embryonic stage

31.1.1 Neurulation

Main article: neurulation

See embryogenesis for understanding the animal development up to this stage.

Neurulation is the formation of the neural tube from the ectoderm of the embryo. It follows gastrulation in all vertebrates.

During gastrulation cells migrate to the interior of embryo, forming three germ layers— the endoderm (the deepest layer), mesoderm and ectoderm (the surface layer)—from which all tissues and organs will arise. In a simplified way, it can be said that the ectoderm gives rise to skin and nervous system, the endoderm to the guts and the mesoderm to the rest of the organs.

After gastrulation the notochord—a flexible, rod-shaped body that runs along the back of the embryo—has been formed from the mesoderm. During the third week of gestation the notochord sends signals to the overlying ectoderm, inducing it to become neuroectoderm. This results in a strip of neuronal stem cells that runs along the back of the embryo. This strip is called the neural plate, and is the origin of the entire nervous system. The neural plate folds outwards to form the neural groove. Beginning in the future neck region, the neural folds of this groove close to create the neural tube (this form of neurulation is called primary neurulation). The ventral (front) part of the neural tube is called the basal plate; the dorsal (rear) part is called the alar plate. The hollow interior is called the neural canal. By the end of the fourth week of gestation, the open ends of the neural tube (the **neuropores**) close off.[1]

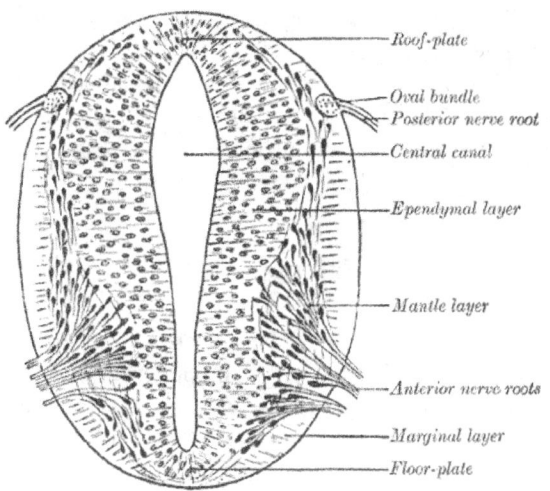

Cross-section of a developing spinal cord at four weeks.

31.1.2 Formation of the spinal cord

The spinal cord forms from the lower part of the neural tube. The wall of the neural tube consists of neuroepithelial cells, which differentiate into neuroblasts, forming the mantle layer (the gray matter). Nerve fibers emerge from these neuroblasts to form the marginal layer (the white matter).

The ventral part of the mantle layer (the basal plates) forms the motor areas of the spinal cord, whilst the dorsal part (the alar plates) forms the sensory areas. Between the basal and alar plates is an intermediate layer that contains neurons of the autonomic nervous system.[2]

31.1.3 Formation of the brain

Late in the fourth week, the superior part of the neural tube flexes at the level of the future midbrain— the mesencephalon. Above the mesencephalon is the prosencephalon (future forebrain) and beneath it is the rhombencephalon (future hindbrain). The optical vesicle (which will eventually become the optic nerve, retina and iris) forms at the basal plate of the prosencephalon.

In the fifth week, the alar plate of the prosencephalon expands to form the cerebral hemispheres (the telencephalon). The basal plate becomes the diencephalon.

The diencephalon, mesencephalon and rhombencephalon constitute the brain stem of the embryo. It continues to flex at the mesencephalon. The rhombencephalon folds posteriorly, which causes its alar plate to flare and form the fourth ventricle of the brain. The pons and the cerebellum form in the upper part of the rhombencephalon, whilst the medulla oblongata forms in the lower part.

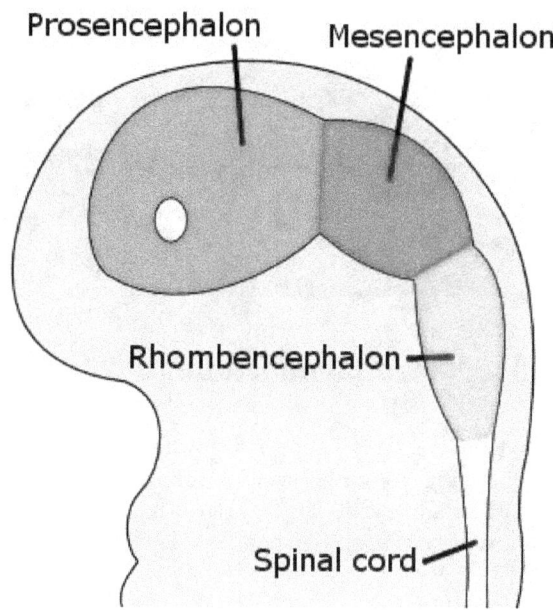

The embryo's brain at four weeks.

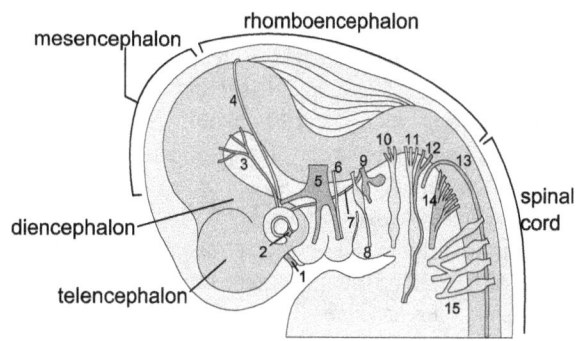

The embryo's nervous system at six weeks.

31.1.4 Evolution of Nervous System

About 400 million years ago, iodine, T4 (and PUFAs) stimulated the evolution of the nervous system in amphibians transforming the aquatic, vegetarian tadpole into the terrestrial, carnivorous frog with better neurological abilities for hunting. Contrary, hypothyroidism in mammals causes cretinism and in adults a neurological regression, as a general slowdown of nervous reflexes with lethargic cerebration, metabolism, digestion, heart rate, hypothermia.[3] [4]

31.2 Human brain development

See also: Human brain development timeline

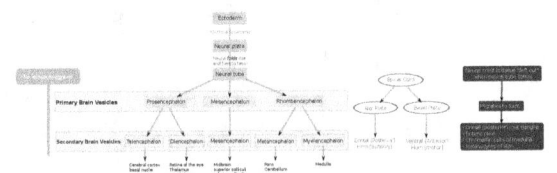

Highly schematic flowchart of human brain development.

31.3 Neuronal migration

Neuronal migration is the method by which neurons travel from their origin or birthplace to their final position in the brain. There are several ways they can do this, e.g. by radial migration or tangential migration.

31.3.1 Radial migration

Neuronal precursor cells proliferate in the ventricular zone of the developing neocortex. The first postmitotic cells to migrate from the preplate which are destined to become Cajal-Retzius cells and subplate neurons. These cells do so by somal translocation. Neurons migrating with this mode of locomotion are bipolar and attach the leading edge of the process to the pia. The soma is then transported to the pial surface by nucleokenisis, a process by which a microtubule "cage" around the nucleus elongates and contracts in association with the centrosome to guide the nucleus to its final destination.[5] Radial fibres (also known as radial glia) can translocate to the cortical plate and differentiate either into astrocytes or neurons.[6] Somal translocation can occur at any time during development.[7]

Subsequent waves of neurons split the preplate by migrating along radial glial fibres to form the cortical plate. Each wave of migrating cells travel past their predecessors forming layers in an inside-out manner, meaning that the youngest neurons are the closest to the surface.[8][9] It is estimated that glial guided migration represents 80-90% of migrating neurons.

31.3.2 Axophilic migration

Many neurons migrating along the anterior-posterior axis of the body use existing axon tracts to migrate along; this is called axophilic migration. An example of this mode of migration is in GnRH-expressing neurons, which make a long journey from their birthplace in the nose, through the forebrain, and into the hypothalamus.[10] Many of the mechanisms of this migration have been worked out, starting with the extracellular guidance cues[11] that trigger intracellular signaling. These intracellular signals, such as calcium signaling, lead to actin [12] and microtubule [13] cytoskeletal dynamics, which produce cellular forces that interact with the extracellular environment through cell adhesion proteins [14] to cause the movement of these cells.

31.3.3 Tangential migration

Most interneurons migrate tangentially through multiple modes of migration to reach their appropriate location in the cortex. An example of tangential migration is the movement of Cajal-Retzius cells from the cortical hem to the superficial part of cortical neuroepithelium.

31.3.4 Others

There is also a method of neuronal migration called **multipolar migration**.[15][16] This is seen in multipolar cells, which are abundantly present in the cortical intermediate zone. They do not resemble the cells migrating by locomotion or somal translocation. Instead these multipolar cells express neuronal markers and extend multiple thin processes in various directions independently of the radial glial fibers.[17]

31.4 Neurotrophic factors

Neurotrophic factors are molecules which promote and regulate neuronal survival in the developing nervous system. They are distinguished from ubiquitous metabolites necessary for cellular maintenance and growth by their specificity; each neurotrophic factor promotes the survival of only certain kinds of neurons during a particular stage of their development. In addition, it had been argued that neurotropihic factors are involved in many other aspects of neuronal development ranging from axonal guidance to regulation of neurotransmitter synthesis. [18]

31.5 Adult neurogenesis

Main article: Adult neurogenesis

Contrary to popular belief, neurogenesis also occurs in specific parts of the adult brain.

31.6 Adult neural development

Main article: Neuroregeneration

Neurodevelopment in the adult nervous system includes mechanisms such as remyelination, generation of new neurons, glia, axons, myelin or synapses. Neuroregeneration differs between the peripheral nervous system (PNS) and the central nervous system (CNS) by the functional mechanisms and especially, the extent and speed.

31.7 See also

- Time lapse sequences of radial migration (also known as glial guidance) and somal translocation.[7]
- Axon guidance
- Neural Darwinism
- Pre- and perinatal psychology
- Neural development

31.8 References

[1] Estomih Mtui; Gregory Gruener (2006). *Clinical Neuroanatomy and Neuroscience*. Philadelphia: Saunders. p. 1. ISBN 1-4160-3445-5.

[2] Atlas of Human Embryology, Chronolab. Last accessed on Oct 30, 2007.

[3] Venturi, Sebastiano (2011). "Evolutionary Significance of Iodine". *Current Chemical Biology-*. **5** (3): 155–162. doi:10.2174/187231311796765012. ISSN 1872-3136.

[4] Venturi, Sebastiano (2014). "Iodine, PUFAs and Iodolipids in Health and Disease: An Evolutionary Perspective". *Human Evolution-*. 29 (1-3): 185–205. ISSN 0393-9375.

[5] Samuels B, Tsai L (2004). "Nucleokinesis illuminated". *Nat Neurosci.* **7** (11): 1169–70. doi:10.1038/nn1104-1169. PMID 15508010.

[6] Campbell K, Götz M (May 2002). "Radial glia: multi-purpose cells for vertebrate brain development". *Trends Neurosci.* **25** (5): 235–8. doi:10.1016/S0166-2236(02)02156-2. PMID 11972958.

[7] Nadarajah B, Brunstrom J, Grutzendler J, Wong R, Pearlman A (2001). "Two modes of radial migration in early development of the cerebral cortex". *Nat Neurosci.* **4** (2): 143–50. doi:10.1038/83967. PMID 11175874.

[8] Nadarajah B, Parnavelas J (2002). "Modes of neuronal migration in the developing cerebral cortex". *Nat Rev Neurosci.* **3** (6): 423–32. doi:10.1038/nrn845. PMID 12042877.

[9] Rakic P (1972). "Mode of cell migration to the superficial layers of fetal monkey neocortex". *J Comp Neurol.* **145** (1): 61–83. doi:10.1002/cne.901450105. PMID 4624784.

[10] Wray S (2010). "From nose to brain: development of gonadotrophin-releasing hormone-1 neurones.". *J Neuroendocrinol.* **22** (7): 743–753. doi:10.1111/j.1365-2826.2010.02034.x. PMC 2919238. PMID 20646175.

[11] Giacobini P, Messina A, Wray S, Giampietro C, Crepaldi T, Carmeliet P, Fasolo A (2007). "Hepatocyte growth factor acts as a motogen and guidance signal for gonadotropin hormone-releasing hormone-1 neuronal migration.". *J Neurosci.* **27** (2): 431–445. doi:10.1523/JNEUROSCI.4979-06.2007. PMID 17215404.

[12] Hutchins BI, Klenke U, Wray S (2013). "Calcium release-dependent actin flow in the leading process mediates axophilic migration.". *J Neurosci.* **33** (28): 11361–71. doi:10.1523/JNEUROSCI.3758-12.2013. PMC 3724331. PMID 23843509.

[13] Hutchins, B. Ian; Wray, Susan (2014). "Capture of microtubule plus-ends at the actin cortex promotes axophilic neuronal migration by enhancing microtubule tension in the leading process.". *Frontiers in Cellular Neuroscience.* **8**: 400. doi:10.3389/fncel.2014.00400. PMC 4245908. PMID 25505874.

[14] Parkash J, Cimino I, Ferraris N, Casoni F, Wray S, Cappy H, Prevot V, Giacobini P (2012). "Suppression of β1-integrin in gonadotropin-releasing hormone cells disrupts migration and axonal extension resulting in severe reproductive alterations.". *J Neurosci.* **32** (47): 16992–7002. doi:10.1523/JNEUROSCI.3057-12.2012. PMID 23175850.

[15] Tabata H, Nakajima K (5 November 2003). "Multipolar migration: the third mode of radial neuronal migration in the developing cerebral cortex". *J Neurosci.* **23** (31): 9996–10001. PMID 14602813.

[16] Nadarajah B, Alifragis P, Wong R, Parnavelas J (2003). "Neuronal migration in the developing cerebral cortex: observations based on real-time imaging". *Cereb Cortex.* **13** (6): 607–11. doi:10.1093/cercor/13.6.607. PMID 12764035.

[17] Tabata H, Nakajima K (5 November 2003). "Multipolar migration: the third mode of radial neuronal migration in the developing cerebral cortex". *J Neurosci.* **23** (31): 9996–10001. PMID 14602813.

[18] Alan M. Davies (1 May 1988)"Trends In Genetics", Volume 4-Issue 5; Department of Anatomy, St George's Hospital Medical School, Cranmer Terrace, Tooting, London SW17 0RE, UK

Chapter 32

Lateralization of brain function

This article is about specialization of function between the left and right hemispheres of the brain. For specialization of brain function generally, see Functional specialization (brain).

The **lateralization of brain function** refers to how some

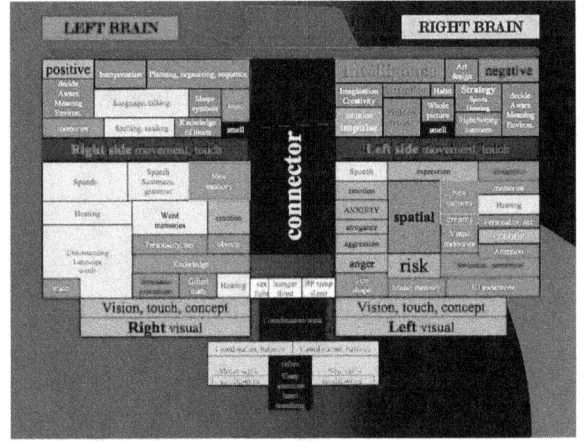

Left and Right Brain

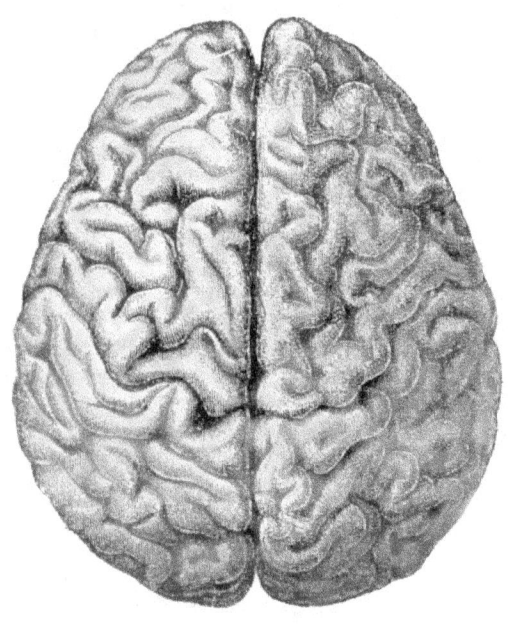

The human brain is divided into two hemispheres–left and right. Scientists continue to explore how some cognitive functions tend to be dominated by one side or the other; that is, how they are lateralized.

neural functions, or cognitive processes tend to be more dominant in one hemisphere than the other. The medial longitudinal fissure separates the human brain into two distinct cerebral hemispheres, connected by the corpus callosum. Although the macrostructure of the two hemispheres appears to be almost identical, different composition of neuronal networks allows for specialized function that is different in each hemisphere.[1]

Lateralization of brain structures is based on general trends expressed in healthy patients; however, there are numerous counterexamples to each generalization. Each human's brain develops differently leading to unique lateralization in individuals. This is different to specialization as lateralization refers only to the function of one structure divided between two hemispheres. Specialization is much easier to observe as a trend since it has a stronger anthropological history.[2] The best example of an established lateralization is that of Broca's and Wernicke's Areas where both are often found exclusively on the left hemisphere. These areas frequently correspond to handedness however, meaning the localization of these areas is regularly found on the hemisphere corresponding to the dominant hand. Function lateralization such as semantics, intonation, accentuation, prosody, etc. has since been called into question and largely been found to have a neuronal basis in both hemispheres.[3][4]

32.1 Interaction and Role

32.1.1 Theme

To get a basic understanding of this complex issue it is easiest to first consider the left (LHS) and right (RHS) hemispheres as distinct but interacting entities. These interactions come in the form of both excitatory and inhibitory signals crossing the corpus callosum and other hemispheric bridges.[5][6] As LHS and RHS each have unique interpretations of situations these signals allow for discussion and an ultimate decision to be made.[7] This interaction is called hemispheric rivalry.[8] This term is apt as both hemispheres are in conflict yet on the same team. In humans the reliance on both hemispheres is the basis of a number of functions including consciousness.[9]

The LHS can be simplified to better understand its role in this rivalry. The LHS is centered around action and is often the driving force behind risky behaviors. This hemisphere heavily relies upon emotional input leading it to make brash and uncalculated decisions. These decisions should not be thought of as ill-conceived, rather illogical and raw.[10]

Similarly the RHS can be brought into understanding by discussing a simple model of its role in the hemispheric interactions. The RHS can be thought of as the opposite of LHS as it relies primarily on critical thinking and calculations to reach its decisions.[11] As such the conclusions reached by the RHS often result in avoidance of risk taking behaviors and overall inaction.[12]

When viewed as action (LHS) and inaction (RHS) the anthropological development of these structures becomes elementary and logical. In environments of scarcity, like those faced by non-human animals, taking risks is the foundational approach to survival. In scarcity it is far more likely to die of starvation than to damaging stimuli from hostile animals or situations. However, in environments of abundance, as humans have observed, it is far more likely to die to damaging stimuli than of starvation. As such the anthropological development of first LHS then much later RHS is understandable.[13]

This also brings clarity to differences observed in modern human brains between environments. In areas of prosperity, where warmth, food, and basic needs for survival are abundant RHS domination is prevalent. Unsurprisingly, in areas of scarcity where cold and limited food are concerns LHS domination is prevalent. This phenomenon has been recorded numerous times when examining LHS dominant cultures, such as those of the Arctic, to RHS dominant cultures, like Africa.[14] Similarly, studies of animals such as the Chickadee have shown similar results. Yukon born chickadees have LHS dominance as compared to Texas born chickadees with RHS dominance. In exchanging Yukon borns for Texas borns it was shown that the action reliant Yukon birds consistently became the alpha of their new environment whereas inaction reliant Texan birds died shortly after their arrival in the Yukon. These studies explain the mass deaths incurred upon large migrations throughout history as people of prosperity attempted and largely failed to survive in scarcity.[15]

When speaking of dominance it is important to recognize that each hemisphere continues to function semi-independently but their interactions become dominated by one side. That is, each hemisphere always provides its input to the decision making process but is drowned out by the other. This occurs as individual decisions are made that biologically alter the state of the brain, changing the weight each hemisphere carries in their rivalry. As a choice of activity or inactivity is made it influences how effectively one hemisphere can inhibit the other and simultaneously teaches the now less effective inhibiting hemisphere to provide more excitatory signals with more frequency. For example: if a child eats a cookie their LHS has successfully inhibited their RHS changing the power dynamic as to make LHS ever so slightly more dominant and RHS more submissive in their rivalry. The reverse is also true should the child not eat the cookie.[16][17]

A lifetime of these anatomically changing decisions, paired with environmental circumstances as discussed above, dictate the structure and plasticity of the human brain. In cases such as common suburban living the LHS has less distinct neural networks and appears significantly blander than the RHS. The opposite asymmetry is observed in individuals such as violent offenders whose LHS is more distinct and pronounced than the RHS. The highest degree of symmetry between the hemispheres has been studied in veteran gang members. These individuals display an amazing propensity to act in extremely risky ways, yet inhibit themselves in the face of provocation to survive.[18][19][20]

32.1.2 Specifics

In gaining a fuller understanding of the interaction and function of cerebral hemispheres one must trace neural networks. The basis of the correlation of LHS with action and emotion is its connectivity with specialized parts throughout both hemispheres that play a role in those behaviors. Most notably LHS has been shown to have integral connections to insula and amygdala. Similarly, the association of RHS with inaction and calculation is tied to its extensive networks connecting to the anterior cingulate, orbitofrontal and prefrontal cortices.[21] These specialized areas are implicated in the calculative and inhibitory processes of inaction and those related to delayed gratification.[22][23] In studies looking at individuals with damaged LHS invariably the temporal lobe, insula and amygdala also sustained connectivity damage.[24] Limiting the connectivity of such areas seriously limits the effectiveness by which they operate. This means that although logistically the LHS is not the 'emotional' hemisphere its connectivity with emotion related areas is crucial to their functioning.[25] This same process implicates the RHS as the 'calculative' hemisphere, though logistically it is not.[26][27]

The role of hemispheric rivalry has become a major talking point for those studying the generation of consciousness within the brain. Some believe the brain's state of conflict is integrally linked to intelligence and genuine free will.[28]

32.2 Failures of lateralization

Lateralization as a concept fails because the brain consistently updates, consolidates, shifts information between the hemispheres.[29] In patients with damage to the LHS, functions traditionally associated with LHS were found predominantly in the RHS, most notably semantic choice.[30] These results hold for those with RHS damage who show a similar amount of functionality in specific motor skills via new neural connections developed in the LHS after RHS damage.[31] Amazingly these shifts continue to maximize brain potential even after extensive damage.[32] In studies where patients incur LHS damage and functionality is shifted to RHS, damage of the same magnitude to the RHS causes RHS functionality to become split once more. This effect has been shown on a neuronal level as the stimulation of specific neurons in the RHS causes similar responses in both hemispheres as neuron clusters on each side enlarge to compensate for the initial stimulation.[33] This points to a sort of equilibrium observed by the hemispheres.[34]

The shift of info and neuron functionality between hemispheres should not be surprising, however, as it has been observed in individuals who have lost a sense. In these individuals, neurons, even those that are specialized, are semi-repurposed to compensate for the loss. For example: those who become blind after years of vision are able to repurpose specialized sections of their brain to have increased visualization and nonspecialization to aid in the efficiency of their other senses.

32.3 History of research on lateralization

32.3.1 Broca

One of the first indications of brain function lateralization resulted from the research of French physician Pierre Paul Broca, in 1861. His research involved the male patient nicknamed "Tan", who suffered a speech deficit (aphasia); "tan" was one of the few words he could articulate, hence his nickname. In Tan's autopsy, Broca determined he had a syphilitic lesion in the left cerebral hemisphere. This left frontal lobe brain area (Broca's area) is an important speech production region. The motor aspects of speech production deficits caused by damage to Broca's area are known as expressive aphasia. In clinical assessment of this aphasia, it is noted that the patient cannot clearly articulate the language being employed.

32.3.2 Wernicke

German physician Karl Wernicke continued in the vein of Broca's research by studying language deficits unlike expressive aphasia. Wernicke noted that not every deficit was in speech production; some were linguistic. He found that damage to the left posterior, superior temporal gyrus (Wernicke's area) caused language comprehension deficits rather than speech production deficits, a syndrome known as receptive aphasia.

32.3.3 Advance in imaging technique

These seminal works on hemispheric specialization were done on patients or postmortem brains, raising questions about the potential impact of pathology on the research findings. New methods permit the *in vivo* comparison of the hemispheres in healthy subjects. Particularly, magnetic resonance imaging (MRI) and positron emission tomography (PET) are important because of their high spatial resolution and ability to image subcortical brain structures.

32.3.4 Movement and sensation

In the 1940s, neurosurgeon Wilder Penfield and his neurologist colleague Herbert Jasper developed a technique of brain mapping to help reduce side effects caused by surgery to treat epilepsy. They stimulated motor and somatosensory cortices of the brain with small electrical currents to activate discrete brain regions. They found that stimulation of one hemisphere's motor cortex produces muscle contraction on the opposite side of the body. Furthermore, the functional map of the motor and sensory cortices is fairly consistent from person to person; Penfield and Jasper's famous pictures of the motor and sensory homunculi were the result.

32.3.5 Split-brain patients

Main article: Split-brain

Research by Michael Gazzaniga and Roger Wolcott Sperry in the 1960s on split-brain patients led to an even greater understanding of functional laterality. Split-brain patients are patients who have undergone corpus callosotomy (usually as a treatment for severe epilepsy), a severing of a large part of the corpus callosum. The corpus callosum connects the two hemispheres of the brain and allows them to communicate. When these connections are cut, the two halves of the brain have a reduced capacity to communicate with each other. This led to many interesting behavioral phenomena that allowed Gazzaniga and Sperry to study the contributions of each hemisphere to various cognitive and perceptual processes. One of their main findings was that the right hemisphere was capable of rudimentary language processing, but often has no lexical or grammatical abilities.[35] Eran Zaidel also studied such patients and found some evidence for the right hemisphere having at least some syntactic ability.

Language is primarily localized in the left hemisphere. One of the experiments carried out by Gazzaniga involved a split-brain patient sitting in front of a computer screen while having words and images presented on either side of the screen and the visual stimuli would go to either the right or left visual field, and thus the left or right brain, respectively. It was observed that if a patient was presented with an image to his left visual field (right brain), he would report not seeing anything. If he was able to feel around for certain objects, he could accurately pick out the correct object, despite not having the ability to verbalize what he saw. This led to confirmation that the left brain is localized for language while the right brain does not have this capability, and when the corpus callosum is cut and the two hemispheres cannot communicate for the speech to be produced.

32.3.6 Pop psychology

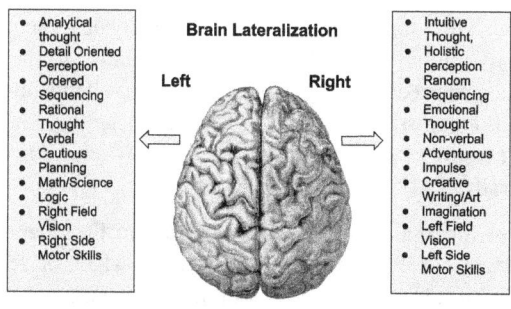

The misrepresentation of lateralization in pop psychology

Popular psychology[36] considers each hemisphere a unique brain that is called upon to serve their lateralized functions. This belief has led to ideas such as 'right-brain and left-brain' dominating popular belief. This stance, however, is rooted in the misinterpretation of studies done in the early 2000s and has since been fully disproven. Ironically these ideas held denote LHS as the 'logical brain' and RHS as the 'creative brain' which, even in the most generalized sense, is opposite the truth.[37][38]

32.4 Sex differences

Significant differences between male and female hemispheric rivalry and dominance have been established. Male brains have significantly better global and rivalry efficiency between the hemispheres, whereas female brains possess considerably better local efficiency within the RHS.[39][40]

32.5 Handedness

Handedness has been implicated in determining which hemisphere is naturally dominant. Due to decussation the dominant hemisphere is opposite to the main hand/foot. Left-handed and ambidextrous individuals have been shown to have more efficient hemispheric interactions.[41][42]

32.6 Self-harm

Damage to the RHS has been shown to drastically increase the likelihood of self-inflicted harm and suicide as calculative ideas such as the prospect of future are lost. RHS damage has also been shown to drastically decrease social performance and appropriateness as these behaviors stem from inhibition of boisterous activities which is no longer possible in these patients.[43]

32.7 Lateralized cognitive processes

For example, structurally, the lateral sulcus generally is longer in the left hemisphere than in the right hemisphere, and functionally, Broca's area and Wernicke's area are located in the left cerebral hemisphere for about 95% of right-handers, but about 70% of left-handers.[44]

Language functions such as grammar, vocabulary and literal meaning[45][46] are typically lateralized to the left hemisphere, especially in right handed individuals.[46] While language production is left-lateralized in up to 90% of right-handed subjects, it is more bilateral, or even right lateralized in approximately 50% of left-handers.[47] In contrast, prosodic language functions, such as intonation and accentuation, often are lateralized to the right hemisphere of the brain.[48][49]

The processing of visual and auditory stimuli, spatial manipulation, facial perception, and artistic ability are represented bilaterally.[47] Numerical estimation, comparison and online calculation depend on bilateral parietal regions[50][51] while exact calculation and fact retrieval are associated with left parietal regions, perhaps due to their ties to linguistic processing.[50][51] Dyscalculia is a neurological syndrome associated with damage to the left temporo-parietal junction.[52] This syndrome is associated with poor numeric manipulation, poor mental arithmetic skill, and the inability to either understand or apply mathematical concepts.[53]

Depression is linked with a hyperactive right hemisphere, with evidence of selective involvement in "processing negative emotions, pessimistic thoughts and unconstructive thinking styles", as well as vigilance, arousal and self-reflection, and a relatively hypoactive left hemisphere, "specifically involved in processing pleasurable experiences" and "relatively more involved in decision-making processes".[54] Additionally, "left hemisphere lesions result in an omissive response bias or error pattern whereas right hemisphere lesions result in a commissive response bias or error pattern."[55] The delusional misidentification syndromes, reduplicative paramnesia and Capgras delusion are also often the result of right hemisphere lesions.[56][57]

32.7.1 Lateralization of language processes

Hemispheric lateralization[58] refers to the distinction of functions of the right and left hemispheres of the brain. If one hemisphere is more heavily involved in a specific function, it is often referred to as being dominant (Bear et al., 2007). Language and speech understanding and function is commonly accepted by linguists and neuroscientists to be a heavily lateralized function.[58] Many specific aspects of language are found to be localized in the left hemisphere, while less so in the right hemisphere as the left hemisphere is most often dominant. This was proposed first through early work in patients with aphasia and language deficits found to have specific areas with lesions and damage.

When looking at patients that have unilateral hemisphere damage, in either the right or left hemisphere their language deficits can be studied. For example; when the left hemisphere has been damaged or lesioned, the right hemisphere is used to take over some functions via brain plasticity, and this damage of the one hemisphere and compensation by the opposite hemisphere creates language understanding and production changes and deficits that can be studied to examine and determine the basis and interaction of brain areas in language processes.

The production of language and language comprehension require the coordination of different subprocesses in time.[58] Though there is debate on how these subprocesses work together and how thinking and comprehending can change, the anatomical basis and role of a loop involving Wernicke's and Broca's area is usually agreed upon.

Neuroscientists generally agree that around the lateral sulcus[59] (or Sylvian Fissure) in the left hemisphere of the brain, there is a neural loop involved both in understanding and producing spoken language. At the front end or beginning of this loop lies Broca's area, which is usually associated with the production of language, or language outputs. At the other end, or specifically in the superior posterior temporal lobe, lies Wernicke's area, which is associated with the processing of words that we hear being spoken, or language inputs. Broca's area and Wernicke's area are connected by a large bundle of nerve fibres called the arcuate fasciculus.[59]

32.7.2 Handedness and language

Broca's area and Wernicke's area are linked by a white matter fiber tract, the arcuate fasciculus. This axonal tract allows the neurons in the two areas to work together in creating vocal language. In more than 95% of right-handed men, and more than 90% of right-handed women, the left hemisphere is dominant in certain aspects of language and speech processing. In left-handed people, the incidence of left-hemisphere language dominance has been reported as 73%[60] and 61%,[61] suggesting left handed people tend to be less lateralized than right-handed people in language function. In general neuroimaging methods, such as functional magnetic resonance imaging and magnetoencephalography, involvement of both hemispheres in many aspects of language processing has been show. The "dominance" discussed in many of these studies simply refers to more brain activation relative to the other hemisphere (or better performance by that hemisphere on psycholinguistic tasks such as dichotic listening); it is not the case that language is "localized" in any one hemisphere laterally.

Brain function lateralization is evident in the phenomena of right- or left-handedness[62] and of right or left ear preference,[63] but a person's preferred hand is not a clear indication of the location of brain function. Although 95% of right-handed people have left-hemisphere dominance for language, 18.8% of left-handed people have right-hemisphere dominance for language function. Additionally, 19.8% of the left-handed have bilateral language functions.[61] Even within various language functions (e.g., semantics, syntax, prosody), degree (and even hemisphere) of dominance may differ.[64]

32.7.3 Methods of study

There are ways of determining whether particular cognitive functions tend to be lateralized to one cerebral hemisphere. The Wada Test introduces an anesthetic to one hemisphere of the brain via one of the two carotid arteries. Once the hemisphere is anesthetized, a neuropsychological examination is effected to determine whether cognitive functions such as language production, language comprehension, verbal memory, or visual memory are retained. Another common way to study neural deficits is to identify the deficits a person exhibits in relation to lesions in different areas of the brain.[65]

Less invasive techniques, such as functional magnetic resonance imaging and transcranial magnetic stimulation may also be used to investigate the role of a particular cerebral hemisphere in a particular task, although these methods may be costly. The divided visual field paradigm is another technique that has contributed to the study of hemispheric specialization. CAT scans, PET scans and EEG are also used to study the brain. CAT scans use tomography to create a 3D image of the brain, which provides insights about neural anatomy, but it is unable to show the brain functioning in real time. PET scans image areas of high metabolic activity and neural activity by scanning for an active substance that has been tagged with positron emitting isotopes, that has been ingested by the patient. Finally, EEGs collect data from the electric fields that are produced by the brain.[66]

32.8 Pathology

32.8.1 Hemisphere damage

Damage to either the right or left hemisphere, and its resulting deficits provide insight into the function of the damaged area. Right hemisphere damage has many effects on language production and perception. Damage or lesions to the right hemisphere can result in a lack of emotional prosody or intonation when speaking. Right hemisphere damage also has monumental effects on understanding discourse. People with damage to the right hemisphere have a reduced ability to generate inferences, comprehend and produce main concepts and a reduced ability to manage alternative meanings. Furthermore, when engaging in discourse people with right hemisphere damage, their discourse is often abrupt and perfunctory or verbose and excessive. They can also have pragmatic deficits in situations of turn taking, topic maintenance and shared knowledge.[67]

Lateral brain damage can also have effects on spatial frequency. People with left hemisphere damage are only able to see low frequency, or big picture, parts of an image. Right hemisphere damage causes damage to low spatial frequency, so people with right hemisphere damage can only see the details of an image, or the high frequency parts of an image.[68]

32.8.2 Plasticity

If a specific region of the brain, or even an entire hemisphere, is injured or destroyed, its functions can sometimes be assumed by a neighboring region in the same hemisphere or the corresponding region in the other hemisphere, depending upon the area damaged and the patient's age.[69] When injury interferes with pathways from one area to another, alternative (indirect) connections may develop to communicate information with detached areas, despite the inefficiencies.

32.8.3 Broca's aphasia

Broca's aphasia is a specific type of expressive aphasia and is so named due to the aphasia that results from damage or lesions to the Broca's area of the brain, that exists most commonly in the left inferior frontal hemisphere. Thus, the aphasia that develops from the lack of functioning of the Broca's area is an expressives and non-fluent aphasia. It is called 'non-fluent' due the issues that arise because Broca's area is critical for language pronunciation and production. The area controls some motor aspects of speech production and articulation of thoughts to words and as such lesions to the area result in the specific non-fluent aphasia.[70]

32.8.4 Wernicke's aphasia

Wernicke's aphasia is the result of damage to the area of the brain that is commonly in the left hemisphere above the sylvian fissure. Damage to this area causes many deficits in language production and cognition. Although the speech produced by a person with Wernicke's aphasia sounds like regular speech, it is riddled with mistakes. They include mild impairments in word selection, grammar, and segmental phonology. Wernicke's aphasia is characterized by phonemic paraphasias, neologism or jargon. Comprehension of spoken language is also mildly impaired in people with Wernicke's aphasia. Another characteristic of a person with Wernicke's aphasia is that they are unconcerned by the mistakes that they are making.[71][72]

32.9 Misapplication of concept

Terence Hines states that the research on brain lateralization is valid as a research program, though commercial promoters have applied it to promote subjects and products far outside the implications of the research.[73] For example, the implications of the research have no bearing on psychological interventions such as EMDR and neurolinguistic programming,[74] brain training equipment, or management training.[75]

32.10 Advantages of brain lateralization

The widespread lateralization of many vertebrate animals indicates an evolutionary advantage associated with the specialization of each hemisphere.[76] In one experiment, baby chicks were lateralized before hatching by exposing their eggs to light.[77] These chicks were set to a task of picking out food from a bed of pebbles. Neither the lateralized, nor the non-lateralized chicks had a problem with this task, but the lateralized chicks only used the eye on the side of which they were lateralized to pick up the pebbles. When presented with a second task of watching for a cutout of a predatory hawk, the discrepancy between lateralized and non-lateralized chicks became evident. Lateralized chicks could pick food out of the pebbles with one eye and one half of the brain[78] while using the other eye and other half of their brain to monitor the skies for predators.[79] Not only could non-lateralized chicks not complete the two tasks simultaneously, but their performance of the single task deteriorated. This suggests that the evolutionary advantage of lateralization comes from the capacity to perform separate parallel tasks in each hemisphere of the brain.[76] It was found in a 2011 study published in the journal of *Brain Behavioral Research* that lateralization of few specific functions as opposed to overall brain lateralization is correlated with parallel tasks efficiency.[80]

32.11 Additional images

- Ventricles of brain and basal ganglia. Superior view. Horizontal section. Deep dissection
- Ventricles of brain and basal ganglia. Superior view. Horizontal section. Deep dissection

32.12 See also

- Ambidexterity
- Bicameralism
- Brain asymmetry
- Cerebral hemisphere
- Chirality
- Cross-dominance
- Dual brain theory
- Emotional lateralization
- Handedness
- Hemispherectomy
- Hemispheres
- Laterality
- Psychoneuroimmunology
- Left brain interpreter
- Ten percent of brain myth

32.13 References

[1] "2. The fields of linguistics — First 1000 ms: Computational neurolinguistics of language 0 documentation". *www.tulane.edu*. Retrieved 4 December 2015.

[2] Boughner, Julie, and Campbell Rolian. "Developmental Approaches to Human Evolution." Google Books. 22 Jan. 2016. Web. 31 Mar. 2016.

[3] Weiss, Peter H., and Simon D. Ubben. "Where Language Meets Meaningful Action: A Combined Behavior and Lesion." Springer. 29 Oct. 2014. Web. 31 Mar. 2016.

[4] Riès, Stephanie K., and Nina F. Dronkers. "Choosing Words: Left Hemisphere, Right Hemisphere, or Both? Perspective on the Lateralization of Word Retrieval." Wiley Online Library. 14 Jan. 2016. Web. 31 Mar. 2016.

[5] ER, Smith-Conway, and Chenery HJ. "A Dual Task Priming Investigation of Right Hemisphere Inhibition for People with Left Hemisphere Lesions." National Center for Biotechnology Information. U.S. National Library of Medicine, 20 Mar. 2012. Web. 29 Mar. 2016.

[6] Garavan, H., T. J. Ross, and E. A. Stein. "Right Hemispheric Dominance of Inhibitory Control: An Event-related Functional MRI Study." Proceedings of the National Academy of Sciences (1999): 8301-306. US National Library of Medicine.

[7] Slagtera, H.A., and S. Prinssena. "Facilitation and Inhibition in Attention: Functional Dissociation of Pre-stimulus Alpha Activity, P1, and N1 Components."Science Direct. 15 Jan. 2016. Web. 31 Mar. 2016.

[8] Miller, Steven M. "Binocular Rivalry and the Cerebral Hemispheres." Brain and Mind. Kluwer Academic Publishers, 14 Mar. 2001. Web. 29 Mar. 2016.

[9] Lindwall, Harry. Knowing Yourself: A Narrative of Accessing the Right Brain Hemisphere. Friesen, 2015. Print.

[10] Grimshaw, Gina M., and David Carmel. "An Asymmetric Inhibition Model of Hemispheric Differences in Emotional Processing." Google Books. 23 May 2014. Web. 29 Mar. 2016.

[11] Harrison, David W. Brain Asymmetry and Neural Systems: Foundations in Clinical Neuroscience and Neuropsychology. Springer International, 2015. Print.

[12] Oosugi, Naoya, and Toru Yanagawa. "Social Suppressive Behavior Is Organized by the Spatiotemporal Integration of Multiple Cortical Regions in the Japanese Macaque." PLOS ONE. 10 Mar. 2016. Web. 31 Mar. 2016.

[13] Sherwood, Chet C., Adam D. Gordon, John S. Allen, Kimberley A. Phillips, Joseph M. Erwin, Patrick R. Hof, and William D. Hopkins. "Aging of the Cerebral Cortex Differs between Humans and Chimpanzees." Proceedings of the National Academy of Sciences of the United States of America. National Academy of Sciences, 25 July 2011. Web. 31 Mar. 2016.

[14] Stopper, Colin M., and Emily B. Green. "Selective Involvement by the Medial Orbitofrontal Cortex in Biasing Risky, But Not Impulsive, Choice." Oxford Journals. 12 Oct. 2012. Web. 31 Mar. 2016.

[15] Brown, John N.A. "The Evolution of Humans and Technology Part 1: Humans."Springer. 17 Mar. 2016. Web. 31 Mar. 2016.

[16] R. J. Morris (2006) Left Brain, Right Brain, Whole Brain? An examination into the theory of brain lateralization, learning styles and the implications for education. PGCE Thesis, Cornwall College St Austell, http://singsurf.org/brain/rightbrain.html

[17] Hatamikia, S., and A. M. Nasrabadi. "Analysis of Inter-hemispheric and Intra-hemispheric Differences of the Correlation Dimension in the Emotional States Based on EEG Signals." IEEE Xplore. IEEE, 27 Nov. 2015. Web. 29 Mar. 2016.

[18] Leutgeba, Verena, and Albert Wabneggera. "Altered Cerebellar-amygdala Connectivity in Violent Offenders: A Resting-state FMRI Study." Science Direct. 01 Jan. 2016. Web. 31 Mar. 2016.

[19] Leutgeba, V., and M. Leitnerb. "Brain Abnormalities in High-risk Violent Offenders and Their Association with Psychopathic Traits and Criminal Recidivism." Science Direct. 12 Nov. 2015. Web. 31 Mar. 2016.

[20] Cristofori, Irene, and Wanting Zhong. "Brain Regions Influencing Implicit Violent Attitudes: A Lesion-Mapping Study." JNeurosci. 02 Mar. 2016. Web. 31 Mar. 2016.

[21] Kumfor, Fiona, and Ramon Landin-Romero. "On the Right Side? A Longitudinal Study of Left- versus Right-lateralized Semantic Dementia." Oxford University Press. 25 Jan. 2016. Web. 31 Mar. 2016.

[22] Eshel, Neir, and Christian C. Ruff. "Effects of Parietal TMS on Somatosensory Judgments Challenge Interhemispheric Rivalry Accounts." Science Direct. Oct. 2010. Web. 29 Mar. 2016.

[23] Wanga, Guangrong, and Jianbiao Lib. "Modulating Activity in the Orbitofrontal Cortex Changes Trustees' Cooperation: A Transcranial Direct Current Stimulation Study." Science Direct. 15 Apr. 2016. Web. 31 Mar. 2016.

[24] Chunha-Bang, Sofi Da, and Liv V. Hjordt. "Serotonin 1B Receptor Binding Is Associated with Trait Anger and Level of Psychopathy in Violent Offenders."Science Direct. 7 Mar. 2016. Web. 31 Mar. 2016.

[25] Brunyé, Tad T., Sarah R. Cavanagh, and Ruth E. Propper. "Hemispheric Bases for Emotion and Memory." Frontiers in Human Neuroscience. Frontiers Media S.A., 05 Dec. 2014. Web. 31 Mar. 2016.

[26] Baldo, Juliania V., and Natalie A. Kacinik. "You May Now Kiss the Bride: Interpretation of Social Situations by Individuals with Right or Left Hemisphere Injury." Science Direct. Elsevier Ltd, 8 Jan. 2016. Web. 29 Mar. 2016.

[27] Mark, Victor W. "Stroke and Behavior." Science Direct. Feb. 2016. Web. 31 Mar. 2016.

[28] Balaban, Noga, Naama Friedmann, and Mira Ariel. "The Effect of Theory of Mind Impairment on Language: Referring after Right-hemisphere Damage." Taylor & Francis. 26 Dec. 2015. Web. 29 Mar. 2016.

[29] Ellefsen, Kai Olav, and Jean-Baptiste Mouret. "Neural Modularity Helps Organisms Evolve to Learn New Skills without Forgetting Old Skills." PLOS Computational Biology:. 02 Apr. 2015. Web. 31 Mar. 2016.

[30] Thompson, Hannah E., and Lauren Henshall. "The Role of the Right Hemisphere in Semantic Control: A Case-series Comparison of Right and Left Hemisphere Stroke." Science Direct. May 2016. Web. 29 Mar. 2016.

[31] Parola, Alberto, and Ilaria Gabbatore. "Assessment of Pragmatic Impairment in Right Hemisphere Damage." Science Direct. Aug. 2016. Web. 31 Mar. 2016.

[32] Save-Pedebosa, Jessica, and Charlotte Pinabiauxe. "The Development of Pragmatic Skills in Children after Hemispherotomy: Contribution from Left and Right Hemispheres." Science Direct. Feb. 2016. Web. 31 Mar. 2016.

[33] Sweatt, J. David. "Neural Plasticity and Behavior – Sixty Years of Conceptual Advances." Wiley Online Library. 10 Mar. 2016. Web. 31 Mar. 2016.

[34] Nakamura, Kimihiro, and Tatsuhide Oga. "Symmetrical Hemispheric Priming in Spatial Neglect: A Hyperactive Left-hemisphere Phenomenon?" Science Direct. 7 Dec. 2010. Web. 29 Mar. 2016.

[35] Kandel E, Schwartz J, Jessel T. *Principles of Neural Science*. 4th ed. p1182. New York: McGraw–Hill; 2000. ISBN 0-8385-7701-6

[36] Nielsen, Jared A., Brandon A. Zielinski, Michael A. Ferguson, Janet E. Lainhart, and Jeffrey S. Anderson. "An Evaluation of the Left-Brain vs. Right-Brain Hypothesis with Resting State Functional Connectivity Magnetic Resonance Imaging." PLOS ONE, 14 August 2013. Web. 30 August 2013.

[37] Westen et al. 2006 *Psychology: Australian and New Zealand edition*. John Wiley p.107

[38] Toga AW, Thompson PM (2003). "Mapping brain asymmetry". *Nature Reviews Neuroscience*. **4** (1): 37–48. doi:10.1038/nrn1009. PMID 12511860.

[39] Jalili, Mahdi. "EEG-BASED FUNCTIONAL BRAIN NETWORKS: HEMISPHERIC DIFFERENCES IN MALES AND FEMALES." EBSCO. Mar. 2015. Web. 29 Mar. 2016.

[40] Cuevas, Kimberly, and Susan D. Calkins. "To Stroop or Not to Stroop: Sex-related Differences in Brain-behavior Associations during Early Childhood." Wiley Online Library. 17 Dec. 2015. Web. 31 Mar. 2016.

[41] McGratha, Robert L., and Shailesh S. Kantak. "Reduced Asymmetry in Motor Skill Learning in Left-handed Compared to Right-handed Individuals." Science Direct. Feb. 2016. Web. 31 Mar. 2016.

[42] N, Cherbuin, and Brinkman C. "Hemispheric Interactions Are Different in Left-handed Individuals." National Center for Biotechnology Information. U.S. National Library of Medicine, 20 Nov. 2006. Web. 31 Mar. 2016.

[43] Borah, Shaina, and Brice McConnell. "Potential Relationship of Self-injurious Behavior to Right Temporo-parietal Lesions." Taylor & Francis. 16 Feb. 2016. Web. 31 Mar. 2016.

[44] Griggs, Richard A. *Psychology: A Concise Introduction*. p. 69.

[45] Boeree, C.G. (2004). "Speech and the Brain". Retrieved 17 February 2012.

[46] Taylor, I.; Taylor, M. M. (1990). *Psycholinguistics: Learning and using Language*. Pearson. ISBN 978-0-13-733817-7. p. 367

[47] Beaumont, J.G. (2008). *Introduction to Neuropsychology, Second Edition*. The Guilford Press. ISBN 978-1-59385-068-5. Chapter 7

[48] Ross ED, Monnot M (January 2008). "Neurology of affective prosody and its functional-anatomic organization in right hemisphere". *Brain Lang*. **104** (1): 51–74. doi:10.1016/j.bandl.2007.04.007. PMID 17537499.

[49] George MS, Parekh PI, Rosinsky N, Ketter TA, Kimbrell TA, Heilman KM, Herscovitch P, Post RM (July 1996). "Understanding Emotional Prosody Activates Right Hemisphere Regions". *Arch Neurol*. **53** (7): 665–670. doi:10.1001/archneur.1996.00550070103017. PMID 8929174.

[50] Dehaene S, Spelke E, Pinel P, Stanescu R, Tsivkin S (May 1999). "Sources of mathematical thinking: behavioral and brain-imaging evidence" (PDF). *Science*. **284** (5416): 970–4. doi:10.1126/science.284.5416.970. PMID 10320379.

[51] Dehaene S, Piazza M, Pinel P, Cohen L (2003). "Three parietal circuits for number processing" (PDF). *Cognitive Neuropsychology*. **20** (3–6): 487–506. doi:10.1080/02643290244000239. PMID 20957581.

[52] Levy LM, Reis IL, Grafman J (August 1999). "Metabolic abnormalities detected by 1H-MRS in dyscalculia and dysgraphia". *Neurology*. **53** (3): 639–41. doi:10.1212/WNL.53.3.639. PMID 10449137.

[53] Dyscalculia Symptoms

[54] Hecht D (October 2010). "Depression and the hyperactive right-hemisphere". *Neurosci. Res*. **68** (2): 77–87. doi:10.1016/j.neures.2010.06.013. PMID 20603163.

[55] Braun CM, Delisle J, Guimond A, Daigneault R (March 2009). "Post unilateral lesion response biases modulate memory: crossed double dissociation of hemispheric specialisations". *Laterality*. **14** (2): 122–64. doi:10.1080/13576500802328613. PMID 18991140.

[56] Devinsky O (January 2009). "Delusional misidentifications and duplications: right brain lesions, left brain delusions". *Neurology*. **72** (1): 80–7. doi:10.1212/01.wnl.0000338625.47892.74. PMID 19122035.

[57] Madoz-Gúrpide A, Hillers-Rodríguez R (April 2010). "[Capgras delusion: a review of aetiological theories]". *Rev Neurol*. **50** (7): 420–30. PMID 20387212.

[58] Friederici, Angela D., and Kai Alter. "Lateralization of auditory language functions: A dynamic dual pathway model." Brain and Language 89.2 (2004). Print.

[59] Dubuc, Bruno. "BROCA'S AREA, WERNICKE'S AREA, AND OTHER LANGUAGE-PROCESSING AREAS IN THE BRAIN." The Brain from Bottom to Top. Ed. Patrick Robert. Douglas Hospital Research Area, Feb. 2004. Web. 4 December 2015.

[60] Knecht S, Dräger B, Deppe M, Bobe L, Lohmann H, Flöel A, Ringelstein EB, Henningsen H (2000). "Handedness and hemispheric language dominance in healthy humans". *Brain*. **123** (12): 2512–2518. doi:10.1093/brain/123.12.2512. PMID 11099452.

[61] Taylor, Insep and Taylor, M. Martin (1990) "Psycholinguistics: Learning and using Language". page 362

[62] Knecht S, Dräger B, Deppe M, Bobe L, Lohmann H, Flöel A, Ringelstein EB, Henningsen H (2000). "Handedness and hemispheric language dominance in healthy humans". *Brain : a journal of neurology.* **123** (12): 2512–2518. doi:10.1093/brain/123.12.2512. PMID 11099452.

[63] Schönwiesner M, Rübsamen R, von Cramon DY (2005). "Hemispheric asymmetry for spectral and temporal processing in the human antero-lateral auditory belt cortex". *European Journal of Neuroscience.* **22** (6): 1521–1528. doi:10.1111/j.1460-9568.2005.04315.x. PMID 16190905.

[64] Regarding different languages: http://www.bbc.co.uk/news/health-11181457

[65] "3. The macrostructure of the brain — First 1000 ms: Computational neurolinguistics of language 0 documentation". *www.tulane.edu.* Retrieved 4 December 2015.

[66] "3. The macrostructure of the brain — First 1000 ms: Computational neurolinguistics of language 0 documentation". *www.tulane.edu.* Retrieved 4 December 2015.

[67] "21. Discourse — First 1000 ms: Computational neurolinguistics of language 0 documentation". *www.tulane.edu.* Retrieved 4 December 2015.

[68] "6. Auditory transduction — First 1000 ms: Computational neurolinguistics of language 0 documentation". *www.tulane.edu.* Retrieved 4 December 2015.

[69] Pulsifer MB, Brandt J, Salorio CF, Vining EP, Carson BS, Freeman JM (2004). "The cognitive outcome of hemispherectomy in 71 children". *Epilepsia.* **45** (3): 243–254. doi:10.1111/j.0013-9580.2004.15303.x. PMID 15009226.

[70] Biopsychology (8th edition), by John J.P. Pinel Pearson 2011

[71] "10. Wernicke's aphasia — First 1000 ms: Computational neurolinguistics of language 0 documentation". *www.tulane.edu.* Retrieved 4 December 2015.

[72] Bogen, J. E., & Bogen, G. M. (1976). Wernicke's region - where is it? Annals of the New York Academy of Sciences, 280, 834-843.

[73] Hines, Terence (1987). "Left Brain/Right Brain Mythology and Implications for Management and Training". *The Academy of Management Review.* **12** (4): 600–606. doi:10.2307/258066. JSTOR 258066.

[74] Drenth JD (2003). "Growing anti-intellectualism in Europe; a menace to science". *Studia Psychologica.* **45** (1): 5–13., available in *ALLEA Annual Report 2003*, pp. 61–72

[75] Sala, Sergio Della (1999). *Mind Myths: Exploring Popular Assumptions about the Mind and Brain.* New York: Wiley. ISBN 0-471-98303-9.

[76] Halpern ME, Güntürkün O, Hopkins WD, Rogers LJ (2005). "Lateralization of the Vertebrate Brain: Taking the Side of Model Systems". *The Journal of Neuroscience.* **25** (45): 10351–10357. doi:10.1523/JNEUROSCI.3439-05.2005. PMC 2654579. PMID 16280571.

[77] Rogers LJ (1990). "Light Input and the Reversal of Functional Lateralization in the Chicken Brain". *Behav Brain Res.* **38** (3): 211–21. doi:10.1016/0166-4328(90)90176-F. PMID 2363841.

[78] Deng C, Rogers LJ (1997). "Differential Contributions of the Two Visual Pathways to Functional Lateralization in Chicks". *Behav Brain Res.* **87** (2): 173–82. doi:10.1016/S0166-4328(97)02276-6. PMID 9331485.

[79] Rogers LJ (2000). "Evolution of Hemispheric Specialization: Advantages and Disadvantages". *Brain Lang.* **73** (2): 236–53. doi:10.1006/brln.2000.2305. PMID 10856176.

[80] Lust, J. M.; Geuze, R. H.; Groothuis, A. G. G.; Bouma, A. (1 March 2011). "Functional cerebral lateralization and dual-task efficiency-testing the function of human brain lateralization using fTCD". *Behavioural Brain Research.* **217** (2): 293–301. doi:10.1016/j.bbr.2010.10.029. ISSN 1872-7549. PMID 21056593.

32.14 Further reading

- Harnad, Stevan; Doty, R.W.; Goldstein, L.; Jaynes, J.; Krauthamer, G. (1977). *Lateralization in the nervous system.* Academic Press. ISBN 978-0-12-325750-5.

- Luria, A. R. (1966). *Higher cortical functions in man.* Basic Books.

- Ornstein, Robert (1998). *The Right Mind: Making Sense of the Hemispheres.* Harcourt Brace International. ISBN 978-0-15-600627-9.

- Drenth, Pieter (2006). *Walks in the Garden of Science: Selected Papers and Lectures* (PDF). Conference *allea*.

- Josse G, Tzourio-Mazoyer N (2003). "Review: Hemispheric specialization for language". *Brain Research Reviews.* **44** (1): 1–12. doi:10.1016/j.brainresrev.2003.10.001. PMID 14739000.

- McGilchrist, Iain (9 October 2009). *The Master and His Emissary: The Divided Brain and the Making of the Western World.* USA: Yale University Press. ISBN 0-300-14878-X. (Hardcover)

Chapter 33

Corpus callosum

For the films, see *Corpus Callosum and Corpus Callosum (2007 film).

The **corpus callosum** (/ˈkɔːrpəs kəˈloʊsəm/; Latin for

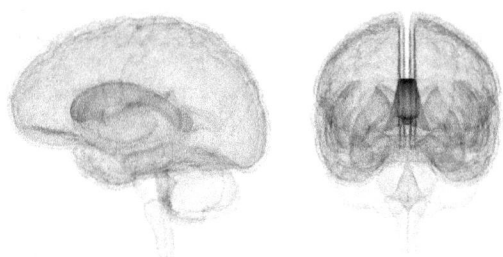

Corpus callosum with Anatomography

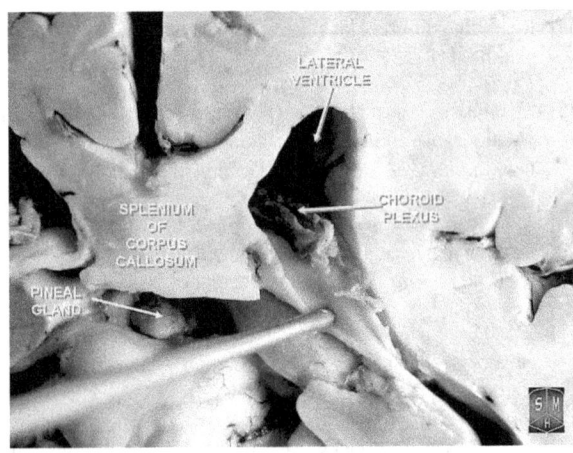

Corpus callosum

"tough body"), also known as the callosal commissure, is a wide, flat bundle of neural fibers about 10 cm long beneath the cortex in the eutherian brain at the longitudinal fissure. It connects the left and right cerebral hemispheres and facilitates interhemispheric communication. It is the largest white matter structure in the brain, consisting of 200–250 million contralateral axonal projections.

33.1 Structure

The posterior (back) portion of the corpus callosum is called the splenium; the anterior (front) is called the genu (or "knee"); between the two is the truncus, or "body", of the corpus callosum. The part between the body and the splenium is often markedly narrowed and thus referred to as the "isthmus". The rostrum is the part of the corpus callosum that projects posteriorly and inferiorly from the anteriormost genu, as can be seen on the sagittal image of the brain displayed on the right. The rostrum is so named for its resemblance to a bird's beak.

On either side of the corpus callosum, the fibers radiate in the white matter and pass to the various parts of the cerebral cortex; those curving forward from the genu into the frontal lobe constitute the *forceps anterior*, and those curving backward into the occipital lobe, the *forceps posterior*. Between these two parts is the main body of the fibers which constitute the **tapetum** and extend laterally on either side into the temporal lobe, and cover in the central part of the lateral ventricle.

Thinner axons in the genu connect the prefrontal cortex between the two halves of the brain; these fibres arise from a fork-like bundle of fibers from the tapetum, the forceps anterior. Thicker axons in the mid body, or trunk of the corpus callosum, interconnect areas of the motor cortex, with proportionately more of the corpus callosum dedicated to supplementary motor regions including Broca's area. The posterior body of the corpus, known as the splenium, communicates somatosensory information between the two halves of the parietal lobe and the visual cortex at the occipital lobe, these are the fibres of the forceps posterior.[1][2]

33.1.1 Variation

Agenesis of the corpus callosum (ACC) is a rare congenital disorder that is one of the most common brain malformations observed in human beings,[3] in which the corpus callosum is partially or completely absent. ACC is usually diagnosed within the first two years of life, and may manifest as a severe syndrome in infancy or childhood, as a milder condition in young adults, or as an asymptomatic incidental finding. Initial symptoms of ACC usually include seizures, which may be followed by feeding problems and delays in holding the head erect, sitting, standing, and walking. Other possible symptoms may include impairments in mental and physical development, hand-eye coordination, and visual and auditory memory. Hydrocephaly may also occur. In mild cases, symptoms such as seizures, repetitive speech, or headaches may not appear for years.

ACC is usually not fatal. Treatment usually involves management of symptoms, such as hydrocephaly and seizures, if they occur. Although many children with the disorder lead normal lives and have average intelligence, careful neuropsychological testing reveals subtle differences in higher cortical function compared to individuals of the same age and education without ACC. Children with ACC accompanied by developmental delay and/or seizure disorders should be screened for metabolic disorders.[4]

In addition to agenesis of the corpus callosum, similar conditions are hypogenesis (partial formation), dysgenesis (malformed), and hypoplasia (underdevelopment, including too thin).

Recent studies have also linked possible correlations between corpus callosum malformation and autism spectrum disorders.[5]

Kim Peek, a savant and the inspiration behind the movie *Rain Man*, was found with agenesis of the corpus callosum.

33.1.2 Sexual dimorphism

The corpus callosum and its relation to sex has been a subject of debate in the scientific and lay communities for over a century. Initial research in the early 20th century claimed the corpus to be different in size between men and women. That research was in turn questioned, and ultimately gave way to more advanced imaging techniques that appeared to refute earlier correlations. However, advanced analytical techniques of computational neuroanatomy developed in the 1990s showed that sex differences were clear but confined to certain parts of the corpus callosum, and that they correlated with cognitive performance in certain tests.[6] One recent study using magnetic resonance imaging (MRI) found that the midsagittal corpus callosum cross-sectional area is, after controlling brain size, on average, proportionately larger in females.[7]

Using diffusion tensor sequences on MRI machines, the rate at which molecules diffuse in and out of a specific area of tissue, anisotropy (directionality), and rates of metabolism can be measured. These sequences have found consistent sex differences in human corpus callosal morphology and microstructure.[8][9][10]

Morphometric analysis has also been used to study specific three-dimensional mathematical relationships with MRIs, and have found consistent and statistically significant differences across genders.[11][12] Specific algorithms have found significant gender differences in over 70% of cases in one review.[13]

33.2 Other correlations

The front portion of the corpus callosum has been reported to be significantly larger in musicians than nonmusicians,[14] and to be 0.75 cm^2[15] or 11% larger in left-handed and ambidextrous people than right-handed people.[15][16] This difference is evident in the anterior and posterior regions of the corpus callosum, but not in the splenium.[15] Other magnetic resonance morphometric study showed corpus callosum size correlates positively with verbal memory capacity and semantic coding test performance.[17] Children with dyslexia tend to have smaller and less-developed corpus callosums than their nondyslexic counterparts.[18][19]

Musical training has shown to increase plasticity of the corpus callosum during a sensitive period of time in development. The implications are an increased coordination of hands, differences in white matter structure, and amplification of plasticity in motor and auditory scaffolding which would serve to aid in future musical training. The study found children who had begun musical training before the age of six (minimum 15 months of training) had an increased volume of their corpus callosum and adults who had begun musical training before the age of 11 also had increased bimanual coordination.[20]

33.3 Clinical significance

33.3.1 Epilepsy

The symptoms of refractory epilepsy can be reduced by cutting the corpus callosum in an operation known as a corpus callosotomy.[21] This is usually reserved for cases in which complex or grand mal seizures are produced by an epileptogenic focus on one side of the brain, causing an interhemispheric electrical storm. The work up for this

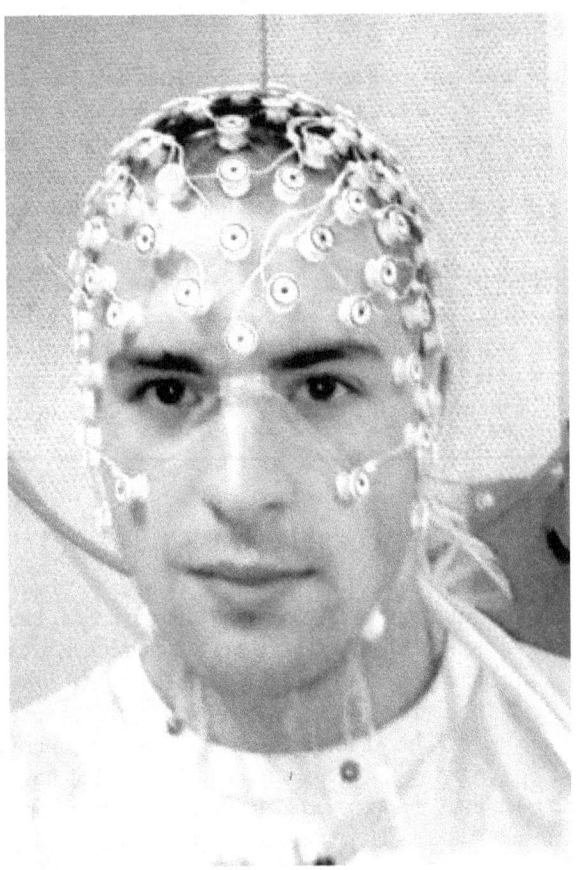

Electroencephalography is used to find the source of electrical activity causing a seizure as part of the surgical evaluation for a corpus callosotomy.

procedure involves an electroencephalogram, MRI, PET scan, and evaluation by a specialized neurologist, neurosurgeon, psychiatrist, and neuroradiologist before surgery can be considered.[22]

33.3.2 Other disease

Anterior corpus callosum lesions may result in akinetic mutism or tactile anomia. Posterior corpus callosum (splenium) lesions may result in alexia (inability to read) without agraphia.

See also:

- Alien hand syndrome
- Alexia without agraphia (seen with damage to splenium of corpus callosum)
- Agenesis of the corpus callosum (also dysgenesis, hypogenesis, hypoplasia), malformations of the corpus callosum
- Split-brain
- Septo-optic dysplasia (deMorsier syndrome)
- Multiple sclerosis with the symptom *Dawson's fingers*
- Mild encephalopathy with a reversible splenial lesion, a rare encephalopathy (or encephalitis) of unknown origin with a transient lesion in the posterior part of the corpus callosum, mostly associated with infectious diseases

33.3.3 Brain split procedure

Main article: Corpus callosotomy

The cerebral cortex is divided into two hemispheres, connected by the corpus callosum. A procedure to help patients alleviate the severity of seizures is called split-brain procedure. As a result, a seizure that starts in one hemisphere is isolated in that hemisphere, since a connection to the other side no longer exists. However, this procedure is dangerous and risky.

33.4 History

The first study of the corpus with relation to gender was by R. B. Bean, a Philadelphia anatomist, who suggested in 1906 that "exceptional size of the corpus callosum may mean exceptional intellectual activity" and that there were measurable differences between men and women. Perhaps reflecting the political climate of the times, he went on to claim differences in the size of the callosum across different races. His research was ultimately refuted by Franklin Mall, the director of his own laboratory.[23]

Of more mainstream impact was a 1982 *Science* article by Holloway and Utamsing that suggested sex difference in human brain morphology, which related to differences in cognitive ability.[24] *Time* published an article in 1992 that suggested that, because the corpus is "often wider in the brains of women than in those of men, it may allow for greater cross-talk between the hemispheres—possibly the basis for women's intuition."[25]

More recent publications in the psychology literature have raised doubt as to whether the anatomic size of the corpus is actually different. A meta-analysis of 49 studies since 1980 found that, contrary to de Lacoste-Utamsing and Holloway, no sex difference could be found in the size of the corpus callosum, whether or not account was taken of larger male brain size.[23] A study in 2006 using thin slice MRI showed no difference in thickness of the corpus when accounting for the size of the subject.[26]

33.5 In other animals

The corpus callosum is found only in placental mammals (the eutherians), while it is absent in monotremes and marsupials,[27] as well as other vertebrates such as birds, reptiles, amphibians and fish.[28] (Other groups do have other brain structures that allow for communication between the two hemispheres, such as the anterior commissure, which serves as the primary mode of interhemispheric communication in marsupials,[29][30] and which carries all the commissural fibers arising from the neocortex (also known as the neopallium), whereas in placental mammals, the anterior commissure carries only some of these fibers.[31]) In primates, the speed of nerve transmission depends on its degree of myelination, or lipid coating. This is reflected by the diameter of the nerve axon. In most primates, axonal diameter increases in proportion to brain size to compensate for the increased distance to travel for neural impulse transmission. This allows the brain to coordinate sensory and motor impulses. However, the scaling of overall brain size and increased myelination have not occurred between chimpanzees and humans. This has resulted in the human corpus callosum's requiring double the time for interhemispheric communication as a macaque's.[1]

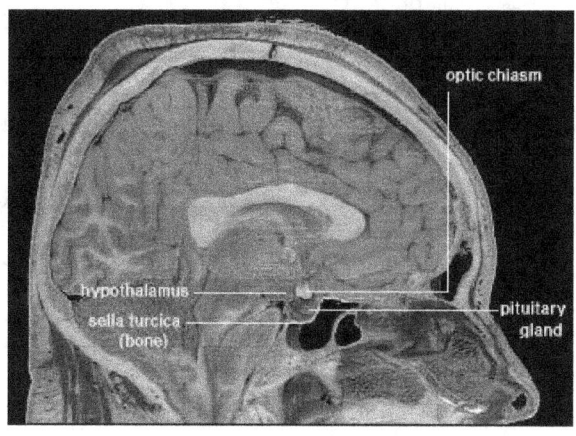

Sagittal post mortem *section through the midline brain. The corpus callosum is the curved band of lighter tissue at the center of the brain above the hypothalamus. Its lighter texture is due to higher myelin content, resulting in faster neuronal impulse transmission.*

The fibrous bundle as which the corpus callosum appears, can and does increase to such an extent in humans that it encroaches upon and wedges apart the hippocampal structures.[32]

33.6 Additional images

- Corpus callosum
- Corpus callosum
- Corpus callosum
- Coronal T2 (grey scale inverted) MRI of the brain at the level of the caudate nuclei emphasizing corpus callosum
- Corpus callosum parts on MRI
- Play media

  DTI Corpus callosum

- Ventricles of brain and basal ganglia.Superior view. Horizontal section.Deep dissection
- Ventricles of brain and basal ganglia.Superior view. Horizontal section.Deep dissection
- Cerebrum. Inferior view.Deep dissection

33.7 References

[1] Caminiti, Roberto; Ghaziri, Hassan; Galuske, Ralf; Hof, Patrick R.; Innocenti, Giorgio M. (2009). "Evolution amplified processing with temporally dispersed slow neuronal connectivity in primates". *Proceedings of the National Academy of Sciences.* **106** (46): 19551–6. Bibcode:2009PNAS..10619551C. doi:10.1073/pnas.0907655106. JSTOR 25593230. PMC 2770441. PMID 19875694.

[2] Hofer, Sabine; Frahm, Jens (2006). "Topography of the human corpus callosum revisited—Comprehensive fiber tractography using diffusion tensor magnetic resonance imaging". *NeuroImage.* **32** (3): 989–94. doi:10.1016/j.neuroimage.2006.05.044. PMID 16854598.

[3] Dobyns, W. B. (1996). "Absence makes the search grow longer". *American Journal of Human Genetics.* **58** (1): 7–16. PMC 1914936. PMID 8554070.

[4] "NINDS Agenesis of the Corpus Callosum Information Page: NINDS". *RightDiagnosis.com.* Retrieved Aug 30, 2011.

[5] "Autism May Involve A Lack Of Connections And Coordination In Separate Areas Of The Brain, Researchers Find". *Medical News Today.*

[6] Davatzikos, C; Resnick, S. M. (1998). "Sex differences in anatomic measures of interhemispheric connectivity: Correlations with cognition in women but not men". *Cerebral Cortex.* **8** (7): 635–40. doi:10.1093/cercor/8.7.635. PMID 9823484.

33.7. REFERENCES

[7] Ardekani, B. A.; Figarsky, K.; Sidtis, J. J. (2012). "Sexual Dimorphism in the Human Corpus Callosum: An MRI Study Using the OASIS Brain Database". *Cerebral Cortex.* **23** (10): 2514–20. doi:10.1093/cercor/bhs253. PMID 22891036.

[8] Dubb, Abraham; Gur, Ruben; Avants, Brian; Gee, James (2003). "Characterization of sexual dimorphism in the human corpus callosum". *NeuroImage.* **20** (1): 512–9. doi:10.1016/S1053-8119(03)00313-6. PMID 14527611.

[9] Westerhausen, René; Kreuder, Frank; Sequeira, Sarah Dos Santos; Walter, Christof; Woerner, Wolfgang; Wittling, Ralf Arne; Schweiger, Elisabeth; Wittling, Werner (2004). "Effects of handedness and gender on macro- and microstructure of the corpus callosum and its subregions: A combined high-resolution and diffusion-tensor MRI study". *Cognitive Brain Research.* **21** (3): 418–26. doi:10.1016/j.cogbrainres.2004.07.002. PMID 15511657.

[10] Shin, Yong-Wook; Jin Kim, Dae; Hyon Ha, Tae; Park, Hae-Jeong; Moon, Won-Jin; Chul Chung, Eun; Min Lee, Jong; Young Kim, In; Kim, Sun I.; et al. (2005). "Sex differences in the human corpus callosum: Diffusion tensor imaging study". *NeuroReport.* **16** (8): 795–8. doi:10.1097/00001756-200505310-00003. PMID 15891572.

[11] Kontos, Despina; Megalooikonomou, Vasileios; Gee, James C. (2009). "Morphometric analysis of brain images with reduced number of statistical tests: A study on the gender-related differentiation of the corpus callosum". *Artificial Intelligence in Medicine.* **47** (1): 75–86. doi:10.1016/j.artmed.2009.05.007. PMC 2732126. PMID 19559582.

[12] Spasojevic, Goran; Stojanovic, Zlatan; Suscevic, Dusan; Malobabic, Slobodan (2006). "Sexual dimorphism of the human corpus callosum: Digital morphometric study". *Vojnosanitetski pregled.* **63** (11): 933. doi:10.2298/VSP0611933S.

[13] Yokota, Y.; Kawamura, Y.; Kameya, Y. (2005). "2005 IEEE Engineering in Medicine and Biology 27th Annual Conference": 3055–8. doi:10.1109/IEMBS.2005.1617119. ISBN 0-7803-8741-4. |chapter= ignored (help)

[14] Levitin, Daniel J. "This is Your Brain on Music", '

[15] Witelson, S. (1985). "The brain connection: The corpus callosum is larger in left-handers". *Science.* **229** (4714): 665–8. Bibcode:1985Sci...229..665W. doi:10.1126/science.4023705. PMID 4023705.

[16] Driesen, Naomi R.; Raz, Naftali (1995). "The influence of sex, age, and handedness on corpus callosum morphology: A meta-analysis". *Psychobiology.* **23** (3): 240–7.

[17] Kozlovskiy, S.A.; Vartanov, A.V.; Pyasik, M.M.; Nikonova, E.Y. (2012). "Functional role of corpus callosum regions in human memory functioning". *International Journal of Psychophysiology.* **85** (3): 396–7. doi:10.1016/j.ijpsycho.2012.07.092.

[18] Hynd, G. W.; Hall, J.; Novey, E. S.; Eliopulos, D.; Black, K.; Gonzalez, J. J.; Edmonds, J. E.; Riccio, C.; Cohen, M. (1995). "Dyslexia and Corpus Callosum Morphology". *Archives of Neurology.* **52** (1): 32–8. doi:10.1001/archneur.1995.00540250036010. PMID 7826273.

[19] Von Plessen, K; Lundervold, A; Duta, N; Heiervang, E; Klauschen, F; Smievoll, AI; Ersland, L; Hugdahl, K (2002). "Less developed corpus callosum in dyslexic subjects—a structural MRI study". *Neuropsychologia.* **40** (7): 1035–44. doi:10.1016/S0028-3932(01)00143-9. PMID 11900755.

[20] Steele, C. J.; Bailey, J. A.; Zatorre, R. J.; Penhune, V. B. (2013). "Early Musical Training and White-Matter Plasticity in the Corpus Callosum: Evidence for a Sensitive Period". *Journal of Neuroscience.* **33** (3): 1282–90. doi:10.1523/JNEUROSCI.3578-12.2013. PMID 23325263.

[21] Clarke, Dave F.; Wheless, James W.; Chacon, Monica M.; Breier, Joshua; Koenig, Mary-Kay; McManis, Mark; Castillo, Edward; Baumgartner, James E. (2007). "Corpus callosotomy: A palliative therapeutic technique may help identify resectable epileptogenic foci". *Seizure.* **16** (6): 545–53. doi:10.1016/j.seizure.2007.04.004. PMID 17521926.

[22] "WebMd Corpus Callotomy". Web MD. July 18, 2010. Retrieved July 18, 2010.

[23] Bishop, Katherine M.; Wahlsten, Douglas (1997). "Sex Differences in the Human Corpus Callosum: Myth or Reality?". *Neuroscience & Biobehavioral Reviews.* **21** (5): 581–601. doi:10.1016/S0149-7634(96)00049-8. PMID 9353793.

[24] Delacoste-Utamsing, C; Holloway, R. (1982). "Sexual dimorphism in the human corpus callosum". *Science.* **216** (4553): 1431–2. Bibcode:1982Sci...216.1431D. doi:10.1126/science.7089533. PMID 7089533.

[25] C Gorman (20 January 1992). "Sizing up the sexes". *Time*: 36–43. As cited by Bishop and Wahlsten.

[26] Luders, Eileen; Narr, Katherine L.; Zaidel, Eran; Thompson, Paul M.; Toga, Arthur W. (2006). "Gender effects on callosal thickness in scaled and unscaled space". *NeuroReport.* **17** (11): 1103–6. doi:10.1097/01.wnr.0000227987.77304.cc. PMID 16837835.

[27] Keeler, Clyde E. (1933). "Absence of the Corpus callosum as a Mendelizing Character in the House Mouse". *Proceedings of the National Academy of Sciences of the United States of America.* **19** (6): 609–11. Bibcode:1933PNAS...19..609K. doi:10.1073/pnas.19.6.609. JSTOR 86284. PMC 1086100. PMID 16587795.

[28] Sarnat, Harvey B., and Paolo Curatolo (2007). *Malformations of the Nervous System: Handbook of Clinical Neurology*, p. 68

[29] Ashwell, Ken (2010). *The Neurobiology of Australian Marsupials: Brain Evolution in the Other Mammalian Radiation*, p. 50

[30] Armati, Patricia J., Chris R. Dickman, and Ian D. Hume (2006). *Marsupials*, p. 175

[31] Butler, Ann B., and William Hodos (2005). *Comparative Vertebrate Neuroanatomy: Evolution and Adaptation*, p. 361

[32] Morris, H., & Schaeffer, J. P. (1953). The Nervous system- The Brain or Encephalon. Human anatomy; a complete systematic treatise. (11th ed., pp. 920–921, 964–965). New York: Blakiston.

33.8 External links

- Stained brain slice images which include the "corpus callosum" at the BrainMaps project
- Comparative Neuroscience at Wikiversity
- NIF Search – Corpus callosum via the Neuroscience Information Framework
- National Organization for Disorders of the Corpus Callosum
- A 3D model of coprus callosum

Chapter 34

Brain mapping

This article is about brain mapping. For broader coverage, see Outline of brain mapping.

Brain mapping is a set of neuroscience techniques predicated on the mapping of (biological) quantities or properties onto spatial representations of the (human or non-human) brain resulting in maps. **Brain mapping** is further defined as the study of the anatomy and function of the brain and spinal cord through the use of imaging (including intra-operative, microscopic, endoscopic and multi-modality imaging), immunohistochemistry, molecular & optogenetics, stem cell and cellular biology, engineering (material, electrical and biomedical), neurophysiology and nanotechnology, according to the definition established in 2013 by Society for Brain Mapping and Therapeutics (SBMT).

34.1 Overview

All neuroimaging can be considered part of brain mapping. Brain mapping can be conceived as a higher form of neuroimaging, producing brain images supplemented by the result of additional (imaging or non-imaging) data processing or analysis, such as maps projecting (measures of) behavior onto brain regions (see fMRI). One such map, called a connectogram, depicts cortical regions around a circle, organized by lobes. Concentric circles within the ring represent various common neurological measurements, such as cortical thickness or curvature. In the center of the circles, lines representing white matter fibers illustrate the connections between cortical regions, weighted by fractional anisotropy and strength of connection.[1]

Brain mapping techniques are constantly evolving, and rely on the development and refinement of image acquisition, representation, analysis, visualization and interpretation techniques. Functional and structural neuroimaging are at the core of the mapping aspect of brain mapping.

Some scientists have criticized the brain image-based claims made in scientific journals and the popular press, like the discovery of "the part of the brain responsible" things like love or musical ability or a specific memory. Many mapping techniques have a relatively low resolution, including hundreds of thousands of neurons in a single voxel. Many functions also involve multiple parts of the brain, meaning that this type of claim is probably both unverifiable with the equipment used, and generally based on an incorrect assumption about how brain functions are divided. It may be that most brain functions will only be described correctly after being measured with much more fine-grained measurements that look not at large regions but instead at a very large number of tiny individual brain circuits. Many of these studies also have technical problems like small sample size or poor equipment calibration which means they cannot be reproduced - considerations which are sometimes ignored to produce a sensational journal article or news headline. In some cases the brain mapping techniques are used for commercial purposes, lie detection, or medical diagnosis in ways which have not been scientifically validated.[2]

34.2 History

In 1962 The origin of Brain Mapping Research was first started in Ohio, and conducted at the Columbus State Hospital. More than 500 subjects were scanned using the US patented Hyper-frequency Electroencephalograph (Hyfreeg) brain scanner for the Brain Mapping Research.[3] A detailed brain mapping report was published by the Battelle Memorial Institute "A New Window into the Human Brain?".[4] The Journal of the American Medical Association also published a report concerning this brain mapping research: "Is Nervous System Amplitude or Frequency Oriented?".[5] JAMA reported: "One of the points on which most neurologist have agreed, is that the nervous system is amplitude oriented. Now a new theory indicates exactly the opposite--that the nervous system actually is frequency oriented." As a result of the brain mapping research, the Psychiatric team members were able to cure: Epilepsy,

Psychomotor Epilepsy, Hallucinations, and Schizophrenia by lowering the neuronal activity in the Reticular Activating System located in the Brain Stem. They also observed the functions of Dreaming and the unique functions of the two Brain Hemispheres that was later confirmed by a girl born with only one Hemisphere.[6] [7]

A book has been published by Kindle Books describing the original Brain Mapping Research project conducted by Battelle Memorial Institute, and identifies a behavioral classification matrix and methods for personality modification.[8]

Victor H. Fischer was the Principal Investigator of the original Psychiatric team incorporating: ten Clinical Psychiatrists, Dr. Paul W. Watkins MD as a member of the Psychiatric staff, Dr. Calvin Baker MD, former commissioner of the Ohio Department of Mental Hygiene, and Neurologist consultant, USAF Colonel Robert F. Hood, MD, Neurology and Psychiatry, Director of Psychiatry, Wright-Patterson Medical Center, USA.

In the late 1980s in the United States, the Institute of Medicine of the National Academy of Science was commissioned to establish a panel to investigate the value of integrating neuroscientific information across a variety of techniques.[9]

Of specific interest is using structural and functional magnetic resonance imaging (fMRI), diffusion MRI (dMRI), magnetoencephalography (MEG), electroencephalography (EEG), positron emission tomography (PET), Near-infrared spectroscopy (NIRS) and other non-invasive scanning techniques to map anatomy, physiology, perfusion, function and phenotypes of the human brain. Both healthy and diseased brains may be mapped to study memory, learning, aging, and drug effects in various populations such as people with schizophrenia, autism, and clinical depression. This led to the establishment of the Human Brain Project.[10] It may also be crucial to understanding traumatic brain injuries (as in the case of Phineas Gage)[11] and improving brain injury treatment.[12]

Following a series of meetings, the International Consortium for Brain Mapping (ICBM) evolved.[13] The ultimate goal is to develop flexible computational brain atlases.

On May 5, 2010 the Supreme Court in India (Smt. Selvi vs. State of Karnataka) declared brain mapping, lie detector tests and narcoanalysis to be unconstitutional, violating Article 20 (3) of Fundamental Rights. These techniques cannot be conducted forcefully on any individual and requires consent for the same. When they are conducted with consent, the material so obtained is regarded as evidence during trial of cases according to Section 27 of the Evidence Act.[14]

34.3 Current Atlas tools

- Talairach Atlas, 1988
- Harvard Whole Brain Atlas, 1995[15]
- MNI Template, 1998 (The standard template of SPM and International Consortium for Brain Mapping)
- Atlas of the Developing Human Brain, 2012[16]

34.4 See also

- Outline of brain mapping
- Outline of the human brain
- Brain Mapping Foundation
- BrainMaps Project
- Center for Computational Biology
- Connectogram
- FreeSurfer
- Human Connectome Project
- IEEE P1906.1
- List of neuroscience databases
- Map projection
- Neuroimaging software
- Whole brain emulation
- Topographic map (neuroanatomy)
- Society for Brain Mapping and Therapeutics

34.5 References

[1] Irimia, Andrei; Chambers, Micah C.; Torgerson, Carinna M.; Horn, John D. (2012). "Circular representation of human cortical networks for subject and population-level connectomic visualization". *NeuroImage*. **60** (2): 1340–51. doi:10.1016/j.neuroimage.2012.01.107. PMC 3594415. PMID 22305988.

[2] Sally Satel; Scott O. Lilienfeld (2015). *Brainwashed: The Seductive Appeal of Mindless Neuroscience*. Basic Books. ISBN 978-0465062911.

[3] Fischer, Victor H. (July 20, 1965). "Detecting Physiological Conditions By Measuring Bioelectric Output Frequency #3,195,533". United States Patent Office.

[4] Fischer, Victor H. (May 1962). "A New Window into the Human Brain?". Battelle Technical Review: 3–9.

[5] Fischer, Victor H. (June 23, 1962). "Is Nervous System Amplitude or Frequency Oriented?". The Journal of the American Medical Association: 30–31.

[6] Fischer, Victor H. (August 4, 2009). "Bilateral visual field maps in a patient with only one hemisphere". **106** (31). PNAS Organization: 13034–13039.

[7] http://www.pnas.org/content/suppl/2009/07/31/0809688106.DCSupplementa

[8] Fischer, Victor H. (January 1, 2013). *Improving Your Thought Process*. Amazon Digital Services, Inc. ASIN B00AW1RZ00.

[9] Pechura, Constance M.; Martin, Joseph B. (1991). *Mapping the Brain and Its Functions: Integrating Enabling Technologies Into Neuroscience Research*. Institute of Medicine (U.S.). Committee on a National Neural Circuitry Database.

[10] Koslow, Stephen H.; Huerta, Michael F. (1997). *Neuroinformatics: An Overview of the Human Brain Project*.

[11] Van Horn, John Darrell; Irimia, Andrei; Torgerson, Carinna M.; Chambers, Micah C.; Kikinis, Ron; Toga, Arthur W. (2012). Sporns, Olaf, ed. "Mapping Connectivity Damage in the Case of Phineas Gage". *PLoS ONE*. **7** (5): e37454. doi:10.1371/journal.pone.0037454. PMC 3353935. PMID 22616011.

[12] Irimia, Andrei; Chambers, Micah C.; Torgerson, Carinna M.; Filippou, Maria; Hovda, David A.; Alger, Jeffry R.; Gerig, Guido; Toga, Arthur W.; Vespa, Paul M.; Kikinis, Ron; Van Horn, John D. (2012). "Patient-Tailored Connectomics Visualization for the Assessment of White Matter Atrophy in Traumatic Brain Injury". *Frontiers in Neurology*. **3**. doi:10.3389/fneur.2012.00010. PMC 3275792. PMID 22363313.

[13] Toga, Arthur W.; Mazziotta, John C., eds. (2002). *Brain Mapping: The Methods*. ISBN 978-0-12-693019-1.

[14] Math, SB (2011). "Supreme Court judgment on polygraph, narco-analysis & brain-mapping: a boon or a bane". *Indian J. Med. Res.* **134**: 4–7. PMC 3171915. PMID 21808125.

[15] Harvard Whole Brain Atlas

[16] Serag, Ahmed; Aljabar, Paul; Ball, Gareth; Counsell, Serena J.; Boardman, James P.; Rutherford, Mary A.; Edwards, A. David; Hajnal, Joseph V.; Rueckert, Daniel (2012). "Construction of a consistent high-definition spatio-temporal atlas of the developing brain using adaptive kernel regression". *NeuroImage*. **59** (3): 2255–65. doi:10.1016/j.neuroimage.2011.09.062. PMID 21985910.

34.6 Further reading

- Rita Carter (1998). *Mapping the Mind*.
- F.J. Chen (2006). *Brain Mapping And Language*
- F.J. Chen (2006). *Focus on Brain Mapping Research*.
- F.J. Chen (2006). *Trends in Brain Mapping Research*.
- F.J. Chen (2006). *Progress in Brain Mapping Research*.
- Koichi Hirata (2002). *Recent Advances in Human Brain Mapping: Proceedings of the 12th World Congress of the International Society for Brain Electromagnetic Topography (ISBET 2001)*.
- Konrad Maurer and Thomas Dierks (1991). *Atlas of Brain Mapping: Topographic Mapping of Eeg and Evoked Potentials*.
- Konrad Maurer (1989). *Topographic Brain Mapping of Eeg and Evoked Potentials*.
- Arthur W. Toga and John C. Mazziotta (2002). *Brain Mapping: The Methods*.
- Tatsuhiko Yuasa, James Prichard and S. Ogawa (1998). *Current Progress in Functional Brain Mapping: Science and Applications*.

34.7 External links

- Epilepsy & Brain Mapping Program
- BrainMapping.org project
- National Centers for Biomedical Computing
- Mapology.org
- Human Brain Mapping
- National Center for Multi-Scale Study of Cellular Networks
- National Center for Biomedical Ontology
- Physics-based Simulation of Biological Structures
- National Alliance for Medical Imaging Computing
- Informatics for Integrating Biology and the Bedside
- National Center for Integrative Biomedical Informatics
- Elekta Neuromag

- Brain Mapping Foundation
- Interactive Brain Map by InformED
- Society for Brain Mapping and Therapeutics

Chapter 35

Outline of brain mapping

"Human Brain Mapping" redirects here. For the scientific journal, see Human Brain Mapping (journal).

The following outline is provided as an overview of and topical guide to brain mapping:

Brain mapping – set of neuroscience techniques predicated on the mapping of (biological) quantities or properties onto spatial representations of the (human or non-human) brain resulting in maps. Brain mapping is further defined as the study of the anatomy and function of the brain and spinal cord through the use of imaging (including intra-operative, microscopic, endoscopic and multi-modality imaging), immunohistochemistry, molecular & optogenetics, stem cell and cellular biology, engineering (material, electrical and biomedical), neurophysiology and nanotechnology.

35.1 Broad scope

- History of neuroscience
- History of neurology
- Brain mapping
- Human brain
- Neuroscience
- Nervous system.

35.1.1 The neuron doctrine

- Neuron doctrine – A set of carefully constructed elementary set of observations regarding neurons. *For more granularity, more current, and more advanced topics, see the cellular level section*
- Asserts that neurons fall under the broader cell theory, which postulates:
  - All living organisms are composed of one or more cells.
  - The cell is the basic unit of structure, function, and organization in all organisms.
  - All cells come from preexisting, living cells.
- The Neuron doctrine postulates several elementary aspects of neurons:
  - The brain is made up of individual cells (neurons) that contain specialized features such as dendrites, a cell body, and an axon.
  - Neurons are cells differentiable from other tissues in the body.
  - Neurons differ in size, shape, and structure according to their location or functional specialization.
  - Every neuron has a nucleus, which is the trophic center of the cell (The part which must have access to nutrition). If the cell is divided, only the portion containing the nucleus will survive.
  - Nerve fibers are the result of cell processes and the outgrowths of nerve cells. (Several axons are bound together to form one nerve fibril. See also: Neurofilament. Several nerve fibrils then form one large nerve fiber. Myelin, an electrical insulator, forms around selected axons.
  - Neurons are generated by cell division.
  - Neurons are connected by sites of contact and not via cytoplasmic continuity. (A cell membrane isolates the inside of the cell from its environment. Neurons do not communicate via direct cytoplasm to cytoplasm contact.)
  - Law of dynamic polarization. Although the axon can conduct in both directions, in tissue there is a preferred *direction* of transmission from cell to cell.
- Elements added later to the initial Neuron doctrine

- A barrier to transmission exists at the site of contact between two neurons that may permit transmission. (Synapse)
- Unity of transmission. If a contact is made between two cells, then that contact can be either *excitatory* or *inhibitory*, but will always be of the same type.
- Dale's law, each nerve terminal releases a *single* type of neurotransmitter.

- Some of the basic postulates in the Neuron doctrine have been subsequently questioned, refuted, or updated. See the cellular level section topics for additional information.

35.1.2 Map, atlas, and database projects

- Brain Activity Map Project 2013 NIH $3 billion project to map every neuron in the human brain in ten years, based upon the Human Genome Project.

  - NIH Brain Research through Advancing Innovative Neurotechnologies (BRAIN) Initiative
  - Community outreach site for above where the public may comment

- Human Brain Project (EU) – 1 billion euro, 10-year project to simulate the human brain with supercomputers.

  - BigBrain A high-resolution 3D atlas of the human brain created as part of the HBP.

- Human Connectome Project – 2009 NIH $30 million project to build a network map of the human brain, including structural (anatomical) and functional elements. Emphasis included research into dyslexia, autism, Alzheimer's disease, and schizophrenia. See also Connectome a, comprehensive map of neural connections in the brain.

- Allen Brain Atlas 2003 $100 million project funded by Paul Allen (Microsoft)

- CONNECT. This project pulls together the EU's world-leading diffusion MRI community to focus on the fundamental advances key to the long-term realisation of microstructure and connectivity mapping of the live human brain as well as exploitation of that information by medical and neuroscience researchers.

- BrainMaps National Institute of Health (NIH) database including 60 terabytes of image scans of primate and non-primates, integrated with information covering structure and function.

- NeuroNames Defines the brain in terms of about 550 *primary* structures (about 850 *unique* structures) to which all other structures, names, and synonyms are related. About 15,000 neuroanatomical terms are cross indexed, including many synonyms in seven languages. Coverage includes the brain and spinal cord of the four species most frequently studied by neuroscientists: human, macaque (monkey), rat and mouse. The controlled, standardized vocabulary for each structure is located in an unambiguous, strict physical hierarchy, and these terms are selected based on ease of pronunciation, mnemonic value, and frequency of use in recent neuroscientific publications. Relation of each structure to its superstructures and substructures is included. The controlled vocabulary is suitable for uniquely indexing neuroanatomical information in digital databases.

- Decade of the Brain 1990-1999 promotion by NIH and the Library of Congress *"to enhance public awareness of the benefits to be derived from brain research"*. Communications targeted Members of Congress, staffs, and the general public to promote funding.

- Talairach Atlas see Jean Talairach

- Harvard Whole Brain Atlas see Human brain

- MNI Template see Medical image computing

- Blue Brain Project and Artificial brain

- International Consortium for Brain Mapping see Brain Mapping

- List of neuroscience databases

- NIH Toolbox National Institute of Health (USA) toolbox for the assessment of neurological and behavioral function

35.2 Imaging and recording systems

This section covers imaging and recording systems. The general section covers history, neuroimaging, and techniques for mapping specific neural connections. The specific systems section covers the various specific technologies, including experimental and widely deployed imaging and recording systems.

35.2.1 General

- Most imaging work to date on individual neurons has been conducted outside the brain, typically on large neurons, and has been most frequently destructive. New techniques are however rapidly emerging. Search on *"Single neuron imaging"* and see related topics: Biological neuron model, Single-unit recording, Neural oscillation#Single neuron model, Computational neuroscience#Single-neuron modeling. dMRI (above) is also promising in nondestructive imaging of single neurons inside the brain.

- History of neuroimaging (redirects from Brain scanner)

- Neuroimaging (redirects from Brain function map)

- Connectomics – mapping technique showing neural connections in a nervous system.

35.2.2 Specific systems

- Cortical stimulation mapping

- Diffusion MRI (dMRI) – includes *diffusion tensor imaging (DTI)* and *diffusion functional MRI (DfMRI)*. dMRI is a recent breakthrough in brain mapping allowing the visualization of cross connections between different anatomical parts of the brain. It allows non-invasive imaging of white matter fiber structure and in addition to mapping can be useful in clinical observations of abnormalities, including damage from stroke.

- Electroencephalography (EEG) Uses electrodes on the scalp and other techniques to detect the electrical flow of currents.

- Electrocorticography intracranial EEG, the practice of using electrodes placed directly on the exposed surface of the brain to record electrical activity from the cerebral cortex.

- Electrophysiological techniques for clinical diagnosis

- Functional magnetic resonance imaging (fMRI)

- Medical image computing (brain research of leads medical and surgical uses of mapping technology)

- Neurostimulation (in research stimulation is frequently used in conjunction with imaging)

- Positron emission tomography (PET) a nuclear medical imaging technique that produces a three-dimensional image or picture of functional processes in the body. The system detects pairs of gamma rays emitted indirectly by a positron-emitting radionuclide (tracer), which is introduced into the body on a biologically active molecule. Three-dimensional images of tracer concentration within the body are then constructed by computer analysis. In modern scanners, three dimensional imaging is often accomplished with the aid of a CT X-ray scan performed on the patient during the same session, in the same machine.

35.2.3 Imaging and recording componentry

Electrochemical

- Haemodynamic response the rapid delivery of blood to active neuronal tissues. Blood Oxygenation Level Dependent signal (BOLD), corresponds to the concentration of deoxyhemoglobin. The BOLD effect is based on the fact that when neuronal activity is increased in one part of the brain, there is also an increased amount of cerebral blood flow to that area. Functional magnetic resonance imaging is enabled by the detection of the BOLD signal.

- Event-related functional magnetic resonance imaging can be used to detect changes in the Blood Oxygen Level Dependent (BOLD) hemodynamic response to neural activity in response to certain events.

Electrical

- Event-related potential positive and negative 10μ to 100μ Volts (μ is millionths) responses, measured via noninvasive electrodes attached to the scalp, that are the reliable and repeatable results of a certain specific sensory, cognitive, or motor event. These are also called *a stereotyped electrophysiological response to a stimulus*. They are called somatosensory evoked potentials when they are elicited by sensory (vs. cognitive or motor) event stimuli. The voltage swing sequences are recorded and broken down by positive and negative, and by how long after the stimulus they are observed. For example, [N100] is a negative swing observed between 80 and 120 milliseconds (100 being the midpoint) after the onset of the stimulus. Alternatively, the voltage swings are labeled based on their order, N1 being the first negative swing observed, N2 the second negative swing, etc. See: N100 (neuroscience), N200 (neuroscience), P300 (neuroscience), N400 (neuroscience), P600 (neuroscience). The first negative and positive swings (see Visual N1, C1 and P1 (neuroscience)) in response to visual stimulation are of particular interest in studying sensitivity and selectiveness of attention.

Electromagnetic

- Magnetoencephalography – a technique for mapping brain activity by recording magnetic fields produced by electrical currents occurring naturally in the brain, using very sensitive magnetometers In research, MEG's primary use is the measurement of *time courses* of activity. MEG can resolve events with a precision of 10 milliseconds or faster, while functional MRI (fMRI), which depends on changes in blood flow, can at best resolve events with a precision of several hundred milliseconds. MEG also accurately pinpoints sources in primary auditory, somatosensory and motor areas. For creating functional maps of human cortex during more complex cognitive tasks, MEG is most often combined with fMRI, as the methods complement each other. Neuronal (MEG) and hemodynamic (fMRI) data do not necessarily agree, in spite of the tight relationship between local field potentials (LFP) and blood oxygenation level dependent (BOLD) signals

Radiological

- Positron-emitting radionuclide (tracer). See Positron emission tomography

- Altanserin a compound that binds to a serotonin receptor. When labeled with the isotope fluorine-18 it is used as a radioligand in positron emission tomography (PET) studies of the brain.

Visual processing and image enhancement

- Scientific visualization an interdisciplinary branch of science primarily concerned with the visualization of three-dimensional phenomena (including medical, biological, and others), where the emphasis is on realistic renderings of volumes, surfaces, illumination sources, and so forth, perhaps with a dynamic (time) component. It is considered a branch of computer science that is a subset of computer graphics. Brain mapping is a leading beneficiary of advances in scientific visualization.

- Blob detection an area in computer vision, A blob is a region of a digital image in which some properties (such as brightness or color, compared to areas surrounding those regions) are constant or vary within a prescribed range of values; all the points in a blob can be considered in some sense to be similar to each other

Information technology

- Determining the number of clusters in a data set A typical application is in data reduction: as the increase in temporal resolution of fMRI experiments routinely yields fMRI sequences containing several hundreds of images, it is sometimes necessary to invoke feature extraction to reduce the dimensionality of the data space.

- Fractional anisotropy a measure often used in diffusion imaging where it is thought to reflect fiber density, axonal diameter, and myelination in white matter. The FA is an extension of the concept of eccentricity of conic sections in 3 dimensions, normalized to the unit range. Anisotropy is the property of being directionally dependent, as opposed to isotropy, which implies identical properties in all directions.

- General linear model – a statistical linear model. It may be written as $Y=XB+U$ where Y is a matrix with series of multivariate measurements, X is a matrix that might be a design matrix, B is a matrix containing parameters that are usually to be estimated, and U is a matrix containing errors or noise. It is frequently used in the analysis of multiple brain scans in scientific experiments where Y contains data from brain scanners, X contains experimental design variables and confounds. See also: statistical parametric mapping

- Resampling (statistics) see section on permutation tests. Nonparametric Permutation Tests are used in fMRI.

Software packages

- Analysis of Functional NeuroImages an open-source environment for processing and displaying functional MRI data

- Cambridge Brain Analysis a software repository developed at University of Cambridge for functional magnetic resonance imaging (fMRI) analysis under the GNU General Public License and runs under Linux.

- Statistical parametric mapping – a statistical technique for examining differences in brain activity recorded during functional neuroimaging experiments using neuroimaging technologies such as fMRI or PET. It may also refer to a specific piece of software created by the Wellcome Department of Imaging Neuroscience (part of University College London) to carry out such analyses.

- ITK-SNAP an interactive software application that allows users to navigate three-dimensional medical images, manually delineate anatomical regions of interest, and perform automatic image segmentation. Its most frequently used to work with magnetic resonance imaging (MRI) and computed tomography (CT) data sets.

35.3 Scientists, academics and researchers

- Mark S. Cohen neuroscientist Professor at the UCLA. Early pioneer of functional brain imaging using magnetic resonance imaging (MRI).

- Anders Dale neuroscientist and Professor University of California, San Diego. He developed FreeSurfer brain imaging analysis software that facilitates the visualization of the functional regions of the highly folded cerebral cortex.

- Pierre Flor-Henry demonstrated in a study of epileptic psychosis, that schizophrenia relates to left and manic-depressive states relate to right hemisphere epilepsies

- Angela D. Friederici director at the Max Planck Institute for Human Cognitive and Brain Sciences in Leipzig, Germany with a specialization in neuropsychology and linguistics.

- Karl J. Friston British neuroscientist and authority on brain imaging. Inventor of statistical parametric mapping

- Isabel Gauthier neuroscientist and head of the Object Perception Lab at Vanderbilt University

- Matthew Howard, III Professor of Neurosurgery at the University of Iowa known for contributions in the field of human brain mapping using intracranial electrophysiology.

- Dr. Surbhi Jain, the first female neurosurgeon from State of Rajasthan. Practices at the Moffitt Cancer Center, Tampa, Florida, and holds world's record for the most number of patients treated by brain mapping guided brain surgery.

- Gitte Moos Knudsen Gitte Moos Knudsen neurobiologist and clinical neurologist professor at Copenhagen University Hospital.

- Kenneth Kwong Scientist at Harvard University known for his work in fMRI

- Robert Livingston (scientist) (October 9, 1918 – April 26, 2002) neuroscientist in 1964 Livingston founded the neuroscience department, the first of its kind in the world, at the newly built University of California, San Diego. His best known research was in the computer mapping and imaging of the human brain. His interest in the brain also extended to questions of cognition, consciousness, emotions, and spirituality.

- Helen S. Mayberg – professor of neurology and psychiatry at Emory University. Specialization includes delineating abnormal brain function in patients with major depression using functional neuroimaging.

- Geraint Rees head of the University College London Faculty of Brain Sciences

- Sidarta Ribeiro neuroscientist and Director of the Brain Institute at Universidade Federal do Rio Grande do Norte

- Perminder Sachdev Neuropsychiatrist Professor at University of New South Wales and director of the Centre for Healthy Brain Ageing

- Pedro Antonio Valdes-Sosa Vice-Director of the Cuban Neuroscience Center which he cofounded in 1990. His specialization includes the statistical analysis of electrophysiological measurements, neuroimaging (fMRI, EEG and MEG tomography), nonlinear dynamical modeling of brain functions including software and electrophysiological equipment development. Member of the Editorial Boards of *NeuroImage*, Medicc, Audioology and Neurotology, *PLosOne*, and Brain Connectivity.

- Robert Turner director at the Max Planck Institute for Human Cognitive and Brain Sciences in Leipzig, Germany with a specialization in brain physics and magnetic resonance imaging (MRI). He is credited with creating the design for the coils found inside every MRI scanner.

- Arno Villringer Director at the Max Planck Institute for Human Cognitive and Brain Sciences in Leipzig, Germany

35.4 Research Institutions

- Laboratory of Neuro Imaging research laboratory within the Department of Neurology at the UCLA School of Medicine. The laboratory conducts a wide variety of brain imaging studies of normal brain anatomy and function, development, aging, and disease.

- University of Texas Health Science Center Department of Radiology – is the second largest academic department in Radiological Sciences in the United States. The department was historically the first program in the United States to establish a Ph.D. program for radiology residents, which is known as the Human Imaging graduate program. See also Stanford Radiology

35.5 Journals

- Behavioral and Brain Sciences
- Developmental Science
- Genes, Brain and Behavior
- Human Brain Mapping (journal)
- Journal of Cerebral Blood Flow & Metabolism
- Journal of Neurochemistry
- Journal of Neurophysiology
- Journal of Neuroscience
- Nature Neuroscience
- Neuroimage
- Neuron
- Trends in Neurosciences

35.6 See also

- Outline of the human brain
- Outline of neuroscience

See also categories

- Category:Brain
- Category:Brain–computer interfacing
- Category:Central nervous system neurons
- Category:Human behavior
- Category:Image processing
- Category:Mind
- Category:Nervous system
- Category:Neural engineering
- Category:Neurobiology
- Category:Neuroimaging journals
- Category:Neurons
- Category:Neuroscience
- Category:Neural coding
- Category:Neuroimaging
- Category:Neuroinformatics
- Category:Neuroscience research centers
- Category:Politics of science – brain research funding issues

35.7 Notes and references

35.8 Text and image sources, contributors, and licenses

35.8.1 Text

- **Human brain** *Source:* https://en.wikipedia.org/wiki/Human_brain?oldid=735474762 *Contributors:* Bryan Derksen, The Anome, Alex.tan, Montrealais, Olivier, Edward, Ubiquity, JohnOwens, Vaughan, Kku, Ixfd64, Ahoerstemeier, Angela, JohnKozak, JWSchmidt, Kingturtle, Julesd, Quickbeam, Ec5618, Timwi, David Newton, RickK, Jay, Andrewman327, Gutza, Jogloran, Dtgm, Maximus Rex, Omegatron, Samsara, Bevo, Raul654, Jerzy, AnthonyQBachler, Hajor, Paranoid, Fredrik, Jredmond, Sverdrup, Hemanshu, Texture, Seth Ilys, Cyrius, Wile E. Heresiarch, Dina, Enochlau, Sethoeph, Centrx, Giftlite, DocWatson42, Washington irving, Nunh-huh, Tom harrison, Brian Kendig, Fastfission, Everyking, Bird, Guanaco, Luigi30, Hazzamon, OldakQuill, Utcursch, Yath, Fangz, Antandrus, Beland, OverlordQ, ClockworkLunch, RaymondByrd, Dubious, YankeeInCA, H Padleckas, AndrewKeenanRichardson, PFHLai, JamesByrd, AOL rules, Icairns, Sayeth, Joyous!, TJSwoboda, Ratiocinate, Demiurge, DMG413, Discospinster, William Pietri, Rich Farmbrough, Chrischan~enwiki, FT2, Vsmith, Dbachmann, Bender235, ESkog, Bcjordan, JustPhil, Ben Webber, Sfahey, Lankiveil, Shanes, Bobo192, Smalljim, Nectarflowed, Arcadian, ParticleMan, Nk, Famousdog, Nsaa, Jjron, Edital, Alansohn, Liao, Eric Kvaalen, Arthena, Jezmck, Sjschen, Lmviterbo, Wouterstomp, Riana, Primalchaos, Bart133, Snowolf, Ombudsman, Velella, Ronark, Egg, Rudresha, DV8 2XL, KTC, Ceyockey, Diogopedrosa, Crosbiesmith, OwenX, Woohookitty, TigerShark, Daniel Case, Uncle G, Kurzon, Zealander, JeremyA, Hdante, Alfakim, CiTrusD, Kmg90, Plegovini, Dolfrog, Dysepsion, Phoenix-forgotten, Phaedrus C, Rjwilmsi, Koavf, Tangotango, SMC, Ekspiulo, Oblivious, Miserlou, Ligulem, St33lbird, Titoxd, Gringo300, Latka, Mister Matt, Crazycomputers, Nivix, Andy85719, RexNL, Gurch, Czar, Alexjohnc3, KFP, Super Sam, Alphachimp, PaulWicks, Gurubrahma, Manufracture, Moocha, DVdm, Satanael, Sceptre, Huw Powell, RussBot, Crazytales, Chuck Carroll, Hydrargyrum, Stephenb, Gaius Cornelius, Cambridge-BayWeather, Pseudomonas, Wimt, Stassats, Irrevenant, NawlinWiki, A314268, Fizan, Wiki alf, BigCow, NickBush24, Nrets, Irishguy, Rjlabs, Nathew, Moe Epsilon, Sallison, Cameronreilly, Gadget850, DeadEyeArrow, Maunus, N. Harmonik, Nick123, Wknight94, Jeremyzone, Ageekgal, Theda, Closedmouth, QmunkE, Spliffy, Allens, Airconswitch, SpLoT, SmackBot, Tomyumgoong, Saravask, KnowledgeOfSelf, NZUlysses, Kmwmtd, Maian, Boris Barowski, Kslays, Dhochron, Mintpieman, TypoDotOrg, Septegram, Commander Keane bot, Yamaguchi??, Gilliam, Skizzik, Bluebot, TimBentley, Persian Poet Gal, SB Johnny, Mdwh, PureRED, Darth Panda, Rlevse, Gracenotes, Audriusa, Muboshgu, Can't sleep, clown will eat me, Laguna77~enwiki, JRPG, A.tenharmsel, Thisisbossi, Rrburke, Joema, Mr.Z-man, Stevenmitchell, Lox, Flyguy649, Khukri, Epachamo, TheLimbicOne, MichaelBillington, Taggart Transcontinental, Lcarscad, Marcus Brute, Daniel.Cardenas, Ligulembot, Jóna Þórunn, Suidafrikaan, Sadi Carnot, Pilotguy, Kukini, Hmoul, Mchavez, Ptpete25, Wvbailey, Sophia, Kuru, WhiteCat, Buchanan-Hermit, Jordanduval, JorisvS, Minna Sora no Shita, IronGargoyle, Grumpyyoungman01, MaximvsDecimvs, Stwalkerster, Optimale, Mr Stephen, Bojan22, Novangelis, Elb2000, Scorpion0422, Cerealkiller13, Kripkenstein, Keitei, KJS77, Fan-1967, Iridescent, TwistOfCain, JoeBot, Wjejskenewr, Blehfu, Robinavery, Courcelles, Tawkerbot2, Dlohcierekim, The Letter J, ChrisCork, Mostly Zen, Dia^, JForget, Mellery, Mattbr, Wafulz, Makeemlighter, CBM, Mcstrother, Qc, KyraVixen, Dshin, 5-HT8, I love editing, Dgw, CCCP, Denstat, MeekMark, Fatalserpent, Jdietsch, Nauticashades, WikiEdit, Rocketboy50, ShoobyD, Naveenbm, Jugad, LouisBB, Anthonyhcole, ST47, Demomoke, Christian75, Colorprobe, Roberta F., DumbBOT, PhineasG, Kozuch, Karuna8, Gerard.percheron, JBrusey, Zalgo, Iss246, FrancoGG, Oleksii0, Epbr123, Lord Hawk, Pstanton, Ucanlookitup, N5iln, Mojo Hand, Anupam, Headbomb, Lmarfell, Marek69, Esowteric, John254, Kathovo, Michael A. White, Klausness, Gossamers, AntiVandalBot, Majorly, Luna Santin, Opelio, Vinigk, KP Botany, SmokeyTheCat, CultureArchitect, Zweifel, Danger, Chill doubt, Lantios, E.James, Hermant patel, Willhaslett, MER-C, The Transhumanist, Gerculanum, PhilKnight, SmokeyTheFatCat, Gert7, WolfmanSF, Bongwarrior, VoABot II, Alexgt, JNW, Rami R, Michele123, Balloonguy, Midgrid, Bubba hotep, Fabricramp, WhatamIdoing, Indon, Ashadeofgrey, DerHexer, JaGa, Esanchez7587, Nevit, Lunakeet, Rustyfence, MartinBot, Grandia01, Whybealive, Audreyt, Munkin3, Wizzywiz, Aleksander.adamowski, Sm8900, Tjamespaul, Xasz, Uriel8, R'n'B, CommonsDelinker, Nono64, EverSince, Nathan412, Etaicq, J.delanoy, Captain panda, Dr LaCombe, CFCF, Trusilver, Tlim7882, Bogey97, Nbauman, Peter Chastain, Cyanolinguophile, Cocoaguy, Octopus-Hands, Kpmiyapuram, Kintakus, Davidm617617, Uparepwe, Mikael Häggström, Psiguy, Dexter prog, Tobias Willmott, Spinach Dip, LittleHow, Belovedfreak, NewEnglandYankee, In Transit, Billybobblue69, Pundit, Angular, BrettAllen, Entropy, Ssault, Jingyangfan, Idiomabot, Waynem37, Semmelweiss, Thedjatclubrock, Temporarily Insane, SleweD, Leebo, Orion99, Jsophrin, 8thstar, Alexkorn, Philip Trueman, AllPurposeGamer, Oshwah, RonSavelo, Philoprof, Vipinhari, Anonymous Dissident, Crohnie, Aymatth2, Corvus cornix, Amaher, Leafyplant, LeaveSleaves, Mannafredo, Cremepuff222, Shadowlapis, Wiae, Romaniandude, Kurowoofwoof111, Comrade Tux, Timothy pp, Laracroft33, Tbtkorg, Lova Falk, Norocron~enwiki, Enviroboy, Prayspot, Morthis, Sapphic, ObjectivismLover, Why Not A Duck, Pimp2323, Vinhtantran, Lando5, Caicai996, Demize, Ronsword, SLarson11, SPQRobin, Kkkkk11111, Markdraper, Tresiden, Frans Fowler, Weeliljimmy, Virtual Cowboy, Sakkura, Dawn Bard, Caltas, Matthew Yeager, Triwbe, Dayacrazy, Yintan, Karonaway, Andersmusician, Smallblueluv, Juoj8~enwiki, Bentogoa, Flyer22 Reborn, Tiptoety, Permacultura, Aravindk editing, Oda Mari, TheGame5050, Oxymoron83, Iain99, Techman224, Dravecky, Promodulus, Perfectapproach, Mike2vil, Monroetransfer, Anchor Link Bot, Mygerardromance, Paulinho28, Ascidian, Levani91, Denisarona, Budhen, Atif.t2, ClueBot, Binksternet, Fyyer, The Thing That Should Not Be, Jagun, R000t, Newone25302, Arakunem, Drmies, Trilobite12, Sanjeev.singh3, LizardJr8, P. S. Burton, Gandaliter, Puchiko, Masterpiece2000, DragonBot, Robert Skyhawk, Excirial, Jusdafax, M4gnum0n, Encyclopedia77, TonyBallioni, WikiNehal, Eeekster, Nadeemj, Gtstricky, Lartoven, Posix memalign, Adrian Comollo, Hazel-roo, NuclearWarfare, Claurie, EhJJ, PeterTheWall, BlueCaper, Tnxman307, OekelWm, Frozen4322, Idontknow610, La Pianista, Bbriggs1, Calor, RenamedUser jaskldjslak901, Thingg, Aitias, PCHS-NJROTC, Sambo1234, Tgruwell, Johnuniq, SoxBot III, Katara4real, DumZiBoT, Wherewithal, BarretB, XLinkBot, Wholebrainer, Staticshakedown, Orcslayer5, PseudoOne, Rror, Vojtěch Dostál, Doc9871, NellieBly, Mifter, FireBrandon, DrAjit-Parkash, P.r.newman, Addbot, Xp54321, Mhines54, Zombie Ah Meng, Willking1979, Eric Drexler, Browlm13, Tcncv, Landon1980, Captaintucker, Otisjimmy1, Crazysane, Binary TSO, DougsTech, Blethering Scot, Fieldday-sunday, Ironholds, Ashton1983, Ka Faraq Gatri, Tills, Looie496, WetSexyLlama, Download, Glane23, Doniago, Humboldtperson97, West.andrew.g, 5 albert square, Burto 19, Lightbot, Ghemachandar, Krano, Jarble, Arbitrarily0, Dwayne Reed, Wagnerax, Legobot, Signus1, Luckas-bot, Yobot, Fraggle81, Esmehwa, JustWong, Shalomamigos, THEN WHO WAS PHONE?, Brougham96, QueenCake, Hamako, Ayrton Prost, Anand011892, Eric-Wester, Tempodivalse, AnomieBOT, Sencinner~enwiki, Tryptofish, Rjanag, Jim1138, IRP, Piano non troppo, AdjustShift, Kingpin13, Law, Flewis, Bluerasberry, Materialscientist, RobertEves92, Citation bot, Flea10, Chahat upreti, Maxis ftw, Intractable, Neurolysis, Dizzyadora, Tekks, Sionus, Apothecia, CptTaco, Addihockey10, Capricorn42, Nasnema, XZeroBot, Ched, Fluke42, Adartsug, Aurelius787, Strossmayer, Xuxa101, Doulos Christos, Trafford09, Sewblon, Gabriel3010, WaysToEscape, Extensor, Žiedas, A.amitkumar, Frozenevolution, FrescoBot, NSH002, Elvato333, Catmannah19, Yijia109, HJ Mitchell, Santiago bronson, Yankeefan95, Gordonlighter, Citation bot 1, Julious me, Shk9664, Pinethicket, Rajveetie, Vicenarian, Plasmanine, 10metreh, Sexypieman, Calmer Waters, Mrwest67, DrFree, Hopeiamfine, Reconsider the static, DeLungMD, Trappist the monk,

Nicholomothy, Tofutwitch11, Dinamik-bot, Vrenator, Contentsaid, January, Defender of torch, Reaper Eternal, Jeffrd10, EMacG, DARTH SIDIOUS 2, Onel5969, Mean as custard, RjwilmsiBot, Wexeb, Sedukai, NerdyScienceDude, Tbaptiste9, Slon02, Skamecrazy123, DASH-Bot, Immunize, Stephanesibani, Ajraddatz, Gcastellanos, Martini999, Anon tan, RA0808, Headcrap101, Halberdmetal, Tommy2010, Dcirovic, Maneganeshpopat, Blaze00, Pancake91, Tmajoor, Imperial Monarch, Pjmurphy81, Jeanpetr, WeijiBaikeBianji, Player-23, Hazard-SJ, Hanjifi, Caiomarinho, AManWithNoPlan, Makecat, Alrik, Augurar, Brandmeister, L Kensington, ନଅର ଠିକଣା, LikeLauren, Donner60, Carmichael, Gooroo72, Elipazooki, AUN4, Petrb, Mikhail Ryazanov, ClueBot NG, RaptorHunter, MelbourneStar, A520, Piast93, Bped1985, SuperKevin48, AveVeritas, Destorywiki, Jj1236, The Master of Mayhem, Intlicious, AnnikaGirlio, Srivastavavishist, Widr, Fish74~enwiki, Helpful Pixie Bot, VanishedUser hjgjktyjhddgf, Electriccatfish2, Kinaro, Wasbeer, Kanadmarick, Niggiecananchez, Snow Rise, Mark Arsten, Exercisephys, Health333, Yowanvista, CitationCleanerBot, Aliiisha, Davefordiscool, P'tit Pierre, Puppypug98, NotWith, Anatomist90, Neel doshi, Klilidiplomus, Gatorfan1983, Biosthmors, Harrypotterrulz123, Alfredpyo1, Iliketacos121, Hghyux, Sonc3924, Dxman345, Carrin19830, Jionpedia, Nathanstheking, Parker poet, Adam26005, Dexbot, G.Kiruthikan, MFSwine, TwoTwoHello, Lugia2453, JakobSteenberg, SFK2, KO6327, Mayank Reeshu, Telfordbuck, SailorMoonLover4Life, Hillbillyholiday, Epicgenius, Fantastic999, Vanamonde93, NMoran0449, Vafaeva, S.Sharma01, Iztwoz, Jakeglass, Dustin V. S., Frederic.Borries, Kelvinwei1, Ugog Nizdast, Spyglasses, LT910001, Seppi333, Mfb, Anrnusna, Meteor sandwich yum, AlexBardoel, Roborule, ???, Rupkatha roy, Concord hioz, Monkbot, Erebusthedark, BethNaught, Rustdustbust, MALHOTRA ABHI, Kollo3250, MicroMacroMania, Multifirefury, Poiuytrewqvtaatv123321, Jaiden-vamp, DCognus, Patel.1340, Solanki.15, Pharoh king, DavidJac, Sarr Cat, Louisbts, Arghyadeep Acharya, Tilifa Ocaufa, Gamingforfun365, Datonewhoknowsstuff, G.tenorio.ruiz, Anpanman, OllieMPS, CAPTAIN RAJU, The Quixotic Potato, Bikram sahoo dilu, Milostelzer103, Felipemossahernandez, Pijushbhatta, Lukethemenace, Foogoofudge, Babyjay1998 and Anonymous: 1321

- **Cerebral cortex** *Source:* https://en.wikipedia.org/wiki/Cerebral_cortex?oldid=734291029 *Contributors:* The Anome, Fnielsen, William Avery, SimonP, Caltrop, Michael Hardy, Angela, Nikai, Evercat, Emperorbma, Novum, Johnmarks, Selket, Dmbowden, Taoster, Robbot, Chris 73, Postdlf, Hadal, Vikreykja, Michael Snow, Cyrius, Pengo, Enochlau, Washington irving, Hokanomono, Electric goat, Michael Devore, Bird, Chinasaur, Jfdwolff, Jabowery, OldakQuill, Gadfium, CryptoDerk, Antandrus, Asbestos, Quota, Cerebral, Danga, Absinf, Cacycle, Bender235, Robert P. O'Shea, Bobo192, Shenme, Arcadian, AnnaP, Katefan0, Super-Magician, Irdepesca572, Freyr, Star Trek Man, Schzmo, Waldir, BD2412, Phillipedison1891, Phaedrus C, Rjwilmsi, Pleiotrop3, Dvulture, Yamamoto Ichiro, FlaBot, Nihiltres, Vsion, Wavelength, RussBot, Archelon, Eleassar, Jehoshaphat, NawlinWiki, A314268, Grafen, Bayle Shanks, Nephron, Daniel Mietchen, Supten, DeadEyeArrow, Phgao, Rto, Superp, Hrvatska, Danny-w, Dontaskme, Limited memory, Caco de vidro, Gaelle Desbordes, SmackBot, Took, Marc Kupper, EncephalonSeven, (boxed), Mike hayes, Chlewbot, Rrburke, Nakon, TheLimbicOne, Drphilharmonic, Clicketyclack, Saerain, Attys, Frozen-Man, Gleng, Riffic, Jcbutler, Iridescent, Aeternus, Benplowman, CmdrObot, Matt1299, BKalesti, RelentlessRecusant, Anthonyhcole, PhineasG, Icehcky8, Gerard.percheron, BCSWowbagger, Thijs!bot, Epbr123, CopperKettle, Edhubbard, AntiVandalBot, Luna Santin, Jj137, Kaobear, Hroðulf, Bongwarrior, VoABot II, Freddyd945, GermanX, MartinBot, ChemNerd, Nikpapag, Pr495du, J.delanoy, CFCF, Chopin-Ate-Liszt!, SteveChervitzTrutane, Lbthrice, LittleHow, Richard D. LeCour, Duras2000, DadaNeem, Paskari, Shoessss, Juliancolton, Guyzero, Raymond-Ferr, Hyteqsystems, Oshwah, Mercy, Ann Stouter, Wagoo, Amaher, Richwil, Lova Falk, RaseaC, Caps tiki, SieBot, K. Annoyomous, Flyer22 Reborn, Tiptoety, Exert, JSpung, Faradayplank, Aspects, OKBot, Mike2vil, Mygerardromance, Gantuya eng, ImageRemovalBot, ClueBot, NickCT, Binksternet, CounterVandalismBot, Thewildtype, Lartoven, RayquazaDialgaWeird2210, M.O.X, Thingg, Llameejones, Cmungall, P.r.newman, Addbot, DOI bot, Elishabet, Averyjack, Looie496, Download, CarsracBot, Maltasanitaeter, Tide rolls, Lightbot, Luckas-bot, Yobot, TaBOT-zerem, Heisenbergthechemist, KamikazeBot, Synchronism, AnomieBOT, Wikijoby, Joule36e5, Piano non troppo, Ulric1313, Bluerasberry, Materialscientist, Citation bot, NinetyNineFennelSeeds, ArthurBot, Tekks, S h i v a (Visnu), Tomdo08, RibotBOT, FrescoBot, Dogposter, NifCurator1, Thanhluan001, DivineAlpha, Citation bot 1, Tom.Reding, Gabrieldavisjones, Shanmugamp7, Hopeiamfine, Юрий Педаченко, 404 page not found, Trappist the monk, Piero le fou, Vrenator, Billyth3pupp3t, Solzhenitsyn1, RjwilmsiBot, Kerrick Staley, EmausBot, Lucien504, Waldheri, Slightsmile, Wikipelli, Dcirovic, Evanh2008, ZéroBot, Tmajoor, Deliciousmonkey, Donner60, Hazard-Bot, Chuispaston-Bot, Woodyim9691, ClueBot NG, Beerse, Wimpus~enwiki, Frietjes, Mesoderm, Skylightleo, Helpful Pixie Bot, BG19bot, Arnavchaudhary, Mishayla, Snow Rise, Cadiomals, Rengekicounter, MrBill3, Tskittles, Thenerdypengwin, Joehill11, Beejayjat, CarrieVS, Sag203, Guptakhy, Aestin, Isarra (HG), Bulba2036, Jamesx12345, ?, Afutureghost, Epicgenius, Wiki naraj, Jht94, Iztwoz, Liberalufp, DavidLeighEllis, Dinisoe, Rohan s pandey, LT910001, Seppi333, Ginsuloft, BruceBlaus, LoSchizzatore, Michael K. Duke, Monkbot, Buggiehuggie, DangerousJXD, Velvel2, Dkeverding, Trick2donging, Neuron Doc, ShaePony, KasparBot, DrJanaOfficial and Anonymous: 272

- **Frontal lobe** *Source:* https://en.wikipedia.org/wiki/Frontal_lobe?oldid=735886544 *Contributors:* Bryan Derksen, LA2, Vaughan, Jovan, ZoeB, JWSchmidt, Emperorbma, Dysprosia, Wik, Selket, Maximus Rex, Robbot, RedWolf, Hadal, Xanzzibar, VanishedUser kfljdfjsg33k, Giftlite, Washington irving, Hokanomono, SoCal, Bird, Jfdwolff, Delta G, Discospinster, Mani1, ESkog, Bcjordan, Arcadian, MPerel, Ral315, Ogress, Alansohn, Jhertel, Arthena, Wtmitchell, Kenyon, 2004-12-29T22:45Z, Commander Keane, JeremyA, Prashanthns, Samvscat, Rjwilmsi, Koavf, Brighterorange, MarnetteD, RexNL, TheDJ, Chobot, YurikBot, Wavelength, Chris Capoccia, Stephenb, Rsrikanth05, Wimt, A314268, Grafen, Robdurbar, Redtailstinger, Sallison, Mysid, Cavan, Zzuuzz, Curpsbot-unicodify, Allens, Kungfuadam, Phl, Eenu, SmackBot, Eaglizard, Yamaguchi??, Cool3, Gilliam, BrotherGeorge, Mordac, Miquonranger03, Darth Panda, Addshore, Flyguy649, Troas, Angela26, IronGargoyle, Camazine, Stwalkerster, Yasha I, Noah Salzman, Big Smooth, Squirepants101, KJS77, Iridescent, Michaelbusch, WU03, Gil Gamesh, Cactus Bob, Ale jrb, Ccie13836, Was a bee, Anthonyhcole, Llort, Fifo, Kiske, Nikopoley, Omicronpersei8, Zalgo, Thijs!bot, RickinBaltimore, Pfranson, Natalie Erin, Danger, Magioladitis, Jmtrivial, WhatamIdoing, Paladin Hammer, STBot, Gaidheal1, Kostisl, Nono64, J.delanoy, Neutron Jack, Rhinestone K, Kpmiyapuram, Katalaveno, Mrs.meganmmc, McSly, Cobi, Elplatt, Xnuala, X!, Deor, Philip Trueman, TXiKiBoT, Oshwah, Mark v1.0, Rcarlosagis, DieBuche, Lova Falk, Why Not A Duck, Schnurrbart, John.n-irl, WereSpielChequers, Dawn Bard, Caltas, Yintan, Flyer22 Reborn, Literaturegeek, ClueBot, Liberfinis, Binksternet, The Thing That Should Not Be, Waterlily16, Wysprgr2005, Myzel, Teleomatic, SoxBot III, Rror, SilvonenBot, Aunt Entropy, Airplaneman, EEng, Addbot, Ardkorjunglist, Wsvlqc, Landon1980, Non-dropframe, Looie496, Download, LaaknorBot, Anonymaus, Bwrs, Tide rolls, Zorrobot, Arbitrarily0, Legobot, Luckas-bot, Yobot, ישראל קרול, Swister-Twister, AnomieBOT, Nutriveg, Weglinton, Materialscientist, Citation bot, Jmarchn, Neurolysis, Xqbot, JimVC3, Capricorn42, Omnipaedista, Musketeer41, Hiersgarr, Shadowjams, Thehelpfulbot, Custoo, Tangent747, Sopheroo, LucienBOT, Remotelysensed, Paine Ellsworth, NifCurator1, BenzolBot, Cannolis, RedBot, Merlion444, Clarkcj12, Bluefist, Onel5969, The Utahraptor, RjwilmsiBot, Viniciusmc, EmausBot, John of Reading, Spunkylyssa, Dcirovic, LWG, Zephyrus Tavvier, ClueBot NG, MitchMcM, Satellizer, Frietjes, Kevin Gorman, Widr, Mtking, Pluma, Daniel Cook00, Titodutta, Rilo.hawn, Regulov, BG19bot, Anatomist90, CeraBot, Millennium bug, Pratyya Ghosh, Joeyman12345, MadGuy7023, Iamfake2662, ABCmouse, Lugia2453, JakobSteenberg, Meddaughc, Iztwoz, ElHef, Dinisoe, LT910001, Booklaunch, Zoey9988, Monkbot, HiYahhFriend, Yurugu, Hugospiers, Crystallizedcarbon, Helloimhungry, Kashish Arora, KasparBot, Dileswar10, Williebeng, Rain-

35.8. TEXT AND IMAGE SOURCES, CONTRIBUTORS, AND LICENSES

Fall and Anonymous: 357

- **Cerebral hemisphere** *Source:* https://en.wikipedia.org/wiki/Cerebral_hemisphere?oldid=733222161 *Contributors:* Bryan Derksen, Stevertigo, Reigh, Lir, Michael Hardy, Gabbe, (, Ahoerstemeier, Evercat, Emperorbma, Selket, Pedant17, Hadal, Giftlite, Washington irving, Pascal666, Khalid hassani, Delta G, Jonel, Sam Hocevar, Anirvan, Cerebral, D6, Mindspillage, Rich Farmbrough, Bender235, ESkog, Rstt, CDN99, Arcadian, NickSchweitzer, Pearle, Gary, Cjthellama, Samohyl Jan, Fivetrees, Freyr, Naveen57, Dolfrog, Jonnabuz, Deltabeignet, Rjwilmsi, Kinu, Salix alba, EBlack, JegaPRIME, Katsuyori, YurikBot, Postglock, Supten, Action potential, Mysid, Tachs, Octavio L~enwiki, Smily, SmackBot, Saravask, Delldot, BrotherGeorge, Bluebot, Rrburke, Stevenmitchell, MrDolomite, Iridescent, Jaksmata, Vega84, Was a bee, Fifo, Mawfive, Pajz, Dasani, AntiVandalBot, Gioto, Jordan Rothstein, Roidroid, Cat Whisperer, Arty4ever, DerHexer, MartinBot, CFCF, Mikael Häggström, Coppertwig, LittleHow, Auditory, VolkovBot, A.Ou, Jmrowland, TXiKiBoT, Zenek.k, SieBot, Gerakibot, Escape Orbit, Wysprgr2005, Mild Bill Hiccup, Auntof6, Vanished user uih38riiw4hjlsd, Addbot, Looie496, AnomieBOT, Ubergeekguy, Citation bot, Jmarchn, ArthurBot, Tekks, Dr.OwenGWilliam, DrFree, Trappist the monk, Angelito7, Waso99, EmausBot, K kisses, ChuispastonBot, ClueBot NG, Wingtipvortex, Wimpus~enwiki, Tideflat, Frietjes, Benmotz, BG19bot, Theobp, MusikAnimal, Ytpete, Anatomist90, HTML2011, Qwe112233, Thirumaran13, Iztwoz, Hira.abbasi, Ahmednawabi, Hira2024, Studentjessica, LT910001, BruceBlaus, Ephelyon, Rsoscia, Neuro101, XenusG, Brianna Stevens and Anonymous: 99

- **Lobe (anatomy)** *Source:* https://en.wikipedia.org/wiki/Lobe_(anatomy)?oldid=723416260 *Contributors:* RoyBoy, Arcadian, The RedBurn, Malcolma, Delldot, John Reaves, COMPFUNK2, Raburton, Dawnseeker2000, MartinBot, VolkovBot, Lova Falk, Rhcastilhos, ClueBot, Pipep-Bot, PixelBot, DumZiBoT, Jmanigold, Addbot, KDS4444, Materialscientist, Hunnjazal, GrouchoBot, RoyGoldsmith, Cypher3c, John Atkins, SporkBot, ClueBot NG, Frietjes, KLBot2, BG19bot, Brest39, Iztwoz, DavidLeighEllis, Akamelfg, LT910001 and Anonymous: 9

- **Parietal lobe** *Source:* https://en.wikipedia.org/wiki/Parietal_lobe?oldid=723736716 *Contributors:* AxelBoldt, Bryan Derksen, Fnielsen, Vaughan, Cyp, Adoarns, Selket, Topbanana, Diberri, Xanzzibar, Giftlite, Washington irving, Bird, OverlordQ, Bumm13, Discospinster, Joshannon, Mani1, Arcadian, Commander Keane, MONGO, JohnJohn, Rjwilmsi, Seraphimblade, Benanhalt, Chobot, Bgwhite, YurikBot, Spaully, Chris Capoccia, NawlinWiki, A314268, Welsh, SigPig, Mysid, Zzuuzz, Sgmanohar, Caliprincess, Reyk, Digfarenough, CWenger, Ffangs, SmackBot, Gilliam, John Reaves, RedHillian, Mr.Z-man, J-Kama-Ka-C, Drphilharmonic, Vina-iwbot~enwiki, Cajolingwilhelm, LaMenta3, Jcbutler, Michaelbusch, George100, Liam Skoda, Badseed, Was a bee, Fifo, Satori Son, Thijs!bot, Shorlin, Karlhahn, Zhang Guo Lao, Bongwarrior, Willpenington, J.delanoy, Kpmiyapuram, Mrs.meganmmc, 83d40m, STBotD, VolkovBot, Oshwah, Lradrama, JhsBot, BotKung, InternetHero, Lova Falk, SieBot, Bentogoa, Fred Birchmore, Hello71, Macy, Martarius, ClueBot, Fyyer, The Thing That Should Not Be, Voxpuppet, Quercus basaseachicensis, Jusdafax, Andy pyro, SchreiberBike, PCHS-NJROTC, SoxBot III, DumZiBoT, SilvonenBot, Addbot, Tanhabot, Diptanshu.D, Looie496, ChenzwBot, West.andrew.g, 84user, Gail, Luckas-bot, ZX81, Yobot, TaBOT-zerem, South Bay, AnomieBOT, Justme89, Somerledi, Citation bot, Jmarchn, 4twenty42o, Jsharpminor, Thehelpfulbot, NifCurator1, A little insignificant, December21st2012Freak, Trappist the monk, Reach Out to the Truth, DARTH SIDIOUS 2, Viniciusmc, Tesseract2, Deagle AP, EmausBot, John of Reading, Wikipelli, Dcirovic, Lucas Thoms, ZéroBot, Donner60, DASHBotAV, Thegoldeneel, ClueBot NG, Kasunchathuranga, Jkwchui, Frietjes, BG19bot, Pastorjamesmiller, Thaki, Anatomist90, Tskittles, Ragas 0214, Dr Bilal Alshareef, Niemiw, Jonadin93, Biosthmors, Tutelary, 331dot, Barndt13, Rachel0821, Jorthi, Gyuwiki, Ruby Murray, Lwsh27, Rahmanmaiesha, Joeshmomo, Iztwoz, LT910001, CdavM, Manul, Neuroit, JaconaFrere, Monkbot, HiYahhFriend, MRD2014, 2gud4you, Miribeth, CAPTAIN RAJU, Flechaveloz 29, HyperKids and Anonymous: 179

- **Temporal lobe** *Source:* https://en.wikipedia.org/wiki/Temporal_lobe?oldid=733947971 *Contributors:* Bryan Derksen, Ewlloyd, Frecklefoot, Michael Hardy, Vaughan, Selket, Chris 73, Puckly, Diberri, Giftlite, StevenS757, Washington irving, SoCal, Bird, SWAdair, Rdsmith4, Sayeth, TiMike, Discospinster, Mani1, Pt, Kime1R, Arcadian, Eje211, Alansohn, Gary, Dismas, Labastar, Consequencefree, Commander Keane, WadeSimMiser, Deepstratagem, Graham87, FreplySpang, Rjwilmsi, Miserlou, Chobot, The Rambling Man, YurikBot, Hawaiian717, RussBot, Wimt, A314268, Rmky87, Moe Epsilon, Mysid, Closedmouth, Ikkyu2, Allens, Junglecat, SmackBot, InvictaHOG, Yamaguchi???, Gilliam, Ian Burnet~enwiki, Vina-iwbot~enwiki, Camazine, Jcbutler, Guillaume777, Michaelbusch, Thinker2006, JForget, Ale jrb, Albia, Pewwer42, Badseed, Was a bee, Anonymi, Anthonyhcole, Thijs!bot, Headbomb, Martin Rizzo, JAnDbot, Bongwarrior, Mbc362, DerHexer, Nikpapag, Bissinger, R'n'B, The Anonymous One, Kpmiyapuram, Eskimospy, Mrs.meganmmc, LordCo Centre, VolkovBot, Philip Trueman, Oshwah, Vanished user ikijeirw34iuaeolaseriffic, Meme92, Psyche825, BotKung, Zachary.nichols, Lova Falk, Semifinalist, Josh830, Foxtrotman, Linuxrules1337, Blamed, ClueBot, Auntof6, Posix memalign, SkyMaja, N.vanstrien, Against the current, XLinkBot, Facts707, Addbot, Diptanshu.D, Looie496, Wolfeye90, Thote, Amirobot, AnomieBOT, Rjanag, Piano non troppo, Materialscientist, Citation bot, Jmarchn, ArthurBot, Xqbot, The sock that should not be, Chickenwarrior22, FrescoBot, عبدالمؤمن, Jonathansuh, NifCurator1, Mightygobbler, I dream of horses, Edderso, Jmsteven, Mean as custard, Wintonian, Dstone66, EmausBot, Outriggr, Dcirovic, ZéroBot, Leverg, Akerans, Cgt, ClueBot NG, Frietjes, Rockingmemusic, Widr, BG19bot, AhMedRMaaty, Snow Rise, Mark Arsten, Anatomist90, Samahm28, Prof. Squirrel, A.naufer, Dexbot, IJimC, Mark Bao, Rimapaul ls, Lasimon1, Nirotify, Wahib zehlawi, Sanafatani, Isaac009988, LT910001, 3AlarmLampscooter, Horseless Headman, HiYahhFriend, Unkitg, TheCognitiveNeuropsychologist, Ifarnswo, Ryubyss, Xenowolf10, KasparBot, CAPTAIN RAJU, Sigma225c and Anonymous: 174

- **Occipital lobe** *Source:* https://en.wikipedia.org/wiki/Occipital_lobe?oldid=735668182 *Contributors:* Bryan Derksen, Arj, Ewen, Charles Matthews, Selket, AaronSw, Hadal, Diberri, Xanzzibar, Giftlite, Washington irving, SoCal, Bird, Luigi30, Wmahan, CryptoDerk, Sootymangabey, Discospinster, Pie4all88, Mani1, Lycurgus, Sietse Snel, Jpgordon, Bobo192, Tronno, R. S. Shaw, Arcadian, Giraffedata, TheProject, Alansohn, ChiBiKi, Wtmitchell, BDD, Bjones, TigerShark, Commander Keane, JeremyA, Rjwilmsi, Boccobrock, The wub, Yamamoto Ichiro, FlaBot, AED, Kolbasz, YurikBot, NawlinWiki, A314268, Ashwinr, Mysid, Closedmouth, Ffangs, Eog1916, SmackBot, Canthusus, Eloy, Bluebot, MalafayaBot, TheFeds, Ctbolt, Ian Burnet~enwiki, Iapetus, Anazem, Drphilharmonic, Wandell, IronGargoyle, Caiaffa, Michaelbusch, NativeForeigner, Tawkerbot2, Ccie13836, Was a bee, Anthonyhcole, Endpoint, Fifo, Thijs!bot, Epbr123, NickRinger, WikiSlasher, Vendettax, Nx01rules, Bongwarrior, VoABot II, Blackpen2008, Esmith512, Roastytoast, Kpmiyapuram, Mrs.meganmmc, Mikael Häggström, Paskari, ItsWoody, The All-Traq, Philip Trueman, WatchAndObserve, Ferengi, JhsBot, Kurowoofwoof111, Acmeinc, Lova Falk, ObjectivismLover, Andrewjlockley, Flyer22 Reborn, Oxymoron83, Myrvin, ClueBot, Niceguyedc, Fluminense, Mspraveen, Excirial, La Pianista, Rankiri, Shunjukun, Addbot, Landon1980, NjardarBot, Looie496, Download, Teles, Ettrig, Luckas-bot, Yobot, Cflm001, Amirobot, Materialscientist, Citation bot, Jmarchn, ArthurBot, Doggy65, FrescoBot, عبدالمؤمن, LucienBOT, NifCurator1, Pinethicket, Jonesey95, Gingermint, Serols, 777sms, Jfmantis, TjBot, John of Reading, Tommy2010, Wikipelli, Dcirovic, K6ka, ZéroBot, H3llBot, ClueBot NG, Frietjes, Twillisjr, Widr, Ngocminh.oss, Cold Season, Yenitha, Bbuunniiee, Anatomist90, Sohrabsalimian, BattyBot, MatthewIreland, Rosemary277, Sarakhany93, EuroCarGT, JurgenNL, Lugia2453, JakobSteenberg, The Anonymouse, Rblrkr, Snakewolf, Tumblinggirl, Jorthi, Dolphins888, Rahmanmaiesha, Iztwoz, Tellthemwhoiwillbe, LT910001, Baconfry, Brainiacal, Monkbot, EdgarCabreraFariña, HiYahhFriend, Praegressus, Swaglord7, Swegfag12 and Anonymous: 158

- **Limbic lobe** *Source:* https://en.wikipedia.org/wiki/Limbic_lobe?oldid=706126602 *Contributors:* Sannse, Bearcat, Pgan002, Arcadian, Wouterstomp, Woohookitty, Rjwilmsi, Was a bee, Anthonyhcole, Allstarecho, Kpmiyapuram, Spiral5800, Beeblebrox, ClueBot, Lenrodman, Baddog144, Beeswaxcandle, AnomieBOT, FrescoBot, NifCurator1, MastiBot, RjwilmsiBot, EmausBot, Frietjes, Helpful Pixie Bot, CitationCleanerBot, JakobSteenberg, LT910001 and Anonymous: 8

- **Insular cortex** *Source:* https://en.wikipedia.org/wiki/Insular_cortex?oldid=732502090 *Contributors:* AxelBoldt, Fnielsen, Michael Hardy, Selket, Alba, Diberri, Pengo, Nmg20, Barbara Shack, Rich Farmbrough, JackWasey, Arcadian, Eric Kvaalen, Dave.Dunford, Ceyockey, Uncle G, Rjwilmsi, PhatRita, JdforresterBot, RussBot, NawlinWiki, Daniel Mietchen, Arch o median, Mdwyer, SmackBot, Domaniac, RDBrown, Zvar, Jwy, Ritafelgate, Giancarlo Rossi, EDUCA33E, Sausagerooster, RichardF, Wikiauthor, CmdrObot, Was a bee, Polomac, Hebrides, Anthonyhcole, Fifo, Thijs!bot, Wikid77, Headbomb, Nasirnaqvi, Cooper24, Ujalm, WLU, Dennisr48, CFCF, Kpmiyapuram, LittleHow, Tulpan, Idioma-bot, VolkovBot, WarddrBOT, WatchAndObserve, Ndaniels, Lamro, Lova Falk, SieBot, Typritc, Mike2vil, ClueBot, Master1228, Excirial, Jgrethe, Addbot, DOI bot, Jncraton, Jwill4, Yobot, Amirobot, Citation bot, Jmarchn, Neurotrip, Tchussle, Makeswell, Xasodfuih, WebCiteBOT, Expsychobabbler, FrescoBot, Citation bot 1, Mathalus, Tom.Reding, Iwakuralain, RjwilmsiBot, Dewritech, Pietac, Dcirovic, AManWithNoPlan, Garyboz, ClueBot NG, Wimpus~enwiki, Frietjes, BG19bot, Joshua Jonathan, BattyBot, Andras jakab, Siuenti, JakobSteenberg, Iztwoz, TaylorTheWiki, Dinisoe, LT910001, Brainiacal, Anrnusna, Monkbot, Rafael1977wfwf, CV9933 and Anonymous: 70

- **Cerebrum** *Source:* https://en.wikipedia.org/wiki/Cerebrum?oldid=736018757 *Contributors:* Bryan Derksen, Alex.tan, Caltrop, Heron, Karada, Angela, Darkwind, Csernica, David Newton, Selket, Saltine, Robbot, Hemanshu, Hadal, Diberri, DocWatson42, Bird, Niteowlneils, Eequor, Edcolins, Slowking Man, Mineminemine, Lesgles, Rfl, Discospinster, Dancxjo, Mani1, Robert P. O'Shea, Kwamikagami, Surachit, Triona, Bobo192, Cmdrjameson, Arcadian, Guidod, Kjkolb, Jhertel, Wouterstomp, Snowolf, Wtmitchell, Oleg Alexandrov, Japanese Searobin, Stemonitis, Commander Keane, WadeSimMiser, JeremyA, Gerbrant, Rjwilmsi, DoubleBlue, Titoxd, Nihiltres, Gurch, Chobot, DVdm, Mercury McKinnon, YurikBot, RobotE, RussBot, Stephenb, Emiellaiendiay, Gustavb, A314268, SAE1962, JShenk, Dogcow, JessicaX, Lipothymia, Mysid, Lt-wiki-bot, Kungfuadam, NeilN, SmackBot, Aflm, Gilliam, Xmahahdu, Postoak, Esteedee, Rrburke, TheLimbicOne, Krashlandon, Gleng, Accurizer, Dcflyer, Nehrams2020, RekishiEJ, JForget, CmdrObot, TheTito, Cydebot, Was a bee, Rcbtdrumwolf, Juansempere, PhineasG, Thebellmaster1x, Epbr123, Qwyrxian, Ghiles, CopperKettle, John254, WillMak050389, E. Ripley, Hmrox, AntiVandalBot, Danger, Res2216firestar, JAnDbot, Dekimasu, Allstarecho, KenyaSong, MartinBot, Nikpapag, Anaxial, Sherybatta, J.delanoy, Captain panda, Sasajid, CFCF, Trusilver, Numbo3, Xris0, Cpiral, AntiSpamBot, Schizoform, FOTEMEH, Evil Egg, StoptheDatabaseState, VolkovBot, TXiKiBoT, Brunton, Lerdthenerd, Dagilson, Alcmaeonid, Ponyo, Malcolmxl5, Caltas, Triwbe, Cooladoola, Keilana, Tombomp, KathrynLybarger, Halcionne, Djadvance, Mygerardromance, Denisarona, ClueBot, Maderchoud447, Niceguyedc, Ledfanatic4, Excirial, Hello Control, Razorflame, Ottawa4ever, Thingg, Coolcat97, Horselover Frost, DumZiBoT, Rror, Little Mountain 5, Facts707, Lab-oratory, Kgibs, Addbot, !Silent, Jncraton, Looie496, Cst17, Ginosbot, Quercus solaris, Tide rolls, Luckas-bot, Yobot, THEN WHO WAS PHONE?, AnomieBOT, Dirjarmocksorz, Tryptofish, AdjustShift, Materialscientist, Citation bot, E2eamon, Tekks, LilHelpa, Skipsscc, Isheden, LucienBOT, WikiDisambiguation, NifCurator1, Cmsrocks, Hyqeom, Suffusion of Yellow, برزکار, فـی‌یزی‌وت‌را‌پیست ابراهیم, Acather96, GoingBatty, Dcirovic, Tiganusi, Deutschgirl, Usb10, ChuispastonBot, NTox, ClueBot NG, Frietjes, Stas000D, BG19bot, Martianme, Anatomist90, Rob Hurt, Ježofska, Biosthmors, Luckydhaliwal, Goldenhuntergow3, Sae Harshberger, Tsuruya, Lugia2453, JakobSteenberg, Frosty, Iztwoz, Dustin V. S., Dinisoe, LT910001, Jianhui67, Monkbot, Buggiehuggie, Սրոսն Սրն, SongofSol, Julietdeltalima, Linda Curttis, ToonLucas22, XxThose ShadesxX, Lehession, Plehreh, Revant Pragadeysh, Swaglord7, Neuron Doc, KasparBot, JJMC89, Wajahath Mohamed, Smartyarpan2000, Qzd, TheoTPV and Anonymous: 271

- **White matter** *Source:* https://en.wikipedia.org/wiki/White_matter?oldid=732848626 *Contributors:* Bryan Derksen, ErdemTuzun, Fnielsen, Gabbe, Delirium, Looix~enwiki, Smack, Selket, Taoster, Robbot, DocWatson42, Michael Devore, Jfdwolff, Gracefool, Junkyardprince, Kaldari, Rich Farmbrough, NeuronExMachina, Bender235, Robert P. O'Shea, Lycurgus, CDN99, Arcadian, Mr2001, Storm Rider, Finduilas, Sjschen, Tycho, Mohawkjohn, Margosbot~enwiki, Brendan Moody, DrFlo1, Celebere, YurikBot, Hairy Dude, Chris Capoccia, Hydrargyrum, A314268, Nephron, Mysid, Modify, Bibliomaniac15, SmackBot, Joconnol, Jfurr1981, Delldot, Jethero, Daniel.Cardenas, Gobonobo, Hermoye~enwiki, Iridescent, Bigmak, JForget, Vaughan Pratt, ShelfSkewed, Outriggr (2006-2009), Asymptote, Was a bee, Juansempere, Alaibot, Thijs!bot, Faigl.ladislav, Louis Waweru, Danger, JAnDbot, Helge Skjeveland, VoABot II, DerHexer, AndoDoug, Anaxial, Nono64, Tgeairn, CFCF, Hahlch, LittleHow, Tagus, James Kipp, Lady Flora, VolkovBot, Vlmastra, Dohgon, Jesus Carp, Lova Falk, SieBot, Toddst1, Flyer22 Reborn, Kareekacha, Asktheneurologist, Mr. Granger, ClueBot, Niceguyedc, Anthonydraco, XLinkBot, AnotherSolipsist, EEng, Addbot, DOI bot, Looie496, Download, Brentdeezee, Tide rolls, Luckas-bot, Yobot, James Cantor, Synchronism, AnomieBOT, Tryptofish, Jim1138, Flopsy Mopsy and Cottonmouth, Citation bot, Jmarchn, GrouchoBot, RibotBOT, January2009, Captain-n00dle, Girlwithgreeneyes, Citation bot 4, Rain drop 45, Angelito7, MollyNYC, RjwilmsiBot, EmausBot, Wikipelli, Dcirovic, K6ka, Serketan, Werieth, Fæ, Hazard-SJ, AManWithNoPlan, ClueBot NG, MelbourneStar, Bndutton, Cntras, Helpful Pixie Bot, Ryker-Smith, Purielku, Glacialfox, Dr Bilal Alshareef, David.moreno72, Darylgolden, Ssscienccce, JakobSteenberg, Me, Myself, and I are Here, Loniucla, LT910001, Sal.ayesa, Monkbot, BiologicalMe, Kashish Arora, M665543, Calebmack, Constantinaki, Francesco Sciacca and Anonymous: 133

- **Grey matter** *Source:* https://en.wikipedia.org/wiki/Grey_matter?oldid=731638026 *Contributors:* Bryan Derksen, Minesweeper, Looix~enwiki, Kingturtle, Selket, Robbot, Ojigiri~enwiki, Hadal, Gracefool, CryptoDerk, Tothebarricades.tk, Achernar7, El C, Lycurgus, Renice, Viriditas, Arcadian, Jumbuck, Interiot, Burn, Cburnett, Jlewis, Woohookitty, Benbest, Rjwilmsi, Margosbot~enwiki, TheMidnighters, RexNL, DrFlo1, Hydrargyrum, Nephron, Mysid, Nicanor5, Theda, Haisook, Skittle, Ramanpotential, Masonbarge, Bibliomaniac15, SmackBot, Triggtay, Jfurr1981, Delldot, Fuzzform, Can't sleep, clown will eat me, Chlewbot, Pwjb, Drphilharmonic, Clicketyclack, Mdrine, Lambiam, Scientizzle, Tim bates, Bobsagat, Shoeofdeath, Brandon.macuser, Outriggr (2006-2009), Neelix, CohibaX, Was a bee, Tawkerbot4, UberScienceNerd, Thijs!bot, Wikid77, Faigl.ladislav, Alangpierce, Natalie Erin, AntiVandalBot, Joehall45, Nthep, Magioladitis, Pedro, Bongwarrior, Mark Lundquist, Robin S, Poeloq, Nikpapag, Anaxial, MerrimacVI, Alro, R'n'B, CFCF, DarkFalls, P4k, LaiPt, Burlywood, Philip Trueman, Anonymous Dissident, Norbu19, Lova Falk, AlleborgoBot, Cowlinator, SieBot, Flyer22 Reborn, EditorInTheRye, Hello71, ShelleyAdams, Mr. Granger, ClueBot, Niceguyedc, Alexbot, Ginbot86, DumZiBoT, XLinkBot, Karimsl89, Addbot, Boomur, Looie496, Numbo3-bot, OlEnglish, Zorrobot, Luckas-bot, Yobot, TaBOT-zerem, Samtar, AnomieBOT, Tryptofish, Killiondude, Piano non troppo, Citation bot, Jmarchn, Naturo13, Sionus, The Great Master DaVinci, Xoxonybabyxoxo, Ruby.red.roses, Bellerophon, I dream of horses, HRoestBot, Trijnstel, Tom.Reding, RedBot, Vrenator, Reaper Eternal, Angelito7, RjwilmsiBot, K6ka, Werieth, JeanneMish, ClueBot NG, Vacation9, Wimpus~enwiki, Razielxbox, Jeremygm, Mtking, Schnupp, Theoldsparkle, Mbfenn, BG19bot, Krenair, BattyBot,

35.8. TEXT AND IMAGE SOURCES, CONTRIBUTORS, AND LICENSES

Acadēmica Orientālis, JakobSteenberg, Graphium, Iztwoz, DavidLeighEllis, LT910001, Mettelrei, Seppi333, Bsridhar6, JaconaFrere, Monkbot, EddieKnight, Joobo, LindaCathy, Sweepy, DatGuy, Edinburgh student 93 and Anonymous: 168

- **Forebrain** *Source:* https://en.wikipedia.org/wiki/Forebrain?oldid=676187665 *Contributors:* AxelBoldt, Bryan Derksen, Emperorbma, David Newton, Hadal, Diberri, Bird, Eequor, Delta G, Beland, Cacycle, Surachit, Jimhutchins, Arcadian, Wouterstomp, Jersyko, FlaBot, Chobot, A314268, Dan Wylie-Sears, Lipothymia, SmackBot, Moshe Constantine Hassan Al-Silverburg, LWF, Gleng, Alaibot, Thijs!bot, Nick Number, Escarbot, Thomasiscool, Schlobb, WarthogDemon, TXiKiBoT, Temporaluser, SieBot, Cooladoola, Drgarden, Sabri76, Addbot, م.اني, Zorrobot, Alessio Facchin, Jmarchn, ArthurBot, NifCurator1, Kelvin Samuel, Stamboltsyan, Onel5969, EmausBot, EME44, ZéroBot, Jenks24, ClueBot NG, Frietjes, Smettems, Dr Bilal Alshareef, Danvasilis, RoniAlyssa, Iztwoz, Dinisoe and Anonymous: 21

- **Midbrain** *Source:* https://en.wikipedia.org/wiki/Midbrain?oldid=731182008 *Contributors:* AxelBoldt, Bryan Derksen, Fnielsen, Kku, Angela, Tristanb, David Newton, Saltine, Pumpie, Vespristiano, Hemanshu, Diberri, Bird, Chinasaur, ELApro, Rich Farmbrough, Kndiaye, Sfahey, Surachit, Arcadian, Hadlock, BRW, Mauvila, Commander Keane, JeremyA, Mreult~enwiki, Tincup~enwiki, Yurik, Rjwilmsi, Vegaswikian, Yamamoto Ichiro, FlaBot, Margosbot~enwiki, Ayla, Choess, Chobot, WriterHound, YurikBot, Arado, Hede2000, Eleassar, A314268, Send513, Lipothymia, Octavio L~enwiki, Lt-wiki-bot, SmackBot, Snowmanradio, Giancarlo Rossi, TenPoundHammer, Gleng, Dicklyon, Geologyguy, Ccie13836, Joelholdsworth, WeggeBot, Was a bee, Anthonyhcole, Fifo, Thijs!bot, Cwtyler, AntiVandalBot, Majorly, Moreza, .anacondabot, VoABot II, JaGa, Schlobb, KylieTastic, VolkovBot, TXiKiBoT, Madhero88, Saif khan33, Tom morse, Cooladoola, Drgarden, The Thing That Should Not Be, Alexbot, Predator47, Subash.chandran007, Cljohns, Vojtěch Dostál, WikHead, Addbot, Aceofhearts1968, Bwrs, Cesiumfrog, Jarble, Bermicourt, Luckas-bot, Yobot, Cflm001, KamikazeBot, AnomieBOT, Jmarchn, ArthurBot, Carturo222, Xqbot, Anna Frodesiak, یوشیچی توویہارا, Erik9bot, Thehelpfulbot, NifCurator1, Adamlankford, Mdstd nyc, Jhbuk, Corinne68, Jonkerz, Callanecc, Reaper Eternal, MrArifnajafov, RjwilmsiBot, EmausBot, WikitanvirBot, Gcastellanos, Niluop, Jdogg182134, Wikipelli, Dcirovic, A930913, Peteb4, ChuispastonBot, ClueBot NG, Frietjes, Widr, Vgross, Jeraphine Gryphon, Anatomist90, Wjdittmar, Dr Bilal Alshareef, BattyBot, Hergilei, Zhaofeng Li, TylerDurden8823, JYBot, Webclient101, Earl Moss, JakobSteenberg, DerekWinters, Iztwoz, BallenaBlanca, LT910001, Nomloc, BruceBlaus, Brainiacal, Rsoscia, KN.Anantha, 3 of Diamonds, Dr-azhar, CAPTAIN RAJU, G-dac and Anonymous: 104

- **Hindbrain** *Source:* https://en.wikipedia.org/wiki/Hindbrain?oldid=733812834 *Contributors:* Pumpie, Diberri, Kndiaye, Surachit, Arcadian, Wouterstomp, NerdyPunk2ML, JeremyA, Cuchullain, Pleiotrop3, FlaBot, Ewlyahoocom, YurikBot, A314268, Nrets, Chakazul, Moe Epsilon, Zwobot, Gadget850, Rto, SmackBot, Jfurr1981, Eurobikermcdog, Gleng, Courcelles, Anthonyhcole, Alaibot, MER-C, Alleborgo, KTo288, Xris0, Aminchar, VolkovBot, Lova Falk, SieBot, The Thing That Should Not Be, Sabri76, DumZiBoT, Monopol, Addbot, Sspeik, Reidlophile, LaaknorBot, م.اني, Ccolen, Luckas-bot, Yobot, Jmarchn, NifCurator1, Rapsar, Bgpaulus, EmausBot, RA0808, Confession0791, Pheobedamsel, ClueBot NG, Frietjes, Nbout, Helpful Pixie Bot, Anatomist90, Smettems, Dr Bilal Alshareef, Ježofska, DerekWinters, Thomasbutts, Iztwoz, Dinisoe, CJ Carter-williams, Jtroiano 2015 and Anonymous: 19

- **Ventricular system** *Source:* https://en.wikipedia.org/wiki/Ventricular_system?oldid=735060133 *Contributors:* Bryan Derksen, The Anome, William Avery, Tristanb, Kylegordon, Mandark~enwiki, David Shay, Robbot, Diberri, Jfdwolff, Rich Farmbrough, Cacycle, Wee Jimmy, Arcadian, Pearle, BD2412, Rjwilmsi, Miserlou, Mohawkjohn, Jakob Suckale, DrFlo1, Bgwhite, Eleassar, Draeco, A314268, Grafen, Deodar~enwiki, Dan Wylie-Sears, Andrewr47, SmackBot, Gilliam, Colonies Chris, Konstable, Snowmanradio, Ortho, Doczilla, Philippschaumann, Was a bee, Wyvyrn, Thijs!bot, Jauntymcd, Nono64, EverSince, Fconaway, CFCF, Mikael Häggström, M-le-mot-dit, Tkenna, Jackfork, Bogwhistle, Peaceful horizon, IngFrancesco, Niceguyedc, Perchy22, DumZiBoT, SilvonenBot, MystBot, Felix Folio Secundus, Addbot, Jv821, Looie496, AndersBot, Taketa, WikiDreamer Bot, Luckas-bot, Yobot, TaBOT-zerem, AnomieBOT, Jo3sampl, Citation bot, ArthurBot, Obersachsebot, Banjaloupe, Gouerouz, Benjybracha, Citation bot 1, Mahnut, RedBot, Trappist the monk, Dcirovic, Tmajoor, 08gymnast12, Frietjes, Rob Hurt, JakobSteenberg, Bulba2036, Iztwoz, Echinacin35, LT910001, BruceBlaus, AlexanderQuent, EivindB, Monkbot, HiYahhFriend, Wanalo, Galorakin, Iair5468, CV9933 and Anonymous: 64

- **Medulla oblongata** *Source:* https://en.wikipedia.org/wiki/Medulla_oblongata?oldid=734811981 *Contributors:* Danny, Liftarn, Cyde, Kosebamse, Ellywa, Tristanb, Selket, Puckly, Hadal, Jfdwolff, Vina, Mike Rosoft, D6, JTN, Cacycle, SocratesJedi, Bender235, Gold Dragon, Wee Jimmy, Bobo192, Smalljim, Malafaya, Arcadian, Mac Davis, Mysdaao, Bart133, Helixblue, Sciurinæ, Japanese Searobin, The JPS, Commander Keane, JeremyA, Tincup~enwiki, BD2412, Canderson7, Miserlou, Marsbound2024, RexNL, Ashwinhgtx, YurikBot, Hede2000, Kirill Lokshin, A314268, Nephron, Gadget850, Cinik, Closedmouth, Moomoomoo, SmackBot, Chazz88, Unyoyega, Eskimbot, IstvanWolf, Oscarthecat, Asclepius, Robth, Rlevse, Rama's Arrow, Drphilharmonic, Doodle77, Acdx, Mattopaedia, Attys, Gleng, Jjz3d83, Noah Salzman, Novangelis, HisSpaceResearch, Iridescent, Supersquid, Ccie13836, Thepatriots, Verdi1, Yaris678, Anthonyhcole, Fifo, Audry2, Thijs!bot, Marek69, A3RO, Nick Number, AntiVandalBot, Danger, Qwerty Binary, JAnDbot, Wasell, WhatamIdoing, A3nm, Brock256, Anaxial, J.delanoy, CFCF, Extransit, Icseaturtles, Mikael Häggström, Ank0ku, Dobber2612, Permafrost, ShinigamiD02, Deor, Philip Trueman, PhilipBembridge, Icemanocp, GcSwRhIc, Lradrama, LeaveSleaves, Raymondwinn, MCTales, DeniabilityPlausible, SieBot, Yerpo, Fimbriata, Manway, DeeSaiD, ClueBot, The Thing That Should Not Be, Razimantv, SuperHamster, Keerthi physio, Excirial, Alexbot, EhJJ, Morel, Jclw1966, 7, Addbot, Captain-tucker, Reidlophile, Looie496, Tide rolls, Carmenyan206, Kurtis, Luckas-bot, Nikosguard, A More Perfect Onion, Rubinbot, IRP, Kingpin13, Materialscientist, Citation bot, ArthurBot, Xqbot, Thehelpfulbot, Jb701926, NifCurator1, Citation bot 1, Pinethicket, Acgator09, Yoyo2os, BogBot, TjBot, Ripchip Bot, LcawteHuggle, Chibby0ne, EmausBot, WikitanvirBot, Lucien504, Dcirovic, Prayerfortheworld, Wackywace, Hrvoje1234, Bmx8295, Donner60, AUN4, ClueBot NG, Harps21, Widr, Helpful Pixie Bot, HMSSolent, Hypoglossal00, BG19bot, Wikichewbaca, Dan653, Gautehuus, Anatomist90, Glacialfox, Gibbja, Dr Bilal Alshareef, ChrisGualtieri, Iztwoz, Koleburgs, LT910001, Ginsuloft, Anon685, AKS.9955, KH-1, LTW83, CyanDye, Jostain12, AlphaAntares, Tilifa Ocaufa, Dr.Flanagan, EremTheVampire and Anonymous: 315

- **Pons** *Source:* https://en.wikipedia.org/wiki/Pons?oldid=729138877 *Contributors:* Vicki Rosenzweig, Fnielsen, Danny, Olivier, Tristanb, Rainer Wasserfuhr~enwiki, Selket, Peak, Hadal, Ancheta Wis, Jfdwolff, Iceberg3k, Roisterer, Indolering, D6, JTN, Bender235, MyNameIsNotBob, Arcadian, Commander Keane, JeremyA, Magister Mathematicae, Ianvitro, Mendaliv, FlaBot, RexNL, Chobot, Celebere, EamonnPKeane, Mercury McKinnon, YurikBot, Drdisque, RobotE, Hede2000, Eleassar, Sanguinity, NawlinWiki, A314268, Nephron, Gadget850, Theda, Moogsi, Maxamegalon2000, KasugaHuang, SmackBot, LordTorgamus, Bobblehead, Kslays, Ohnosjamie, Eug, Shicoco, AKMask, Mike hayes, Muboshgu, Emt14, Nick125, JakGd1, BryanG, Gleng, Avedomni, IFinishWhatIStar, Michaelbusch, Tawkerbot2, Ccie13836, Vladaig, Mcstrother, Gogo Dodo, Was a bee, Anthonyhcole, Fifo, Chrislk02, Epbr123, AntiVandalBot, Danger, JAnDbot, Deflective, WhatamIdoing, Animum, R'n'B, Mrs.meganmmc, Mikael Häggström, Flyingidiot, TXiKiBoT, Plot9098, EnJx, BotMultichill, Flyer22 Reborn, Fimbriata, DerRoteBaron, The-Catalyst31, ImageRemovalBot, Martarius, Sfan00 IMG, Icarusgeek, Ghefley, Trivialist, DragonBot, BOTarate, Thingg, McLondon, Jendarrow, Addbot, Shivaniakshya, NjardarBot, Looie496, Guffydrawers, Dayewalker, OlEnglish, Luckas-bot, Yobot, MonaLynne, QueenCake, Tryptofish,

Materialscientist, Citation bot, Jmarchn, Thomas.ades, ArthurBot, Xqbot, BlackTooth93, Gigemag76, Medic2008, LucienBOT, NifCurator1, RedBot, Keepstherainoff, Jordgette, Fama Clamosa, Tbhotch, DARTH SIDIOUS 2, Ripchip Bot, Evjason, Sarakathryn, EmausBot, Dcirovic, ClueBot NG, Frietjes, Helpful Pixie Bot, Anatomist90, Dr Bilal Alshareef, I am One of Many, Iztwoz, Drbemore, LT910001, BLMac, Crystallizedcarbon, Iwilsonp and Anonymous: 118

- **Brainstem** *Source:* https://en.wikipedia.org/wiki/Brainstem?oldid=732039581 *Contributors:* AxelBoldt, Eloquence, Fnielsen, Edward, (, Ellywa, Julesd, Habj, Tristanb, Selket, Tpbradbury, Kaare, Hadal, Everyking, Bird, Jfdwolff, Jpkoester1, Antandrus, Bumm13, Mike Rosoft, Chris Howard, JTN, Smyth, Bobo192, Brim, Arcadian, Giraffedata, Ogress, Anthony Appleyard, Jared81, Atomicthumbs, DSatz, Ceyockey, Commander Keane, JeremyA, GregorB, Macaddct1984, Tincup~enwiki, BD2412, Ligulem, Boccobrock, RexNL, Gurch, Pgiii, Alphachimp, Chobot, Aethralis, YurikBot, Wavelength, Eleassar, A314268, JSLR, SmackBot, Hydrogen Iodide, Hraefen, TimBentley, Telempe, King Arthur6687, Tdivala, Rlevse, Tsca.bot, Rrburke, Maratanos, Decltype, JakGd1, Drphilharmonic, Ligulembot, Gleng, JorisvS, IronGargoyle, MTSbot~enwiki, Hu12, Iridescent, Michaelbusch, Tucaaue, Hydra Rider, Courcelles, Orangutan, Kitra101, Myasuda, Was a bee, Danorton, Anthonyhcole, Thijs!bot, Epbr123, Paragon12321, Marek69, Nick Number, Mentifisto, AntiVandalBot, LibLord, Joehall45, JAnDbot, Robina Fox, Karlhahn, VoABot II, JNW, Richrobison, Ksanyi, Stupefaction, Kevin Dufendach, Anaxial, R'n'B, Nono64, CFCF, Kpmiyapuram, Kataiveno, Mrs.meganmmc, Jayden54, HiEv, Funandtrvl, VolkovBot, Philip Trueman, Kww, GDonato, Cremepuff222, Epgui, Ziphon, Lova Falk, Doc James, Medcin, SieBot, Mikemoral, Vandy27, Doctorfluffy, Oxymoron83, Asktheneurologist, AnonGuy, Waves00, WikiLaurent, Martarius, ClueBot, The Thing That Should Not Be, Osm agha, VandalCruncher, Excirial, Alexbot, Juddlawr, Some student, Some student2, Earcatching, Attaboy, DumZiBoT, Staticshakedown, Nixpix99, McLondon, Sophie1979, Vojtěch Dostál, Sgpsaros, Addbot, Ronhjones, Looie496, MrOllie, Download, Chamal N, CarsracBot, Quercus solaris, Tassedethe, Tide rolls, Luckas-bot, Yobot, Senator Palpatine, Fraggle81, KamikazeBot, AnomieBOT, Tryptofish, IRP, Kingpin13, JonathanWilliford, Sir Pinade, Reggiecarey, La comadreja, Jmarchn, Xqbot, Mattashner, SassoBot, Gouerouz, NifCurator1, Xbcj0843hck3, Pinethicket, MastiBot, Jonkerz, Leondumontfollower, برزکار ابراهیم راپیست‌یوتزیوفی, Salvio giuliano, EmausBot, WikitanvirBot, Immunize, Gcastellanos, Alysakow, Tommy2010, Wikipelli, Jasonanaggie, Mz7, Ebrambot, AbdulKareem92, GrayFullbuster, ClueBot NG, Wimpus~enwiki, Frietjes, Genericuser744, Allardo, Anatomist90, Vashjanus, Enchanted 2 meet you, Ježofska, Luckydhaliwal, TwoTwoHello, Fox2k11, Jamesx12345, Zziccardi, Iztwoz, Hoskatti, Dinisoe, ï¿½, LT910001, Rondodude2000, Seppi333, BruceBlaus, Brainiacal, James Gabriel II, Oxford Professor 001, Gordon S. Johnson, Jr. Esq, Lemniscus, Devwebtel, KasparBot, Qzd, Anupruj, LCSW2020 and Anonymous: 212

- **Superior colliculus** *Source:* https://en.wikipedia.org/wiki/Superior_colliculus?oldid=732353645 *Contributors:* William Avery, Edward, Selket, Penboy07, Pengo, Rdsmith4, ELApro, Rich Farmbrough, Rspeer, Wee Jimmy, Kghose, Brim, Arcadian, Puzzle123, Yogi de, Jlewis, JeremyA, Rjwilmsi, AED, Nihiltres, Who, Tillmo, RussBot, A314268, Nrets, Iamnotanorange~enwiki, Tony1, Gadget850, Ikkyu2, Jfurr1981, Bluebot, Stevenmitchell, Nuklear, Drphilharmonic, Dr.saptarshi, Gleng, Sohale, InedibleHulk, Mcstrother, Scott Coleman, Was a bee, Anthonyhcole, CopperKettle, Headbomb, Issueskid, CFCF, Mikael Häggström, VolkovBot, Ninjawarriordex, Gerakibot, Fluminense, Choney, Heyzeuss, Sgpsaros, Jonsarsiat, Addbot, Looie496, Yobot, Bogey4, AnomieBOT, Citation bot, Cureden, LucienBOT, NifCurator1, Citation bot 1, Keepstherainoff, Haaninjo, EmausBot, Dcirovic, Wimpus~enwiki, Frietjes, Helpful Pixie Bot, PhnomPencil, Anatomist90, Rob Hurt, JakobSteenberg, Cfgranda, Iztwoz, LT910001, Seppi333, Anrnusna, Monkbot, Sarake leeg, TrishApps, Vivek.bekhabar and Anonymous: 37

- **Thalamus** *Source:* https://en.wikipedia.org/wiki/Thalamus?oldid=723929352 *Contributors:* AxelBoldt, Bryan Derksen, Alex.tan, Fnielsen, Youssefsan, Ixfd64, Tristanb, Emperorbma, Wikiborg, Selket, Robbot, Modulatum, Webhat, Diberri, Wile E. Heresiarch, Ancheta Wis, DocWatson42, Lupin, Everyking, Bird, Niteowlneils, Jfdwolff, Sonance, Gravy, Cerebral, Pyramidal, Maikel, DanielCD, Sfeldman, Discospinster, NeuronExMachina, Cacycle, Bender235, Bobo192, Arcadian, Alansohn, Babajobu, Goldom, Velella, Danntm, Sciurinæ, Japanese Searobin, Crosbiesmith, Nuggetboy, Brendanconway, Douzzer, Rjwilmsi, NatusRoma, Carfro, FlaBot, Chanting Fox, B44H, Salvadorjo~enwiki, Chobot, YurikBot, Chris Capoccia, Gaius Cornelius, A314268, Nrets, Bayle Shanks, Dppowell, Richardcavell, Arthur Rubin, Frenkmelk, Caco de vidro, SmackBot, RDBury, Ericbateson, Stepa, Delldot, Ichbinkerl, Kslays, Gilliam, Marc Kupper, Sviemeister, Kazkaskazkasako, Bluebot, AlexDitto, Kittybrewster, Lathal, Pazda, Roger.lee, Spiritia, WhiteCat, FrozenMan, Sir marek, Gleng, Slakr, Ryulong, Nismo3112, Iridescent, Michaelbusch, Lanem, Sophomoric, Beno1000, CmdrObot, Mcstrother, Christophersnelson, Was a bee, JFreeman, Brain-mapper, Fifo, Gerard.percheron, Casliber, Epbr123, CopperKettle, Mojo Hand, Skovorodkin, Grayshi, Fireice, Greensburger, SiobhanHansa, Magioladitis, Catgut, Fallschirmjäger, Ksanyi, Freddyd945, Schlobb, Nikpapag, R'n'B, J.delanoy, Mrs.meganmmc, Joncaplan, SJP, Michael Angelkovich, Lrunge, VMesc5er, VolkovBot, TXiKiBoT, Wettingthebedsince1956, Xwdl~enwiki, Madhero88, @pple, Sevela.p, Bortain, Doc James, Dan Polansky, Dmdstudent, Rhcastilhos, Mike2vil, ClueBot, GorillaWarfare, Wwheaton, Master1228, Niceguyedc, Rodolfo Llinas, Wouldeven9, DumZiBoT, Ptsdprof, Addbot, DOI bot, Looie496, LaaknorBot, CarsracBot, Chzz, Luckas-bot, Yobot, Dede2008, TaBOT-zerem, Untrue Believer, AnomieBOT, Enlightenmentreloaded, Zxabot, Materialscientist, Citation bot, Xqbot, TheAMmollusc, DSisyphBot, GrouchoBot, Omnipaedista, Shirik, Stealthaxe, FrescoBot, Ionutzmovie, NifCurator1, Argumzio, Éder Santos, Citation bot 1, Dinamik-bot, EmausBot, Rami radwan, Dcirovic, Serketan, SherwoodB, Dgwingert, Xiutwel-0003, Mikhail Ryazanov, ClueBot NG, MelbourneStar, Medbenmedben, 10k, Jj1236, Helpful Pixie Bot, Allardo, Drift chambers, Mejoribus, Anatomist90, Rococo1700, Jonadin93, ChrisGualtieri, Ducknish, Ssnn, Suchithra.ravi, JakobSteenberg, I am One of Many, Iztwoz, ArmbrustBot, Ginsuloft, JaconaFrere, Monkbot, HiYahhFriend, Anonyneuroscience, PErdos and Anonymous: 226

- **Hypothalamus** *Source:* https://en.wikipedia.org/wiki/Hypothalamus?oldid=732572676 *Contributors:* Bryan Derksen, Youssefsan, Netesq, Someone else, Frecklefoot, Vaughan, Karada, Looxix~enwiki, Mkweise, Buckwad, Robbot, Kd4ttc, Fuelbottle, Diberri, Dratman, Alison, Michael Devore, Varlaam, Thpn, Harikishore~enwiki, Antandrus, Mrtrey99, Discospinster, Bender235, Chairboy, Matteh, Brim, Tritonal, Dungodung, Larsie, Arcadian, Haham hanuka, Silver hr, Atlant, Wouterstomp, Super-Magician, LFaraone, Redvers, MartinSpacek, Jersyko, Mark K. Jensen, JeremyA, Bipedal, Eras-mus, SDC, Dysepsion, Mandarax, BD2412, Rjwilmsi, Nightscream, Kinu, Kasha.re, Margosbot~enwiki, Nihiltres, Pete.Hurd, Srleffler, Jpfagerback, Stephenb, Slodave, A314268, Роман Беккер, Nrets, Dan Wylie-Sears, Lipothymia, DGJM, DeadEyeArrow, Xabian40409, Arthur Rubin, Josh3580, Allens, David.hillshafer, SmackBot, Methoxyroxy~enwiki, Bluebot, MalafayaBot, Snowmanradio, Ww2censor, TheLimbicOne, Pissant, RandomP, Drphilharmonic, Jklin, Gleng, JorisvS, RomanSpa, Ben Moore, Hikoto, Stwalkerster, Hnc, Muadd, Mr Stephen, Optakeover, DabMachine, Iridescent, Michaelbusch, DreamsReign, AGK, Bulldozzer, Tawkerbot2, Atomobot, JForget, CmdrObot, Cogpsych, Ccie13836, Nikolabc, Coldbringer, Corpx, Adolphus79, Fifo, B, DKVII, Thijs!bot, Sugarcream, Mercury~enwiki, Classof2006smr, CopperKettle, Mojo Hand, Eb.eric, Mentifisto, Hmrox, Jack D. Zwemer, Avdignan, Hermant patel, Joehall45, MikeLynch, JAnDbot, Michaelkemp, VoABot II, WhatamIdoing, MartinBot, Jack007, Kostisl, CommonsDelinker, Nono64, J.delanoy, Pharaoh of the Wizards, Rhinestone K, Ohnm, Mrs.meganmmc, Mikael Häggström, Belovedfreak, Vegetent, Wingnut99, Lanternix, Placidstorms, CardinalDan, VolkovBot, Philip Trueman, TXiKiBoT, Oshwah, Zidonuke, Zamphuor, Godingo, Guillaume2303, Greggogil, Hopcraft, Clajef, AlleborgoBot,

35.8. TEXT AND IMAGE SOURCES, CONTRIBUTORS, AND LICENSES

Wikiscottcha, SieBot, Winchelsea, Phe-bot, Da Joe, AlbertHall, Oxymoron83, Iain99, Mike2vil, Huku-chan, Hordaland, Denisarona, Drgarden, Explicit, Myth010101, Animeronin, ClueBot, Bwfrank, The Thing That Should Not Be, Lastbetrayal, Master1228, Mild Bill Hiccup, Eod79, Niceguyedc, Excirial, Alexbot, Jusdafax, RTaptap, Redpriest187, Sun Creator, Peter.C, Kikos, AC+79 3888, DumZiBoT, Chymæra, Roxy the dog, Teh Rote~enwiki, Jamesinspace, Cminard, Rungladwin, Avoided, Cjurkoshek, Addbot, DOI bot, Captain-tucker, AkhtaBot, Looie496, Favonian, Terrillja, Tide rolls, Bfigura's puppy, Legobot, Luckas-bot, Jbm377, Amirobot, Tjkinsey, AnomieBOT, Tryptofish, ESCapade, Kingpin13, Citation bot, Quebec99, LilHelpa, Marshallsumter, Xqbot, GrouchoBot, Omnipaedista, Brandon5485, RibotBOT, Joaquin008, A.amitkumar, Prari, FrescoBot, Aelindor, NifCurator1, Citation bot 1, Variasveces, Pinethicket, Rainbowofknowledge, Tom.Reding, Mjs1991, VMHman, Trappist the monk, Dinamik-bot, Inferior Olive, RjwilmsiBot, Slon02, Lehacarpenter, EmausBot, Combee123, Immunize, Haon 2.0, Ado2013, Delilah fitzgerald, Dcirovic, ZéroBot, Josve05a, MithrandirAgain, ArizonaLifeScience, Hazard-SJ, Lji1942, Feral mage, Donner60, Harpern1, ClueBot NG, Jamesdpalmer, Mesoderm, Widr, Helpful Pixie Bot, BG19bot, Esoteric10, Vokesk, Rwb594, Piguy101, Jialuzeng, BattyBot, Angelaquency, Luckydhaliwal, Jorgelopest, Miguelrangeljr, Enterprisey, Suchithra.ravi, Earl Moss, JakobSteenberg, WroteOddly, Tony Mach, SFK2, Iztwoz, Shelbystripes, RileyBot, Tentinator, Hazelares, NYBrook098, Musthu mueen, Dinisoe, Nilufar Paseban, LT910001, Seppi333, BruceBlaus, JWNoctis, Alexpiet, Brainiacal, Cyrillec, Monkbot, Shelbyknorr23, HiYahhFriend, BethNaught, Skollu, 12alteration21, Pablothepenguin, Meep366, رضا جواد, Sumaira Naz, Tilifa Ocaufa, DominicJohnsonStudent, Wkenkel, BornNHawaii, Barbara (WVS) and Anonymous: 378

- **Basal ganglia** *Source:* https://en.wikipedia.org/wiki/Basal_ganglia?oldid=724219477 *Contributors:* Bryan Derksen, The Anome, Alex.tan, Andre Engels, Fnielsen, Someone else, Ram-Man, Vaughan, Ec5618, Silvonen, Braininfo.rprc.washington.edu, Hadal, Diberri, Adam78, Ancheta Wis, Tarek, Everyking, Chinasaur, Jfdwolff, OldakQuill, Sonjaaa, Dan aka jack, Hehelol, Icairns, Mike Rosoft, Ulflarsen, DanielCD, Discospinster, Rich Farmbrough, Cacycle, MarkS, Bender235, BjarteSorensen, CanisRufus, Pjf, Gilgamesh he, Wisdom89, Arcadian, Pearle, Mdd, Alansohn, Eric Kvaalen, AzadMashari, Scottishmatt, Japanese Searobin, Jackhynes, Kelly Martin, Commander Keane, JeremyA, Anaru, Rjwilmsi, Uwe Gille, RexNL, Bgwhite, WriterHound, Bubbachuck, YurikBot, Chris Capoccia, Yakuzai, Eleassar, A314268, Nrets, Xabian40409, Octavio L~enwiki, Arthur Rubin, Modify, Ulmo~enwiki, Superp, Digfarenough, SmackBot, Jfurr1981, Eskimbot, Amatulic, Bluebot, Torzsmokus, Huji, Drphilharmonic, Clicketyclack, Gary.goldberg.md, Squiggle, Attys, FrozenMan, Gleng, Hu12, DabMachine, Tomwood0, Tawkerbot2, Makeemlighter, General Tojo, Leevanjackson, WeggeBot, Was a bee, Anthonyhcole, D666D, Brain-mapper, Gerard.percheron, Niubrad, Thijs!bot, Carpentc, JustAGal, Nick Number, Hermant patel, Schlegel, Ph.eyes, Igodard, Gfmer, Nucleophilic, Jackfirst, R.Tempest, Mmoneypenny, JustVisiting, CFCF, Xris0, Kpmiyapuram, Gbin2000, Mikael Häggström, (jarbarf), Per Alm~enwiki, STBotD, Kelapstick, Butseriouslyfolks, Viral T, Lova Falk, CMBJ, Sema nini, SieBot, Gerakibot, Danierrr, Sergeyf1, Scubert1, Ludovich-eng, David Buglar, Jaksmith, ClueBot, EoGuy, William Ortiz, Iohannes Animosus, Janelle91165, Manco Capac, Drisar, Qradua, Addbot, DOI bot, Jncraton, Diptanshu.D, Looie496, Glane23, SamatBot, LinkFA-Bot, Lightbot, Luckas-bot, R500Mom, AnomieBOT, Tryptofish, Materialscientist, Citation bot, Xqbot, Mattashner, RibotBOT, SassoBot, NifCurator1, Citation bot 1, Jonesey95, RedBot, JohnnyTremolo, Trappist the monk, DixonDBot, Jonkerz, Jkwascom, فی‌زیوتراپیست ابراهیم بزرگار, Sulbactam, DASHBot, EmausBot, John of Reading, WikitanvirBot, Shameerbabu986, Benlansdell, Dcirovic, Choosebrad, Shuipzv3, AUN4, ClueBot NG, Dr. Persi, Frietjes, Widr, Gmzh, Charles800, Helpful Pixie Bot, Iman Kamali Sarvestani, Wikisian, BG19bot, MrBill3, Aquacryst, Anatomist90, BattyBot, Luckydhaliwal, ChrisGualtieri, Dexbot, CuriousMind01, ققنوس, Joshtaco, Jmhl83, Randykitty, Iztwoz, OhGodItsSoAmazing, Andrey.a.mitin, Christez, Clr324, Babitaarora, Dinisoe, LT910001, Seppi333, BruceBlaus, Mizrahi58, SkateTier, Filedelinkerbot, U08kk12, JaunJimenez, HiYahhFriend, PurplePrawn97, Soonkeun-khu, Abnihil, FourViolas and Anonymous: 182

- **Olfactory bulb** *Source:* https://en.wikipedia.org/wiki/Olfactory_bulb?oldid=725355790 *Contributors:* Bryan Derksen, Edward, Wikiborg, Dysprosia, Selket, Fuelbottle, Pengo, Ancheta Wis, FleaPlus, Bird, Neilc, OldakQuill, Sayeth, Discospinster, Rich Farmbrough, Kndiaye, Cmdrjameson, Arcadian, Galaxiaad, JeremyA, Dolfrog, Male1979, Marudubshinki, Rjwilmsi, FlaBot, AJR, YurikBot, Wavelength, Gaius Cornelius, Chefyinghi, SmackBot, Gilliam, Dlohcierekim's sock, Mattv, RomanSpa, SandyGeorgia, Edwin, Was a bee, Wikid77, Headbomb, Nick Number, Escarbot, Equinexus, Xb2u7Zjzc32, Appraiser, Srice13, MartinBot, Anaxial, Petter Bøckman, Katharineamy, Jeff G., Dudanunielul, Spinningspark, NB-NB, Mimihitam, DoNNNald, Dabomb87, ClueBot, Leppac, Arjayay, Dthomsen8, NellieBly, Tameamseo, Addbot, Fyrael, Dwesson, Luckas-bot, Yobot, Ptbotgourou, AnomieBOT, Justme89, Citation bot, ArthurBot, LilHelpa, John Bessa, The Wiki ghost, Strawbaby, NifCurator1, Jonesey95, Trappist the monk, Bearboat, Gould363, Azzurro2882, Dcirovic, MendicantBias1, Rcsprinter123, ClueBot NG, Albaalba, MerscratianAce, BG19bot, Snow Rise, Anatomist90, BattyBot, Jimw338, ChrisGualtieri, ProBonoPublicoA90, JakobSteenberg, Faizan, LT910001, Emmaskyewilkinson, Vmanjarrez, Kpsychas, Wiki at Royal Society John, Monkbot, Lıovioboti, Erebusthedark, HiYahhFriend, Frschu, Hgwartney and Anonymous: 74

- **Hippocampus** *Source:* https://en.wikipedia.org/wiki/Hippocampus?oldid=730438806 *Contributors:* AxelBoldt, Bryan Derksen, Robert Merkel, Alex.tan, Fnielsen, SimonP, Riptor~enwiki, Tucci528, Edward, Michael Hardy, JWSchmidt, Darkwind, Korpo~enwiki, Pedant17, Tpbradbury, Owen, Sewing, Chuunen Baka, Robbot, Arseni, Premeditated Chaos, Odin.de, Diberri, Dave6, Ancheta Wis, Washington irving, Michael Devore, Bird, Niteowlneils, Jfdwolff, AJim, Wikiwikifast, Jackol, Dan aka jack, Tsemii, ELApro, Freakofnurture, DanielCD, Rich Farmbrough, Cacycle, Vsmith, Dave souza, Bender235, Kelvinc, Art LaPella, Peter M Gerdes, FG~enwiki, Robotje, Arcadian, Famousdog, Hajenso, Hagerman, Jumbuck, Alansohn, Enirac Sum, Eric Kvaalen, Wouterstomp, Angelic Wraith, Vedantm, Japanese Searobin, Benbest, Urod, JeremyA, Hdante, Rachel1, Ianvitro, Mendaliv, Rjwilmsi, TBHecht, FayssalF, FlaBot, RobertG, Joe07734, Kri, Chobot, Jlam4911, Gwernol, YurikBot, Borgx, Butsuri, Sceptre, Jimp, Pacaro, Artur Lion~enwiki, Wimt, NawlinWiki, A314268, Nrets, Rbarreira, Nephron, Entilword, Epipelagic, Action potential, Gadget850, Nikkimaria, Closedmouth, Colin, Digfarenough, Dando~enwiki, SmackBot, Reedy, Jfurr1981, Delldot, Eskimbot, Iph, Kintetsubuffalo, JRGL~enwiki, Chris the speller, Skater11091, Jprg1966, Miquonranger03, Danielkueh, Stevenmitchell, TheLimbicOne, Duckbill, RandomP, Drphilharmonic, Ligulembot, Cheekywee~enwiki, Mgiganteus1, Invisifan, SandyGeorgia, Ryulong, Sausagerooster, Kurtle, Sasata, Michaelbusch, Neuromusic, Beno1000, Chirality, Tawkerbot2, Fvasconcellos, Ccie13836, Juhachi, Outriggr (2006-2009), Cydebot, Was a bee, Anthonyhcole, Brain-mapper, Casliber, Thijs!bot, Epbr123, Mercury~enwiki, CopperKettle, Headbomb, Drmslam, John254, Lewallen, George dubya Bush, Mfirbank, Hmrox, AntiVandalBot, Gioto, Music&Medicine, Danger, Kaini, IvanLitvinov, East718, Stuartlayton, VoABot II, Dekimasu, Farquaadhnchmn, Hekerui, Presearch, WhatamIdoing, Thuglas, David Eppstein, Kineticsicentist, Brockston, Misarxist, Amaranth12498, CommonsDelinker, Nono64, LedgendGamer, DrKay, CFCF, Nbauman, Boghog, Ginsengbomb, Kpmiyapuram, Mrs.meganmmc, Mikael Häggström, LittleHow, Belovedfreak, NewEnglandYankee, Paskari, Jrolston, DorganBot, Xenonice, Dogerty12, Elikarag, VolkovBot, Mbmaciver, ABF, GimmeBot, Guillaume2303, James Baraldi, Jr., Anna Lincoln, Martin451, Rcarlosagis, BotKung, Autodidactyl, Spiral5800, Lova Falk, Doc James, Quietbritishjim, Elbo821, Oda Mari, Arbor to SJ, Faradayplank, Ealdgyth, Psychosomatic Tumor, Kudret abi, Altzinn, Dabomb87, Tal Celes, Myrvin, ClueBot, PipepBot, Pdmckinley, RODERICKMOLASAR, Spikey1973, Mild Bill Hiccup, Piledhigheranddeeper, Alexbot, PsychoProf, I might be batman, Leonard^Bloom, Redpriest187, SpikeToronto, Bucketoftruth, Navicular, Vegetator, Aitias, N.vanstrien, Dana boomer, Canihaveacookie, Johnuniq, Avoided, Rreagan007, Tameamseo, WikiDao, Sgpsaros, Wythy, Addbot, DOI bot, DougsTech, Cricinfouser, AkhtaBot, D0762, CanadianLinuxUser, Fluffernutter, Looie496, Cst17, Najmakb, Skamnelis, McMannDavid, Favonian, ChenzwBot, LinkFA-Bot, Tide rolls, Cesiumfrog, Merlin the Wizard, Legobot, Yobot, Vcmartin, KamikazeBot, A More Perfect Onion, Götz, Jean-Francois Gariepy, Materialscientist, Citation bot, Techdoctor, Jmarchn, UltraBibendum, Xqbot, Loliveke, 5glacieres, BenTheMen, Omnipaedista, Rurigok, NifCurator1, Thanhluan001, Dtone157, HamburgerRadio, Citation bot 1, Keenan ahern, Pinethicket, Jordan-ITP, Lars Washington, MastiBot, Ongar the World-Weary, 404 page not found, Cocacola456123, GoodOlRickyTicky3, Jonkerz, MrX, Weedwhacker128, MBVECO, RjwilmsiBot, Gunderberg, EmausBot, WikitanvirBot, Andykolandy, ScottyBerg, Moswento, Dcirovic, K6ka, ZéroBot, Jenks24, AManWithNoPlan, Wingman4l7, Koala0090, Thine Antique Pen, Cit helper, Donner60, Gordon1104, Shirleybayer, Bhappylots, Wcfios, SYTYCSM, ClueBot NG, Rezabot, Widr, Lanoitanretni, PhD4NRG, Zibart, Bibcode Bot, Turf Einar, MusikAnimal, Phs951, Geoff B Hall, Brenp12, Wahewila, Rob Hurt, Biosthmors, Salem79, Prison gates open, YFdyh-bot, Guywholikesca2+, Wpd0001, Dexbot, Bhbuehler, Mr. Guye, Wayneandkarl, ResearchRules, SassyLilNugget, TruthProf, ECRobertson90, Sacliff, Iztwoz, Caiaphodus, Poodlelover18, Dinisoe, LT910001, Seppi333, Neurosciency, Philgalinsky, Brainiacal, Anrnusna, Bahar.rahsepar, JaconaFrere, Minecwaflova69, ص‌ع‌ب, Monkbot, Anon376, HiYahhFriend, GinAndChronically, BethNaught, Chickaehrtyejhjte, Derdikman, Velvel2, Spizaetus, Allenwookie, Jameshawes2008, Adebowaleareo, Tilifa Ocaufa, KasparBot, Liechtenstein96, ComicBookGuy1221, Slwa357 and Anonymous: 331

- **Amygdala** *Source:* https://en.wikipedia.org/wiki/Amygdala?oldid=735801508 *Contributors:* AxelBoldt, Derek Ross, Bryan Derksen, Taw, Ap, Fnielsen, PierreAbbat, Kosebamse, Ellywa, Notheruser, JWSchmidt, Schneelocke, Emperorbma, Bemoeial, Selket, Wiwaxia, Sander123, Lowellian, Odin.de, João Sousa, Giftlite, Washington irving, Jfdwolff, Cyberied, Wikiwikifast, AlistairMcMillan, Jabowery, LiDaobing, Beland, Piotrus, Sayeth, Bobhearn, Lumidek, DanielCD, Cfailde, Kndiaye, Bender235, Techtoucian, Kwamikagami, Arcadian, Googie man, Alansohn, Andrewpmk, Apoc2400, Wdfarmer, Ombudsman, Angelic Wraith, BRW, Oghmoir, Bacteria, Before My Ken, FoxInShoes, Male1979, Rjwilmsi, Nightscream, Stardust8212, Sdornan, Yamamoto Ichiro, RexNL, Gurch, BMF81, Chobot, Jared Preston, WriterHound, Wjfox2005, YurikBot, Borgx, RussBot, Chris Capoccia, Stephenb, Eleassar, Pseudomonas, Artur Lion~enwiki, A314268, Anomo, Grafen, Nrets, Welsh, Thesloth, Mistercow, Elkman, Richardcavell, Where next Columbus?, Fram, KnightRider~enwiki, SmackBot, Bobet, Phi-Gastrein~enwiki, Jfurr1981, Kslays, Mifren, Gilliam, BrotherGeorge, Dov-El, Chris the speller, Bluebot, Mikepurvis, Dlohcierekim's sock, Liontooth, Max David, Scarletsmith, MeekSaffron, Rogerrp, Pazda, TheLimbicOne, Drphilharmonic, Michalchik, Ohconfucius, SashatoBot, Tarcieri, Rlalumiere, Kyoko, Mfourman, Nismo3112, Iridescent, Michaelbusch, Twas Now, Octane, Chirality, Tawkerbot2, Marilyn.hanson, JForget, Mellery, Wolfdog, CmdrObot, Cogpsych, Unionhawk, Searles2sels, Dgw, Juhachi, Penbat, Cydebot, Ubiq, Was a bee, Karafias, Casliber, EvocativeIntrigue, Mattisse, Thijs!bot, Sugarcream, CopperKettle, Headbomb, AntiVandalBot, Majorly, Nipisiquit, Verizone, MoniqueRN, Trendline, JAnDbot, Deflective, Greensburger, Canjth, VoABot II, Nyq, JamesBWatson, Caesarjbsquitti, Animum, Allstarecho, LaMona, Zip123, SuziAnvin, Maurice Carbonaro, Kpmiyapuram, Davidm617617, Mrs.megannmc, 1000Faces, Nicotinamide, SteveChervitzTrutane, LittleHow, MegaDaytime, Juliancolton, Plindenbaum, Philip Trueman, TXiKiBoT, Oshwah, WatchAndObserve, Rcarlosagis, Cremepuff222, Eubulides, Lova Falk, AlleborgoBot, Michelleem, Logan, SieBot, Eren01, Robotchampion, Lightmouse, Martarius, ClueBot, Binksternet, Isuperhero, Dreamback1116, Sun Creator, Aitias, DumZiBoT, XLinkBot, Earcanal, Addbot, DOI bot, Timothydanaos, CanadianLinuxUser, Fluffernutter, Looie496, Favonian, Willondon, Legobot, Luckas-bot, Yobot, Choms, KamikazeBot, AnomieBOT, Tryptofish, Lynntyler, Jim1138, Royote, Prudhommei1, Stefanculumov, Beatnik86, Citation bot, Jmarchn, Quebec99, Termininja, Fair-normal, XZeroBot, Makeswell, GrouchoBot, Omnipaedista, RibotBOT, Musketeer41, FrescoBot, Paine Ellsworth, D'ohBot, Steve Quinn, NifCurator1, Citation bot 1, Aldy, Chaiten1, Abductive, Hessamnia, Hopeiamfine, Trappist the monk, Ansumang, RjwilmsiBot, NB32, DASHBot, EmausBot, Dewritech, Rarevogel, Slightsmile, Dsanchez89, Dennis714, Cit helper, Shirleybayer, Nirakka, ClueBot NG, Hamlet 2010a, Gameguy118, Frietjes, Qyqyqy1, Mrbazoun, Kshobe, Cyono, Da5id403, BG19bot, Nardzom, Mukeshsamani, Geoff B Hall, AnieHall, Brad7777, Souvhsduhfu, Rob Hurt, BattyBot, HectorMoffet, ChrisGualtieri, JY-Bot, SamLinscho, Dexbot, Sae Harshberger, Sarkisoo, Ix3chocobos, CuriousMind01, Jakesamazing, Bhiron, ECRobertson90, Slt47, B14709, Dinisoe, LT910001, Liz, Anrnusna, NeuroGlossary, Justin8785, Rsoscia, Wiki at Royal Society John, Schoemann, Monkbot, Stevemaren, Tkennedy239, Swege, Archiloc, Medgirl131, CardinalsFan06, Julietdeltalima, DavidJac, Jamienh123, Mahfuzur rahman shourov, Snazzywiki, Emiba13, Rkentkirkwood and Anonymous: 244

- **Pallium (neuroanatomy)** *Source:* https://en.wikipedia.org/wiki/Pallium_(neuroanatomy)?oldid=706324815 *Contributors:* Jhertel, MrTree, Umofomia, Boghog, Bobber0001, VolkovBot, WereSpielChequers, Invertzoo, Addbot, TutterMouse, Looie496, AnomieBOT, Bluerasberry, Sideways713, MrArifnajafov, Gorthian, Iztwoz, Robevans123, James Gabriel II and Anonymous: 8

- **Gyrus** *Source:* https://en.wikipedia.org/wiki/Gyrus?oldid=728458295 *Contributors:* Alex.tan, Emperorbma, Ancheta Wis, Arcadian, TaintedMustard, Rjwilmsi, PhatRita, RussBot, Reo On, A314268, Daniel Mietchen, Zwobot, Mysid, JoanneB, Superp, Ut2491, EncycloPetey, Stevage, Kuru, Gleng, Falk Lieder, Badseed, Was a bee, Alaibot, Casliber, THEMlCK, Jairuscobb, Equazcion, VolkovBot, Wiae, Lova Falk, Caltas, Exert, Martarius, PipepBot, Elenaschifirnet, Addbot, Looie496, Tassedethe, AnomieBOT, Lynntyler, Xqbot, FrescoBot, Salvidrim!, EmausBot, ZéroBot, Ὁ οἶστρος, ClueBot NG, Wimpus~enwiki, GoShow, Chuntuk, SteenthIWbot, Neuron Doc, Unenthusiastic and Anonymous: 12

- **Sulcus (neuroanatomy)** *Source:* https://en.wikipedia.org/wiki/Sulcus_(neuroanatomy)?oldid=704978510 *Contributors:* Fnielsen, Gzuckier, Arcadian, Alansohn, TaintedMustard, Thiseye, Daniel Mietchen, Xaje, Mysid, Caerwine, SmackBot, Giancarlo Rossi, Falk Lieder, Was a bee, Alaibot, Nick Number, LittleHow, Ltnature, Elenaschifirnet, Addbot, Lightbot, Yobot, Amirobot, Lynntyler, Xqbot, Mlpearc, Seventeensquared, D'ohBot, Wikipelli, Donner60, ClueBot NG, Frietjes, Wbm1058, JakobSteenberg, SteenthIWbot, Iztwoz, LT910001, BruceBlaus, Derek hord, Janfreyberg, Unenthusiastic and Anonymous: 15

- **Development of the nervous system in humans** *Source:* https://en.wikipedia.org/wiki/Development_of_the_nervous_system_in_humans?oldid=732040971 *Contributors:* Beland, Billymac00, Arcadian, Kurzon, Rjwilmsi, Bgwhite, SmackBot, RDBrown, Niubrad, CopperKettle, Headbomb, Coyets, Mild Bill Hiccup, DOI bot, Diptanshu.D, Looie496, Glass Sword, AnomieBOT, Tryptofish, Zawer, Citation bot, Jonathansuh, I dream of horses, Tauiris, Pile-Up, Klbrain, Dcirovic, Jenks24, ISpamThisSite3, ClueBot NG, Jjbraun, BG19bot, Khazar2, Neuroes, Iztwoz, Altojells, Monkbot, Neuroez and Anonymous: 12

- **Lateralization of brain function** *Source:* https://en.wikipedia.org/wiki/Lateralization_of_brain_function?oldid=735654572 *Contributors:* Ed Poor, Fnielsen, William Avery, Reigh, Michael Hardy, Lousyd, Julesd, Dwo, Ec5618, Shavenwarthog, Topbanana, Moink, Gwalla, JamesMLane, Mboverload, Gracefool, Aecarol, Histrion, ELApro, D6, Discospinster, FT2, Bender235, *drew, GTubio, Giraffedata, Tgr, Edital, Alansohn, Dowcet, Bookandcoffee, Natalya, Dolfrog, AdinaBob, Mandarax, Rjwilmsi, Miserlou, Brighterorange, DoubleBlue, Ewlyahoocom, A314268,

Action potential, Charlie Wiederhold, Closedmouth, Donald Albury, Ikkyu2, Loginer, JLaTondre, Yaco, Smily, SmackBot, KnowledgeOfSelf, Chris the speller, Tjrudebeck, Skatche, Nick Levine, Harnad, PPBlais, Clicketyclack, Ceoil, John, Gleng, Extremophile, Yms, Doczilla, Lamshuwing, Iridescent, Michaelbusch, Grothmag, CzarB, S-Ranger, Adriatikus, DavidHOzAu, Vaughan Pratt, Lighthead, Nczempin, ArmyOfFluoride, Lentower, Myocastor, Gmcomp, Cydebot, Peterdjones, Bridgecross, Jayen466, Quibik, Sp, PizzaMan, Headbomb, Sobreira, Marek69, Esowteric, Bobblehead, Edhubbard, AntiVandalBot, Ferg1986, Tyco.skinner, Atrax~enwiki, Jordan Rothstein, Roman à clef, Matlee, Roidroid, VoABot II, Zenomax, Arty4ever, ExplicitImplicity, Lifelike, Tarheel blue, CommonsDelinker, LedgendGamer, RockMFR, J.delanoy, Boghog, KMQ0729, Cpiral, Mrs.meganmmc, Notreallydavid, LittleHow, DadaNeem, Silentlambs, Buakawnage, Jmrowland, Deleet, Slysplace, Wiae, InternetHero, Lova Falk, VanishedUserABC, HiDrNick, Justinhaynes, Banjotime, TTLLOGIC, LeadSongDog, Oxymoron83, Hello71, Karl2620, Elassint, ClueBot, Jbening, The Thing That Should Not Be, Icarusgeek, EoGuy, Jonathansand, LeftBrainforLife, Razorflame, Egmontaz, XLinkBot, Balance Terence, WikHead, Mifter, Aunt Entropy, EEng, Addbot, ERK, Tutonite, Willking1979, Simplicissimo, Wikimichael22, Diptanshu.D, Looie496, SamatBot, Lightbot, Yobot, Lacrymocéphale, Becky Sayles, Kailas007, AnomieBOT, DemocraticLuntz, Götz, Rjanag, Materialscientist, Citation bot, Beserkyourmerk, LilHelpa, Tomwsulcer, Jrulya12, Mikeoo17, Mmmeg, FrescoBot, Citation bot 1, Laemonicus, Patafisik, Elockid, MichaelExe, Trappist the monk, Scire9, Dinamik-bot, Gardrek, 6golf9, RjwilmsiBot, Cstanford.math, WikitanvirBot, UsüF, Leftyluv, Tisane, Peter K. Livingston, Robert P. Miller IV, Dcirovic, Lucas Thoms, Josve05a, AManWithNoPlan, EvroAziya, Madhatter198, Kashuhivorodu, Rmashhadi, Skoenig3, ClueBot NG, CocuBot, Frietjes, Widr, Stephanie Lahey, Benmotz, Helpful Pixie Bot, BG19bot, Episeda, Witch Hazell, Spacemountainmike, CitationCleanerBot, Anatomist90, Iamnotfreud, Aisteco, Rob Hurt, ChrisGualtieri, Countered, Mr. Guye, Lamarkfonda, Dfarrell007, Epicgenius, UofTJonii, Iztwoz, Psychscholar29, DavidLeighEllis, Jdis1110, Chickensaresocute, NottNott, YiFeiBot, Mr. George D Patnoe Jr, Monkbot, JustClix, HiYahhFriend, Happy Attack Dog, BiologicalMe, Kanyeweston, Rennertoad24, Constantinaki, Shootingstar88, Kduncan100, Charlie Oh, XenusG, Vensco and Anonymous: 238

- **Corpus callosum** *Source:* https://en.wikipedia.org/wiki/Corpus_callosum?oldid=732598165 *Contributors:* AxelBoldt, Bryan Derksen, Fnielsen, PierreAbbat, Michael Hardy, (, Kosebamse, Looxix~enwiki, William M. Connolley, Jeandré du Toit, Mrehker, Timwi, Selket, Maximus Rex, Furrykef, Ke4roh, Diberri, Pengo, DragonflySixtyseven, Apalsola, Mattman723, Discospinster, Rich Farmbrough, Bender235, HiddenInPlainSight, Arcadian, Querent, Keenan Pepper, Wouterstomp, SidP, Knowledge Seeker, Dirac1933, Js229, Urod, Huhsunqu, Liface, Cod, Pentawing, Rjwilmsi, FlaBot, Bgwhite, YurikBot, Huw Powell, Blackworm, Chris Capoccia, Hydrargyrum, Gaius Cornelius, Sekhui, Carl Daniels, Kewp, Limetom, Lt-wiki-bot, Ninly, Mike Dillon, Terry Longbaugh, Allens, D Monack, SmackBot, Amcbride, Mvanderw, Delldot, Rmx256, TheWoodsman, Persian Poet Gal, Jprg1966, Cregox, Grover cleveland, Drphilharmonic, Nyvhek, Mgiganteus1, Hypnosifl, DabMachine, Go rators, Michaelbusch, George100, Vanisaac, Gveret Tered, CmdrObot, John M Baker, Memills, JVinocur, Asymptote, Gogo Dodo, Was a bee, Lugnuts, LastBall, Drewerd, I do not exist, Nick Number, Noclevername, Widefox, Awien, Alastair Haines, Lenny Kaufman, JamesBWatson, Bahhasg, Catgut, Bticho, MartinBot, CommonsDelinker, J.delanoy, CFCF, Hans Dunkelberg, Mrs.meganmmc, Jefferson61345, LittleHow, KylieTastic, Bovineboy2008, Naohiro19 revertvandal, Xwdl~enwiki, Msjaitly, Falcon8765, AlleborgoBot, Thx1200, SieBot, Tylers75, Flyer22 Reborn, Halcionne, Drgarden, Myrvin, Governator~enwiki, ClueBot, Dylan McCall, Mild Bill Hiccup, Wikijens, Thewildtype, Arjayay, Muro Bot, Dark Mage, Psychonautic, Facts707, Addbot, DOI bot, Looie496, Rj.amdphreak, Moroderen, Tassedethe, Lightbot, Jarble, Wing, Yobot, AnomieBOT, Jim1138, Bluerasberry, Materialscientist, Citation bot, La comadreja, Maxis ftw, Jmarchn, Control.valve, GrouchoBot, Omnipaedista, Aldo samulo, Tylerritchie, Alicein123456land, Sagahryan, Gouerouz, NifCurator1, Argumzio, Citation bot 1, Citation bot 4, I dream of horses, Jonesey95, RedBot, Timyrlan, Trappist the monk, Goldy419, Sideways713, RjwilmsiBot, EmausBot, John of Reading, WikitanvirBot, Theus PR, GoingBatty, Klbrain, ZéroBot, Donner60, Kippelboy, FeatherPluma, Dixwerr, ClueBot NG, TehGrauniad, Dyerdata, Frietjes, Helpful Pixie Bot, Bibcode Bot, Finanbryan, Milner syndrome, Amriya.naufer, Ruby.pennell, MrBill3, Zimpoo, Jlaing17, Anatomist90, Ntsingley, BattyBot, GoShow, TheJJJunk, TylerDurden8823, Ducknish, FoCuSandLeArN, Jason.straathof, TAbildskov, Scizor 99, Faizan, Liannahu07, Iztwoz, Asm233, Raymattwiki, Rileyandotherguy, Moroplogo, Dinisoe, LT910001, Jackmcbarn, ScotXW, Zinedine Socrates, Icecreamcooper, HiYahhFriend, Andrew Macallister, Proff125, Anonyneuroscience, BiologicalMe, Skyexers, Okman8392, Constantinaki and Anonymous: 180

- **Brain mapping** *Source:* https://en.wikipedia.org/wiki/Brain_mapping?oldid=730725312 *Contributors:* Michael Hardy, Sannse, Beland, Xezbeth, Kndiaye, Arcadian, Mdd, Woohookitty, Rjwilmsi, Afterwriting, Chris Capoccia, Grubber, A314268, Daniel Mietchen, Rjlabs, Jogers, Crystallina, SmackBot, Kslays, Richmeister, Iwaterpolo, Radagast83, Clicketyclack, Bushsf, Aeternus, CmdrObot, John.d.van.horn, PKT, Thijs!bot, Alphachimpbot, Minerva2, The Transhumanist, Brewhaha@edmc.net, Indon, Just H, Flowanda, CommonsDelinker, Mange01, DigitalCatalyst, Kpmiyapuram, Louislemieux, Mikael Häggström, Medlat, Jponnoly, WereSpielChequers, FSHL, Jazzmen301, Hordaland, ClueBot, Staticshakedown, MarkSCohen, Roryethanr, Addbot, Looie496, Bowenseeg, Leonidas from XIV, Yobot, AnomieBOT, Citation bot, HamburgerRadio, Robert.Baruch, Trappist the monk, InMktgWeTrust, Whisky drinker, Mentibot, ClueBot NG, Helpful Pixie Bot, David.moreno72, Babak Kateb, Tentinator, Loniucla, Afhse, Comp.arch, Kokapellimt, Pvpoodle, BruceBlaus, Choupz, Yukio22, HakanIST, Vicxtor H. Fischer and Anonymous: 37

- **Outline of brain mapping** *Source:* https://en.wikipedia.org/wiki/Outline_of_brain_mapping?oldid=727482764 *Contributors:* Bearcat, Giraffedata, LtNOWIS, Wtmitchell, Dolfrog, BD2412, Wavelength, Rjlabs, Chris the speller, The Transhumanist, Magioladitis, Zoofroot, JL-Bot, Keysanger, Dthomsen8, Looie496, Yobot, AnomieBOT, Tryptofish, LilHelpa, FrescoBot, John of Reading, Gcastellanos, GoingBatty, ClueBot NG, Snotbot, Cncmaster, Mogism, Me, Myself, and I are Here, Jodosma, Seppi333, MrScorch6200, Choupz, Sîc S. Drêwalla, Vieque and Anonymous: 4

35.8.2 Images

- **File:1311_Brain_Stem.jpg** *Source:* https://upload.wikimedia.org/wikipedia/commons/6/69/1311_Brain_Stem.jpg *License:* CC BY 4.0 *Contributors:* https://cnx.org/contents/FPtK1zmh@8.25:fEI3C8Ot@10/Preface *Original artist:* OpenStax

- **File:1315_Brain_Sinuses.jpg** *Source:* https://upload.wikimedia.org/wikipedia/commons/a/a2/1315_Brain_Sinuses.jpg *License:* CC BY 4.0 *Contributors:* https://cnx.org/contents/FPtK1zmh@8.25:fEI3C8Ot@10/Preface *Original artist:* OpenStax

- **File:1316_Meningeal_LayersN.jpg** *Source:* https://upload.wikimedia.org/wikipedia/commons/4/4c/1316_Meningeal_LayersN.jpg *License:* CC BY 4.0 *Contributors:* https://cnx.org/contents/FPtK1zmh@8.25:fEI3C8Ot@10/Preface *Original artist:* OpenStax

- **File:1511_The_Limbic_Lobe.jpg** *Source:* https://upload.wikimedia.org/wikipedia/commons/7/7a/1511_The_Limbic_Lobe.jpg *License:* CC BY 3.0 *Contributors:* Anatomy & Physiology, Connexions Web site. http://cnx.org/content/col11496/1.6/, Jun 19, 2013. *Original artist:* OpenStax College

- **File:1604_Types_of_Cortical_Areas-02.jpg** *Source:* https://upload.wikimedia.org/wikipedia/commons/a/aa/1604_Types_of_Cortical_Areas-02.jpg *License:* CC BY 3.0 *Contributors:* Anatomy & Physiology, Connexions Web site. http://cnx.org/content/col11496/1.6/, Jun 19, 2013. *Original artist:* OpenStax College

- **File:3DSlicer-KubickiJPR2007-fig6.jpg** *Source:* https://upload.wikimedia.org/wikipedia/commons/b/b6/3DSlicer-KubickiJPR2007-fig6.jpg *License:* CC-BY-SA-3.0 *Contributors:* Kubicki M., McCarley R.W., Westin C-F., Park H-J., Maier S.E., Kikinis R., Jolesz F.A., Shenton M.E. A review of diffusion tensor imaging studies in schizophrenia. J Psychiatr Res. 2007 Jan-Feb;41(1-2):15-30. PMID: 16023676. PMCID: PMC2768134. *Original artist:* ?

- **File:4_week_embryo_brain.jpg** *Source:* https://upload.wikimedia.org/wikipedia/commons/4/4c/4_week_embryo_brain.jpg *License:* Public domain *Contributors:* Own work *Original artist:* Kurzon

- **File:6_week_human_embryo_nervous_system.svg** *Source:* https://upload.wikimedia.org/wikipedia/commons/3/33/6_week_human_embryo_nervous_system.svg *License:* Public domain *Contributors:* Own work *Original artist:* Kurzon

- **File:Alzheimer'{}s_Disease.gif** *Source:* https://upload.wikimedia.org/wikipedia/commons/f/f4/Alzheimer%27s_Disease.gif *License:* CC BY 4.0 *Contributors:* http://docjana.com/#/alzheimers; https://www.patreon.com/posts/4380136 *Original artist:* Doctor Jana

- **File:Ambox_contradict.svg** *Source:* https://upload.wikimedia.org/wikipedia/commons/2/2e/Ambox_contradict.svg *License:* Public domain *Contributors:* self-made using Image:Emblem-contradict.svg *Original artist:* penubag, Rugby471

- **File:Amigdale1.jpg** *Source:* https://upload.wikimedia.org/wikipedia/commons/c/c3/Amigdale1.jpg *License:* CC-BY-SA-3.0 *Contributors:* ? *Original artist:* ?

- **File:Amyg.png** *Source:* https://upload.wikimedia.org/wikipedia/commons/8/8b/Amyg.png *License:* CC-BY-SA-3.0 *Contributors:* Originally from en.wikipedia; description page is (was) here * 14:35, 13 February 2004 [[:en:User:Washington irving|Washington irving]] 189×230 (22,159 bytes) <span class="comment">(Location of the Amygdala in the Human Brain)</span> *Original artist:* User Washington irving on en.wikipedia

- **File:Basal-ganglia-classic.png** *Source:* https://upload.wikimedia.org/wikipedia/commons/4/45/Basal-ganglia-classic.png *License:* CC-BY-SA-3.0 *Contributors:* ? *Original artist:* Created by Andrew Gillies

- **File:Basal-ganglia-coronal-sections-large.png** *Source:* https://upload.wikimedia.org/wikipedia/commons/3/33/Basal-ganglia-coronal-sections-large.png *License:* CC-BY-SA-3.0 *Contributors:* Own work *Original artist:* Andrew Gillies (User:Anaru)

- **File:Basal_ganglia_2.jpg** *Source:* https://upload.wikimedia.org/wikipedia/commons/a/af/Basal_ganglia_2.jpg *License:* CC BY-SA 3.0 *Contributors:* Own work *Original artist:* Anatomist90

- **File:Basal_ganglia_circuits.svg** *Source:* https://upload.wikimedia.org/wikipedia/commons/9/9e/Basal_ganglia_circuits.svg *License:* CC BY-SA 3.0 *Contributors:* Introduction to Parkinson's Disease By Zaneta Navratilova. Last updated: May 10, 2004. Modified from Nestler et. al, 2001, p. 306 and Ottley et. al, 1999.

Original artist: Mikael Häggström, based on images by Andrew Gillies/User:Anaru and Patrick J. Lynch

- **File:Blausen_0076_BasalGanglia.png** *Source:* https://upload.wikimedia.org/wikipedia/commons/5/55/Blausen_0076_BasalGanglia.png *License:* CC BY 3.0 *Contributors:* Own work *Original artist:* BruceBlaus. When using this image in external sources it can be cited as:

- **File:Blausen_0114_BrainstemAnatomy.png** *Source:* https://upload.wikimedia.org/wikipedia/commons/e/e0/Blausen_0114_BrainstemAnatomy.png *License:* CC BY 3.0 *Contributors:* Own work *Original artist:* BruceBlaus. When using this image in external sources it can be cited as:

- **File:Blausen_0115_BrainStructures.png** *Source:* https://upload.wikimedia.org/wikipedia/commons/1/1a/Blausen_0115_BrainStructures.png *License:* CC BY 3.0 *Contributors:* Own work *Original artist:* BruceBlaus. When using this image in external sources it can be cited as:

- **File:Blausen_0896_Ventricles_Brain.png** *Source:* https://upload.wikimedia.org/wikipedia/commons/d/d4/Blausen_0896_Ventricles_Brain.png *License:* CC BY 3.0 *Contributors:* Own work *Original artist:* BruceBlaus

- **File:Bone.png** *Source:* https://upload.wikimedia.org/wikipedia/commons/2/2c/Bone.png *License:* CC BY 3.0 *Contributors:* [1] [2] *Original artist:* BanzaiTokyo

- **File:Brain_Lateralization.svg** *Source:* https://upload.wikimedia.org/wikipedia/commons/7/78/Brain_Lateralization.svg *License:* CC BY-SA 3.0 *Contributors:* Own work *Original artist:* Chickensaresocute

- **File:Brainmaps-macaque-hippocampus.jpg** *Source:* https://upload.wikimedia.org/wikipedia/commons/9/9a/Brainmaps-macaque-hippocampus.jpg *License:* CC BY 3.0 *Contributors:* http://brainmaps.org/index.php?p=screenshots *Original artist:* brainmaps.org

- **File:Brodmann-areas.png** *Source:* https://upload.wikimedia.org/wikipedia/commons/0/09/Brodmann-areas.png *License:* Public domain *Contributors:* This image was made by modifying a scan of p 288 of the book "Anatomy of the Nervous System", by Stephen Walter Ranson, W. B. Saunders, 1920. *Original artist:* modified by User:Looie496, original artist unknown but probably Brodmann

- **File:CajalHippocampus_(modified).png** *Source:* https://upload.wikimedia.org/wikipedia/commons/2/25/CajalHippocampus_%28modified%29.png *License:* Public domain *Contributors:* File:CajalHippocampus.jpeg from: Santiago Ramón y Cajal (1911) [1909] *Histologie du Système nerveux de l'Homme et des Vertébrés*, Paris: A. Maloine *Original artist:*

- original: Santiago Ramón y Cajal (1852–1934)

- **File:Cajal_cortex_drawings.png** *Source:* https://upload.wikimedia.org/wikipedia/commons/5/5b/Cajal_cortex_drawings.png *License:* Public domain *Contributors:* "Comparative study of the sensory areas of the human cortex" by Santiago Ramon y Cajal, published 1899, ISBN 9781458821898 *Original artist:* User:Looie496 created file, Santiago Ramon y Cajal created artwork

- **File:Cerebral_lobes.png** *Source:* https://upload.wikimedia.org/wikipedia/commons/8/85/Cerebral_lobes.png *License:* CC-BY-SA-3.0 *Contributors:* derivative work of this - Gutenberg Encyclopedia *Original artist:* ?

35.8. TEXT AND IMAGE SOURCES, CONTRIBUTORS, AND LICENSES

- **File:Cerebral_vascular_territories.jpg** *Source:* https://upload.wikimedia.org/wikipedia/commons/4/4a/Cerebral_vascular_territories.jpg *License:* CC BY 2.5 *Contributors:*
- Brain_stem_normal_human.svg *Original artist:*
- derivative work: Frank Gaillard (<a href='//commons.wikimedia.org/wiki/User_talk:Frank_Gaillard' title='User talk:Frank Gaillard'>talk</a>)
- **File:Cerebrum_animation_small.gif** *Source:* https://upload.wikimedia.org/wikipedia/commons/f/fa/Cerebrum_animation_small.gif *License:* CC BY-SA 2.1 jp *Contributors:* Polygon data are from BodyParts3D[1] *Original artist:* Polygon data were generated by Database Center for Life Science(DBCLS)[2].
- **File:Choroid_plexus.jpg** *Source:* https://upload.wikimedia.org/wikipedia/commons/c/c6/Choroid_plexus.jpg *License:* CC BY-SA 3.0 *Contributors:* Own work *Original artist:* Anatomist90
- **File:Cod_brain_showing_tectum.png** *Source:* https://upload.wikimedia.org/wikipedia/commons/2/27/Cod_brain_showing_tectum.png *License:* Public domain *Contributors:* This image was scanned from p 112 of the book "The Anatomy of the central nervous system of man and of vertebrates", by Ludwig Edinger, Davis, 1899, and then modified using Gimp. *Original artist:* original artist unknown, modified by user:Looie496
- **File:Commons-logo.svg** *Source:* https://upload.wikimedia.org/wikipedia/en/4/4a/Commons-logo.svg *License:* CC-BY-SA-3.0 *Contributors:* ? *Original artist:* ?
- **File:Computed_tomography_of_human_brain_-_large.png** *Source:* https://upload.wikimedia.org/wikipedia/commons/5/50/Computed_tomography_of_human_brain_-_large.png *License:* CC0 *Contributors:* Radiology, Uppsala University Hospital. Uploaded by Mikael Häggström. *Original artist:* Department of Radiology, Uppsala University Hospital. Uploaded by Mikael Häggström.
- **File:Constudoverbrain.png** *Source:* https://upload.wikimedia.org/wikipedia/commons/b/bc/Constudoverbrain.png *License:* CC-BY-SA-3.0 *Contributors:* this is Image:Constudoverbrain.gif from Commons, cropped and resaved in PNG format *Original artist:* Original uploader of gif version was RobinH at en.wikibooks
- **File:Corpus_callosum.jpg** *Source:* https://upload.wikimedia.org/wikipedia/commons/b/b6/Corpus_callosum.jpg *License:* CC-BY-SA-3.0 *Contributors:* This image was made by combining two images from Wikimedia Commons, Image:Human brain midsagittal cut .JPG and Image:Human brain frontal (coronal) section.JPG *Original artist:* modified image created by user:Looie496. original images created by John A. Beal, Ph.D.
- **File:Corpus_callosum.png** *Source:* https://upload.wikimedia.org/wikipedia/commons/6/60/Corpus_callosum.png *License:* CC BY-SA 2.1 jp *Contributors:* from Anatomography[1] website maintained by Life Science Databases(LSDB). *Original artist:* Images are generated by Life Science Databases(LSDB).
- **File:Development_of_nervous_system.svg** *Source:* https://upload.wikimedia.org/wikipedia/commons/6/67/Development_of_nervous_system.svg *License:* Public domain *Contributors:* Own work, based on :File:Development_of_nervous_system.png *Original artist:* Tauiris
- **File:Diagram_showing_some_of_the_main_areas_of_the_brain_CRUK_188.svg** *Source:* https://upload.wikimedia.org/wikipedia/commons/d/d1/Diagram_showing_some_of_the_main_areas_of_the_brain_CRUK_188.svg *License:* CC BY-SA 4.0 *Contributors:* Original email from CRUK *Original artist:* Cancer Research UK
- **File:EEG_cap.jpg** *Source:* https://upload.wikimedia.org/wikipedia/commons/b/bf/EEG_cap.jpg *License:* Public domain *Contributors:* Transferred from en.wikipedia to Commons by Sreejithk2000 using CommonsHelper. *Original artist:* Thuglas at English Wikipedia
- **File:Early_Olfactory_System.svg** *Source:* https://upload.wikimedia.org/wikipedia/commons/0/07/Early_Olfactory_System.svg *License:* CC BY 3.0 *Contributors:* http://www.frontiersin.org/systems_neuroscience/10.3389/fnsys.2011.00084/full . doi:10.3389/fnsys.2011.00084 *Original artist:* Benjamin Auffarth, Bernhard Kaplan, Anders Lansner
- **File:Edit-clear.svg** *Source:* https://upload.wikimedia.org/wikipedia/en/f/f2/Edit-clear.svg *License:* Public domain *Contributors:* The *Tango! Desktop Project*. *Original artist:*

  The people from the Tango! project. And according to the meta-data in the file, specifically: "Andreas Nilsson, and Jakub Steiner (although minimally)."
- **File:Endocrine_central_nervous_en.svg** *Source:* https://upload.wikimedia.org/wikipedia/commons/9/96/Endocrine_central_nervous_en.svg *License:* Public domain *Contributors:* made myself based on the information found on the wikipedia article *Original artist:* LadyofHats
- **File:Free-to-read_lock_75.svg** *Source:* https://upload.wikimedia.org/wikipedia/commons/8/80/Free-to-read_lock_75.svg *License:* CC0 *Contributors:*

  Adapted from <a href='//en.wikipedia.org/wiki/File:Open_Access_logo_PLoS_white_green.svg' class='image' title='Open_Access_logo_PLoS_white_green.svg'><img alt='Open_Access_logo_PLoS_white_green.svg' src='//upload.wikimedia.org/wikipedia/commons/thumb/9/90/Open_Access_logo_PLoS_white_green.svg/9px-Open_Access_logo_PLoS_white_green.svg.png' width='9' height='14' srcset='//upload.wikimedia.org/wikipedia/commons/thumb/9/90/Open_Access_logo_PLoS_white_green.svg/14px-Open_Access_logo_PLoS_white_green.svg.png 1.5x, //upload.wikimedia.org/wikipedia/commons/thumb/9/90/Open_Access_logo_PLoS_white_green.svg/18px-Open_Access_logo_PLoS_white_green.svg.png 2x' data-file-width='640' data-file-height='1000' /></a>

  Original artist:

  This version:Trappist_the_monk (talk) (Uploads)
- **File:Frontal_lobe_animation.gif** *Source:* https://upload.wikimedia.org/wikipedia/commons/5/54/Frontal_lobe_animation.gif *License:* CC BY-SA 2.1 jp *Contributors:* Polygon data are from BodyParts3D.[11] *Original artist:* Polygon data were generated by Life Science Databases(LSDB).
- **File:Functional_magnetic_resonance_imaging.jpg** *Source:* https://upload.wikimedia.org/wikipedia/commons/8/87/Functional_magnetic_resonance_imaging.jpg *License:* Public domain *Contributors:* ? *Original artist:* ?
- **File:Golgi_Hippocampus.jpg** *Source:* https://upload.wikimedia.org/wikipedia/commons/5/5e/Golgi_Hippocampus.jpg *License:* Public domain *Contributors:* ? *Original artist:* ?

- **File:Gray640.png** *Source:* https://upload.wikimedia.org/wikipedia/commons/2/20/Gray640.png *License:* Public domain *Contributors:* Henry Gray (1918) *Anatomy of the Human Body* (See "Book" section below)

  Original artist: Henry Vandyke Carter

- **File:Gray651.png** *Source:* https://upload.wikimedia.org/wikipedia/commons/c/ce/Gray651.png *License:* Public domain *Contributors:* Henry Gray (1918) *Anatomy of the Human Body* (See "Book" section below)

  Original artist: Henry Vandyke Carter

- **File:Gray653.png** *Source:* https://upload.wikimedia.org/wikipedia/commons/a/a1/Gray653.png *License:* Public domain *Contributors:* Henry Gray (1918) *Anatomy of the Human Body* (See "Book" section below)

  Original artist: Henry Vandyke Carter

- **File:Gray654.png** *Source:* https://upload.wikimedia.org/wikipedia/commons/7/75/Gray654.png *License:* Public domain *Contributors:* Henry Gray (1918) *Anatomy of the Human Body* (See "Book" section below)

  Original artist: Henry Vandyke Carter

- **File:Gray694.png** *Source:* https://upload.wikimedia.org/wikipedia/commons/9/94/Gray694.png *License:* Public domain *Contributors:* Henry Gray (1918) *Anatomy of the Human Body* (See "Book" section below)

  Original artist: Henry Vandyke Carter

- **File:Gray719.png** *Source:* https://upload.wikimedia.org/wikipedia/commons/0/0b/Gray719.png *License:* Public domain *Contributors:* Henry Gray (1918) *Anatomy of the Human Body* (See "Book" section below)

  Original artist: Henry Vandyke Carter

- **File:Gray722.png** *Source:* https://upload.wikimedia.org/wikipedia/commons/c/c0/Gray722.png *License:* Public domain *Contributors:* Henry Gray (1918) *Anatomy of the Human Body* (See "Book" section below)

  Original artist: Henry Vandyke Carter

- **File:Gray726.png** *Source:* https://upload.wikimedia.org/wikipedia/commons/3/35/Gray726.png *License:* Public domain *Contributors:* Henry Gray (1918) *Anatomy of the Human Body* (See "Book" section below)

  Original artist: Henry Vandyke Carter

- **File:Gray727.svg** *Source:* https://upload.wikimedia.org/wikipedia/commons/f/fe/Gray727.svg *License:* Public domain *Contributors:* Henry Gray (1918) *Anatomy of the Human Body* (See "Book" section below)

  Original artist: Henry Vandyke Carter

- **File:Gray739-emphasizing-hippocampus.png** *Source:* https://upload.wikimedia.org/wikipedia/commons/2/2e/Gray739-emphasizing-hippocampus.png *License:* Public domain *Contributors:* Henry Gray (1918) *Anatomy of the Human Body* (See "Book" section below)

  Original artist: Henry Vandyke Carter

- **File:Gray754.png** *Source:* https://upload.wikimedia.org/wikipedia/commons/8/89/Gray754.png *License:* Public domain *Contributors:* Henry Gray (1918) *Anatomy of the Human Body* (See "Book" section below)

  Original artist: Henry Vandyke Carter

- **File:Gray_718-amygdala.png** *Source:* https://upload.wikimedia.org/wikipedia/commons/f/f4/Gray_718-amygdala.png *License:* Public domain *Contributors:* Henry Gray (1918) *Anatomy of the Human Body* (See "Book" section below)

  Original artist: Henry Vandyke Carter

- **File:HE_stain_murine_optic_tectum.jpg** *Source:* https://upload.wikimedia.org/wikipedia/commons/3/37/HE_stain_murine_optic_tectum.jpg *License:* CC BY-SA 2.0 *Contributors:*

- NOGO-A_expression_during_cell_migration.jpg *Original artist:* NOGO-A_expression_during_cell_migration.jpg: Caltharp SA, Pira CU, Mishima N, Youngdale EN, McNeill DS, Liwnicz BH, Oberg KC

- **File:Hippocampus_(brain).jpg** *Source:* https://upload.wikimedia.org/wikipedia/commons/3/39/Hippocampus_%28brain%29.jpg *License:* CC-BY-SA-3.0 *Contributors:* Original image donated by Frank Gaillard Designs. First uploaded to the English Wikipedia as Hippocampus_(brain).jpg. *Original artist:* Fg, contact:frank.gaillard@gmail.com

- **File:Hippocampus_and_seahorse_cropped.JPG** *Source:* https://upload.wikimedia.org/wikipedia/commons/5/5b/Hippocampus_and_seahorse_cropped.JPG *License:* CC BY-SA 3.0 *Contributors:*

- Hippocampus_and_seahorse.JPG *Original artist:* Hippocampus_and_seahorse.JPG: Professor Laszlo Seress

- **File:Hippocampus_small.gif** *Source:* https://upload.wikimedia.org/wikipedia/commons/f/ff/Hippocampus_small.gif *License:* CC BY-SA 2.1 jp *Contributors:* URL. *Original artist:* Images are generated by Life Science Databases(LSDB).

- **File:Hirnventrikel_mittelgroß.gif** *Source:* https://upload.wikimedia.org/wikipedia/commons/2/29/Hirnventrikel_mittelgro%C3%9F.gif *License:* CC BY-SA 2.5 *Contributors:* Polygon data are from BodyParts3D. *Original artist:* Polygon data were generated by Life Science Databases(LSDB).

- **File:Homo_habilis-2.JPG** *Source:* https://upload.wikimedia.org/wikipedia/commons/6/61/Homo_habilis-2.JPG *License:* CC-BY-SA-3.0 *Contributors:*

- Homo_habilis.JPG *Original artist:*

- unknown (klimaundmensch.de)

35.8. TEXT AND IMAGE SOURCES, CONTRIBUTORS, AND LICENSES

- **File:Human_Cortical_Development.png** *Source:* https://upload.wikimedia.org/wikipedia/commons/f/fd/Human_Cortical_Development.png *License:* CC BY 2.5 *Contributors:* Kapellou O, Counsell SJ, Kennea N, Dyet L, Saeed N, et al. (2006) Abnormal Cortical Development after Premature Birth Shown by Altered Allometric Scaling of Brain Growth. PLoS Med 3(8): e265. doi:10.1371/journal.pmed.0030265 *Original artist:* Kapellou O, Counsell SJ, Kennea N, Dyet L, Saeed N, et al.

- **File:Human_brainstem_anterior_view_description.JPG** *Source:* https://upload.wikimedia.org/wikipedia/commons/6/6e/Human_brainstem_anterior_view_description.JPG *License:* CC BY 2.5 *Contributors:* http://www.healcentral.org/healapp/showMetadata?metadataId=40566 (Internet Archive of file description page) *Original artist:* John A Beal, PhD Dep't. of Cellular Biology & Anatomy, Louisiana State University Health Sciences Center Shreveport

- **File:Human_embryo_8_weeks_4.JPG** *Source:* https://upload.wikimedia.org/wikipedia/commons/7/77/Human_embryo_8_weeks_4.JPG *License:* CC BY-SA 3.0 *Contributors:* Own work *Original artist:* Anatomist90

- **File:Human_motor_cortex_topography.png** *Source:* https://upload.wikimedia.org/wikipedia/commons/0/0b/Human_motor_cortex_topography.png *License:* Public domain *Contributors:* This file is modified from a scan of p 315 of the book "Anatomy of the Nervous System", by Stephen Walter Ranson, WB Saunders, 1920. *Original artist:* original artist unknown; modified by user:Looie496

- **File:Hypothalamus_small.gif** *Source:* https://upload.wikimedia.org/wikipedia/commons/6/67/Hypothalamus_small.gif *License:* CC BY-SA 2.1 jp *Contributors:* URL. *Original artist:* Images are generated by Life Science Databases(LSDB).

- **File:Isthmii_circuit.svg** *Source:* https://upload.wikimedia.org/wikipedia/commons/6/67/Isthmii_circuit.svg *License:* CC BY-SA 3.0 *Contributors:* Own work *Original artist:* Looie496

- **File:Lateral_sulcus.gif** *Source:* https://upload.wikimedia.org/wikipedia/commons/5/58/Lateral_sulcus.gif *License:* CC BY-SA 2.1 jp *Contributors:* vector data is by BodyParts3D[1]. Coloring and animation are by was_a_bee. *Original artist:* Vector data were generated by Life Science Databases(LSDB). Computer animation was generated by was_a_bee.

- **File:Lateral_surface_of_cerebral_cortex_-_gyri.png** *Source:* https://upload.wikimedia.org/wikipedia/commons/3/35/Lateral_surface_of_cerebral_cortex_-_gyri.png *License:* CC BY 2.5 *Contributors:* Hagmann P, Cammoun L, Gigandet X, Meuli R, Honey CJ, et al. (2008) Mapping the Structural Core of Human Cerebral Cortex. PLoS Biol 6(7): e159. doi:10.1371/journal.pbio.0060159 *Original artist:* Patric Hagmann et.al.

- **File:Left_and_Right_Brain.jpg** *Source:* https://upload.wikimedia.org/wikipedia/commons/6/67/Left_and_Right_Brain.jpg *License:* Public domain *Contributors:* No machine-readable source provided. Own work assumed (based on copyright claims). *Original artist:* No machine-readable author provided. Webber assumed (based on copyright claims).

- **File:Lobes_of_the_brain_NL.svg** *Source:* https://upload.wikimedia.org/wikipedia/commons/0/0e/Lobes_of_the_brain_NL.svg *License:* Public domain *Contributors:* Henry Gray (1918) *Anatomy of the Human Body* (See "Book" section below)

  Original artist: Henry Vandyke Carter

- **File:LocationOfHypothalamus.jpg** *Source:* https://upload.wikimedia.org/wikipedia/commons/9/9f/LocationOfHypothalamus.jpg *License:* Public domain *Contributors:* ? *Original artist:* ?

- **File:MRI_Location_Amygdala_up.png** *Source:* https://upload.wikimedia.org/wikipedia/commons/2/2f/MRI_Location_Amygdala_up.png *License:* CC0 *Contributors:* Own work *Original artist:* Amber Rieder, Jenna Traynor, Geoffrey B Hall

- **File:MRI_Location_Hippocampus_up..png** *Source:* https://upload.wikimedia.org/wikipedia/commons/3/3c/MRI_Location_Hippocampus_up..png *License:* CC0 *Contributors:* Own work *Original artist:* Amber Rieder, Jenna Traynor

- **File:Medial_surface_of_cerebral_cortex_-_entorhinal_cortex.png** *Source:* https://upload.wikimedia.org/wikipedia/commons/1/15/Medial_surface_of_cerebral_cortex_-_entorhinal_cortex.png *License:* CC BY 2.5 *Contributors:* File:Medial surface of cerebral cortex - gyri.png *Original artist:* Hagmann P, Cammoun L, Gigandet X, Meuli R, Honey CJ, et al.

- **File:Medulla_-_Inferior_level_cross_section.svg** *Source:* https://upload.wikimedia.org/wikipedia/commons/8/8a/Medulla_-_Inferior_level_cross_section.svg *License:* CC BY-SA 3.0 *Contributors:* Own work *Original artist:* Kevin Dufendach

- **File:Medulla_-_Middle_level_cross_section.svg** *Source:* https://upload.wikimedia.org/wikipedia/commons/8/8e/Medulla_-_Middle_level_cross_section.svg *License:* CC BY-SA 3.0 *Contributors:* Own work *Original artist:* Kevin Dufendach

- **File:Medulla_-_Rostral_level_cross_section.svg** *Source:* https://upload.wikimedia.org/wikipedia/commons/3/38/Medulla_-_Rostral_level_cross_section.svg *License:* CC BY-SA 3.0 *Contributors:* Own work *Original artist:* Kevin Dufendach

- **File:Medulla_oblongata_and_foramen_magnum_animation_small.gif** *Source:* https://upload.wikimedia.org/wikipedia/commons/2/2a/Medulla_oblongata_and_foramen_magnum_animation_small.gif *License:* CC BY-SA 2.1 jp *Contributors:* Anatomography (setting page of this image.) *Original artist:* Anatomography

- **File:Medulla_oblongata_small.gif** *Source:* https://upload.wikimedia.org/wikipedia/commons/0/01/Medulla_oblongata_small.gif *License:* CC BY-SA 2.1 jp *Contributors:* URL. *Original artist:* Images are generated by Life Science Databases(LSDB).

- **File:Mergefrom.svg** *Source:* https://upload.wikimedia.org/wikipedia/commons/0/0f/Mergefrom.svg *License:* Public domain *Contributors:* ? *Original artist:* ?

- **File:Midbrain_-_inferior_colliculus.svg** *Source:* https://upload.wikimedia.org/wikipedia/commons/5/5e/Midbrain_-_inferior_colliculus.svg *License:* CC BY-SA 3.0 *Contributors:* Own work *Original artist:* Kevin Dufendach

- **File:Midbrain_-_superior_colliculus.svg** *Source:* https://upload.wikimedia.org/wikipedia/commons/f/fa/Midbrain_-_superior_colliculus.svg *License:* CC BY-SA 3.0 *Contributors:* Own work *Original artist:* Kevin Dufendach

- **File:Midbrainsuperiorcolliculus.png** *Source:* https://upload.wikimedia.org/wikipedia/commons/f/f7/Midbrainsuperiorcolliculus.png *License:* CC-BY-SA-3.0 *Contributors:* ? *Original artist:* ?

- **File:Mouse_Amygdala.pdf** *Source:* https://upload.wikimedia.org/wikipedia/commons/8/84/Mouse_Amygdala.pdf *License:* CC BY-SA 4.0 *Contributors:* Own work *Original artist:* Rob Hurt

- **File:Mouse_MOB_three_color.jpg** *Source:* https://upload.wikimedia.org/wikipedia/commons/1/10/Mouse_MOB_three_color.jpg *License:* Public domain *Contributors:* Released by author *Original artist:* Matt Valley
- **File:NPH_MRI_272_GILD.gif** *Source:* https://upload.wikimedia.org/wikipedia/commons/3/38/NPH_MRI_272_GILD.gif *License:* CC BY-SA 3.0 *Contributors:* Own work *Original artist:* Nevit Dilmen (<a href='//commons.wikimedia.org/wiki/User_talk:Nevit' title='User talk: Nevit'>talk</a>)
- **File:Neuro_logo.png** *Source:* https://upload.wikimedia.org/wikipedia/commons/f/f8/Neuro_logo.png *License:* Public domain *Contributors:* The PNG crusade bot automatically converted this image to the more efficient PNG format. ; 19:29:54, 18 April 2007 (UTC) Remember the dot (*ShouldBePNG*) *Original artist:* The PNG crusade bot automatically converted this image to the more efficient PNG format. 19:29:54, 18 April 2007 (UTC) Remember the dot (*ShouldBePNG*)
- **File:NeuronGolgi.png** *Source:* https://upload.wikimedia.org/wikipedia/commons/8/88/NeuronGolgi.png *License:* CC BY 2.5 *Contributors:* Transferred from en.wikipedia to Commons. *Original artist:* The original uploader was Pr495du at English Wikipedia
- **File:Nicolas_P._Rougier'{}s_rendering_of_the_human_brain.png** *Source:* https://upload.wikimedia.org/wikipedia/commons/7/73/Nicolas_P._Rougier%27s_rendering_of_the_human_brain.png *License:* GPL *Contributors:* http://www.loria.fr/~{}rougier *Original artist:* Nicolas Rougier
- **File:Occipital_lobe_animation_small.gif** *Source:* https://upload.wikimedia.org/wikipedia/commons/8/8f/Occipital_lobe_animation_small.gif *License:* CC BY-SA 2.1 jp *Contributors:* Polygon data are from BodyParts3D[1] *Original artist:* Polygon data were generated by Database Center for Life Science(DBCLS)[2].
- **File:PET-image.jpg** *Source:* https://upload.wikimedia.org/wikipedia/commons/c/c6/PET-image.jpg *License:* Public domain *Contributors:* Own work *Original artist:* Jens Maus (http://jens-maus.de/)
- **File:Parietal_lobe_animation_small.gif** *Source:* https://upload.wikimedia.org/wikipedia/commons/3/32/Parietal_lobe_animation_small.gif *License:* CC BY-SA 2.1 jp *Contributors:* Polygon data are from BodyParts3D[1] *Original artist:* Polygon data were generated by Database Center for Life Science(DBCLS)[2].
- **File:Place_Cell_Spiking_Activity_Example.png** *Source:* https://upload.wikimedia.org/wikipedia/commons/5/5e/Place_Cell_Spiking_Activity_Example.png *License:* CC BY-SA 3.0 *Contributors:* I personally created this image using data I collected during my graduate studies. *Original artist:* Stuartlayton
- **File:Pons_-_Inferior.svg** *Source:* https://upload.wikimedia.org/wikipedia/commons/7/72/Pons_-_Inferior.svg *License:* CC BY-SA 3.0 *Contributors:* Own work *Original artist:* Kevin Dufendach
- **File:Pons_-_Middle.svg** *Source:* https://upload.wikimedia.org/wikipedia/commons/e/e9/Pons_-_Middle.svg *License:* CC BY-SA 3.0 *Contributors:* Own work *Original artist:* Kevin Dufendach
- **File:Portal-puzzle.svg** *Source:* https://upload.wikimedia.org/wikipedia/en/f/fd/Portal-puzzle.svg *License:* Public domain *Contributors:* ? *Original artist:* ?
- **File:Question_book-new.svg** *Source:* https://upload.wikimedia.org/wikipedia/en/9/99/Question_book-new.svg *License:* Cc-by-sa-3.0 *Contributors:*

  Created from scratch in Adobe Illustrator. Based on Image:Question book.png created by User:Equazcion *Original artist:* Tkgd2007
- **File:Rat-hippocampal-activity-modes.png** *Source:* https://upload.wikimedia.org/wikipedia/commons/3/3f/Rat-hippocampal-activity-modes.png *License:* Public domain *Contributors:* Own work *Original artist:* Looie496
- **File:Rhombencephalon.jpg** *Source:* https://upload.wikimedia.org/wikipedia/commons/4/4f/Rhombencephalon.jpg *License:* CC BY-SA 3.0 *Contributors:* Own work *Original artist:* Adrian Halga
- **File:Rorschach_blot_03.jpg** *Source:* https://upload.wikimedia.org/wikipedia/commons/8/82/Rorschach_blot_03.jpg *License:* Public domain *Contributors:* http://www.pasarelrorschach.com/en/inkblots.htm *Original artist:* Hermann Rorschach (died 1922)
- **File:Schematic_illustration_of_differences_in_neuronal_specification_and_migration_patterns_between_the_mammalian_and_avian_pallium.png** *Source:* https://upload.wikimedia.org/wikipedia/commons/e/e9/Schematic_illustration_of_differences_in_neuronal_specification_and_migration_patterns_between_the_mammalian_and_avian_pallium.png *License:* CC BY 2.5 *Contributors:* Nomura T, Takahashi M, Hara Y, Osumi N. Patterns of neurogenesis and amplitude of Reelin expression are essential for making a mammalian-type cortex. PLoS ONE. 2008 Jan 16;3(1):e1454. PMID 18197264 doi: 10.1371/journal.pone.0001454 http://www.pubmedcentral.nih.gov/articlerender.fcgi?artid=2175532&rendertype=figure&id=pone-0001454-g005 *Original artist:* Nomura T et al
- **File:Skull_and_brain_normal_human.svg** *Source:* https://upload.wikimedia.org/wikipedia/commons/e/ec/Skull_and_brain_normal_human.svg *License:* CC BY 2.5 *Contributors:* Patrick J. Lynch, medical illustrator *Original artist:* Patrick J. Lynch, medical illustrator
- **File:Sobo_1909_623.png** *Source:* https://upload.wikimedia.org/wikipedia/commons/1/1e/Sobo_1909_623.png *License:* Public domain *Contributors:* Atlas and Text-book of Human Anatomy Volume III Vascular System, Lymphatic system, Nervous system and Sense Organs *Original artist:* Dr. Johannes Sobotta
- **File:Sobo_1909_624.png** *Source:* https://upload.wikimedia.org/wikipedia/commons/e/ea/Sobo_1909_624.png *License:* Public domain *Contributors:* Atlas and Text-book of Human Anatomy Volume III Vascular System, Lymphatic system, Nervous system and Sense Organs *Original artist:* Dr. Johannes Sobotta
- **File:SparrowTectum.jpg** *Source:* https://upload.wikimedia.org/wikipedia/commons/0/09/SparrowTectum.jpg *License:* Public domain *Contributors:* ? *Original artist:* ?

35.8. TEXT AND IMAGE SOURCES, CONTRIBUTORS, AND LICENSES

- **File:Spinal_cord_tracts_-_English.svg** *Source:* https://upload.wikimedia.org/wikipedia/commons/b/b2/Spinal_cord_tracts_-_English.svg *License:* CC BY-SA 3.0 *Contributors:*
  Original artist: Polarlys and Mikael Häggström
- **File:Temporal_lobe_animation.gif** *Source:* https://upload.wikimedia.org/wikipedia/commons/1/1c/Temporal_lobe_animation.gif *License:* CC BY-SA 2.1 jp *Contributors:* Polygon data are from BodyParts3D[1] *Original artist:* Polygon data were generated by Database Center for Life Science(DBCLS)[2].
- **File:Thalamus_small.gif** *Source:* https://upload.wikimedia.org/wikipedia/commons/7/78/Thalamus_small.gif *License:* CC BY-SA 2.1 jp *Contributors:* URL. *Original artist:* Images are generated by Life Science Databases(LSDB).
- **File:Thalmus.png** *Source:* https://upload.wikimedia.org/wikipedia/commons/0/0b/Thalmus.png *License:* CC BY-SA 3.0 *Contributors:* Own work by uploader, sources [1] [2] [3] [4] [5] [6] [7] *Original artist:* Madhero88
- **File:Tyrannosaurus_brain_aus.jpg** *Source:* https://upload.wikimedia.org/wikipedia/commons/b/be/Tyrannosaurus_brain_aus.jpg *License:* GFDL *Contributors:* Own work *Original artist:* Matt Martyniuk (Dinoguy2)
- **File:Vertebrate_pallium.svg** *Source:* https://upload.wikimedia.org/wikipedia/commons/7/73/Vertebrate_pallium.svg *License:* Public domain *Contributors:* Own work *Original artist:* Looie496
- **File:Visible_Human_head_slice.jpg** *Source:* https://upload.wikimedia.org/wikipedia/commons/d/de/Visible_Human_head_slice.jpg *License:* Public domain *Contributors:* http://erie.nlm.nih.gov/~{}dave/vh/avf1067a.png *Original artist:* This image was created by a US government project in the National Library of Medicine, a branch of NIH. As government work, it is in the public domain. The original image was modified by user:Looie496
- **File:Visual_cortex_-_low_mag.jpg** *Source:* https://upload.wikimedia.org/wikipedia/commons/4/47/Visual_cortex_-_low_mag.jpg *License:* CC BY-SA 3.0 *Contributors:* Own work *Original artist:* Nephron
- **File:Wiki_letter_w_cropped.svg** *Source:* https://upload.wikimedia.org/wikipedia/commons/1/1c/Wiki_letter_w_cropped.svg *License:* CC-BY-SA-3.0 *Contributors:* This file was derived from Wiki letter w.svg: <a href='//commons.wikimedia.org/wiki/File:Wiki_letter_w.svg' class='image'><img alt='Wiki letter w.svg' src='https://upload.wikimedia.org/wikipedia/commons/thumb/6/6c/Wiki_letter_w.svg/50px-Wiki_letter_w.svg.png' width='50' height='50' srcset='https://upload.wikimedia.org/wikipedia/commons/thumb/6/6c/Wiki_letter_w.svg/75px-Wiki_letter_w.svg.png 1.5x, https://upload.wikimedia.org/wikipedia/commons/thumb/6/6c/Wiki_letter_w.svg/100px-Wiki_letter_w.svg.png 2x' data-file-width='44' data-file-height='44' /></a>
  Original artist: Derivative work by Thumperward
- **File:Wiktionary-logo-v2.svg** *Source:* https://upload.wikimedia.org/wikipedia/commons/0/06/Wiktionary-logo-v2.svg *License:* CC BY-SA 4.0 *Contributors:* Own work *Original artist:* Dan Polansky based on work currently attributed to Wikimedia Foundation but originally created by Smurrayinchester

35.8.3 Content license

- Creative Commons Attribution-Share Alike 3.0

www.ingramcontent.com/pod-product-compliance
Lightning Source LLC
Chambersburg PA
CBHW082326220526
45470CB00008B/2412